X-RAY DIFFRACTION

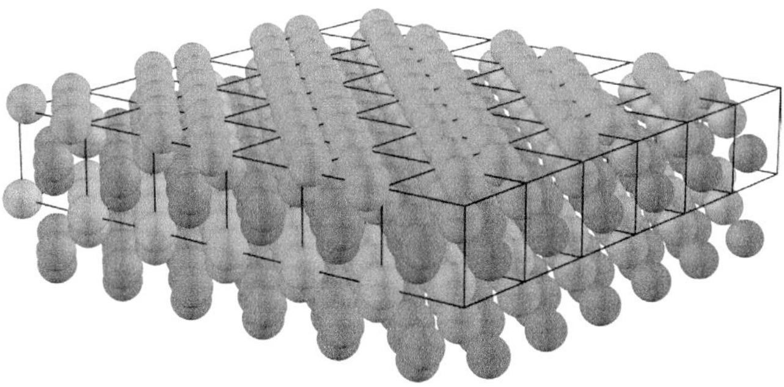

X-RAY DIFFRACTION

Modern Experimental Techniques

edited by

OLIVER H. SEECK
BRIDGET M. MURPHY

PAN STANFORD PUBLISHING

Published by

Pan Stanford Publishing Pte. Ltd.
Penthouse Level, Suntec Tower 3
8 Temasek Boulevard
Singapore 038988

Email: editorial@panstanford.com
Web: www.panstanford.com

British Library Cataloguing-in-Publication Data
A catalogue record for this book is available from the British Library.

X-Ray Diffraction: Modern Experimental Techniques

ISBN 978-981-4303-59-0 (Hardcover)
ISBN 978-981-4303-60-6 (eBook)

Printed in the USA

Contents

Preface

The development of human culture accompanies the progressive understanding of nature. In the last few centuries, the progress was tremendous, especially upon realization that nature is based on complex interplay between interactions on microscopic and macroscopic scale. Regarding the properties of matter, microscopic interactions, in particular between the atoms, are of eminent importance and they basically determine all characteristics. Even fully macroscopic properties such as melting point, viscosity, and stiffness are based on interatomic and intermolecular interaction parameters. To accomplish detailed understanding of the microscopic aspects of nature, science fields such as atomic physics, materials science, chemistry, and theoretical biology have been established. During the past decade, the focus has been additionally put on engineering and technology applications, resulting in the so-called micro- and nanotechnology. In the 21st century, the miniaturization and use of nanomaterials is omnipresent, e.g., in computer and sensor technology and in optics, medicine, and cosmetics; the future potentials are huge.

For further advancements in micro- and nanotechnology, profound knowledge of the interatomic and intermolecular interaction parameters is essential. On the one hand, this is challenging for theoretical science groups that develop mathematical tools to understand nature. On the other hand, experimental tools have to be designed and utilized to actually probe the interactions on the atomic scale. Therefore, scientific instruments with methods based on electrons, ions, or photons have been designed. Some of them are available as (more or less) inexpensive laboratory equipment. However, for high-end applications, they can be very costly and complex with a need of well-trained personnel for operation.

Tremendous progress has been achieved in the development of tools based on X-radiation. During the past 50 years, the evolution went from laboratory sources, so-called X-ray tubes, which are still available today, to parasitic use of synchrotron radiation from particle physics experiments, dedicated storage rings for X-radiation, and finally to X-ray lasers. The latter two are large-scale facilities with construction costs of several hundred million US dollars up to USD 1 billion and significant manpower with hundreds of FTEs to run the experiments. All over the world, approximately 20 modern sources are available, of which four deliver high-energy photons and two X-ray lasers. Synchrotron radiation sources offer extraordinary high X-ray beam quality for high-precision measurements on the atomic or molecular scale with accessible time scales from seconds down to femtoseconds (in the case of X-ray lasers). At each of the sources, a large number of experimental stations have been accommodated, which are specialized on certain X-ray methods, such as micro-diffraction, small-angle scattering, X-ray photo emission, fluorescence spectroscopy, tomography, and many more.

Modern synchrotron radiation sources are available for the general scientific and industrial community. Users are mostly from fields in physics, chemistry, geoscience, materials science, biology, archeology, and related fields. Usually, beam time is distributed on a proposal-based system with external referees. For this, an applicant has to define the science case and to choose an experimental station that fits his purpose best. At this point, a potential user should be able to evaluate the capabilities of the experimental stations at the synchrotron radiation sources and to identify the X-ray methods that he wants to apply. Aside from the experimental station, the X-ray photon flux and energy, the beam size and the divergence, the coherence and timing are properties of eminent importance.

In this book, the most important X-ray scattering and diffraction methods are introduced along with some aspects about the production of X-radiation at synchrotrons. In the first two chapters, the basics of X-ray diffraction and scattering methods and an overview of the characteristics of synchrotron radiation are presented. Also, the X-ray optics of a synchrotron radiation experiment are explained, which enables the reader to estimate the flux and the other beam parameters at the sample. In the later chapters, experts explain the different scattering and diffraction techniques.

The chapters on micro-diffraction and small-angle scattering give insights into the research of macromolecular samples, crystalline or amorphous. For both methods, focusing of the beam is of eminent importance; therefore, in the micro-diffraction section, focusing techniques are introduced. The following two chapters focus on inelastic scattering and X-ray standing waves, which are widely used to investigate phonon- and electron-density distribution in hard condensed matter. The next two chapters are devoted to magnetism. Two fully different X-ray methods are applicable: Magnetic scattering, which is a diffraction method based on magnetic interaction with the X-rays, and nuclear scattering, which monitors changes in the hyperfine field of the nuclei induced by magnetism in the sample.

The three following chapters deal with special topics: scattering at liquid interfaces, extreme condition science with X-rays, and tomography. The first is demanding as many chemical and biological reactions appear at liquid interfaces. Extreme condition science (high temperature and high pressure) relies on well-established X-rays powder diffraction methods; however, the experimental setup is very complex and the present status is explained in the book. Tomography is also introduced, though it is not a particular scattering or diffraction method. In many cases, such as metallic sintered powders, tomography and scattering methods are complementary.

The last two chapters describe applications of coherent X-rays. The so-called speckle pattern that arises from scattering of coherent beams at disordered samples contains more information than standard scattering data and can be used to do imaging or time-resolved studies. The experimental techniques and the rather complex theory are introduced in these chapters.

This book gives an insight into the up-to-date X-ray scattering methods that are available at modern synchrotron radiation sources. It enables the reader to understand the basic concept behind the methods and therefore to plan an appropriate, synchrotron radiation–based experiment.

Chapter 1

Overview of X-Ray Scattering and Diffraction Theory and Techniques

Oliver H. Seeck
Deutsches Elektronen Synchrotron (DESY), Notkestraße 85, 22607 Hamburg, Germany
oliver.seeck@desy.de

Based on the Maxwell's equations, which are the fundamental equations of electrodynamics, electromagnetic waves interact with electric and magnetic potentials. Interaction in this context means that the waves can locally change the potential and the potential may be able to modify the waves. An example is the interaction of high-field electromagnetic waves with plasmas, which leads to the so-called parametric coupling [see 1,2].

We are concentrating here on the interaction of electromagnetic waves with potentials depending on time or space. On an atomic scale, electric and magnetic potentials with spatial dependence are always present in any kind of matter: Electrons are negatively charged and have a magnetic spin. The cores are positively charged and can also exhibit a spin. On an atomic scale, the electric potential in matter is

X-Ray Diffraction: Modern Experimental Techniques
Edited by Oliver H. Seeck and Bridget M. Murphy

ISBN 978-981-4303-59-0 (Hardcover), 978-981-4303-60-6 (eBook)
www.panstanford.com

very strong as it is inversely proportional to the distance between the charges, which is about 1 Å (1 Å = 0.1 nm) in matter. Therefore, the electric potential of matter cannot be easily altered by an externally applied electromagnetic field, but electromagnetic waves can be modified by the atomic potential. This means that electromagnetic waves can be used to investigate the atomic structure of matter without affecting the atomic potentials.

X-rays are such electromagnetic waves with wave lengths in the same order as the typical atomic distances in matter, namely, 0.1...10 Å. Therefore, X-rays are especially sensitive to the atomic potential. X-ray scattering and diffraction means that the incoming electromagnetic wave field is coherently deflected at the electric or magnetic potential of sample [see 3]. The full dynamical X-ray scattering theory, which includes multiple scattering effects, is rather complex [see 4–6]. Fortunately, in many cases approximations can be applied, in particular the kinematical approximation or Born approximation for elastic scattering. The Born approximation neglects any kind of multiple scattering effects or energy gain or loss and gives a very descriptive insight into scattering theory. It can be used to explain a large number of different scattering experiments [7].

In this chapter, the fundamental terms of X-ray diffraction and scattering are introduced, and the Born approximation is explained. Specific examples are presented for which the Born approximation yields good results. The conditions where the Born approximation fails are also shown. For these cases, methods from the dynamical scattering theory are presented. Special techniques such as coherent scattering and tomography are discussed in following chapters.

1.1 Scattering at Single Electrons

The scattering process of X-rays at single electrons has been described in text books [see 3,7,8] and is not explained here in detail. Instead, the scattering will be introduced in simple terms so that in the following the Born approximation of X-ray scattering can be introduced.

For this, it is assumed that the incident electromagnetic wave is a plane wave with wave length λ. The wave length and the photon energy of electromagnetic waves are connected by $E = \hbar\omega$. Here, $\omega = ck$ is the frequency with the speed of light c and the wave number, $k = 2\pi/\lambda$. The usual unit for the photon energy is electron volts [eV], whereas for the wavelength it is angstrom [Å] with E [eV] = 12398.4244 λ[Å] and 1 Å = 0.1 nm. The complex amplitude of a plane wave can be written as follows:

$$A(\mathbf{k}) \propto \exp(i\mathbf{k}\cdot\mathbf{r}). \tag{1.1}$$

Here, the wave number is the modulus of the wave vector **k** which points along the traveling direction.

If the incident electromagnetic wave illuminates a single electron this wave excites the electron, which as a result radiates a spherical wave with the same wave length as the incident beam.[1] Consequently, this wave length conserving scattering is called elastic scattering.

In the Born approximation, the scattering of an ensemble of electrons, called sample, is simply calculated by adding all corresponding spherical waves. In the following, it is assumed that the distance of the observer from the sample is much larger than the sample size (far-field approximation [9]). Then the curvature of the spherical waves is negligible, and they can be approximated by plane waves. When adding the plane waves scattered from each electron, the correct phase difference of the waves has to be taken into consideration. The phase difference results from the different paths along which the waves have travelled. The path difference has to be normalized to the wave length. This is depicted in Fig. 1.1 which shows the incoming wave $\mathbf{k}_i$ and the outgoing wave $\mathbf{k}_f$. The reflected wave from the particle at $\mathbf{r}_0$ is behind the wave reflected from the origin. Therefore, the phase difference between the two waves is $\mathbf{k}_f\cdot \mathbf{r}_0 - \mathbf{k}_i \cdot \mathbf{r}_0 = (\mathbf{k}_f - \mathbf{k}_i)\cdot \mathbf{r}_0 = \mathbf{q}\cdot \mathbf{r}_0$. As the result, in the Born approximation, the scattered amplitude of a single particle is given by the following expression:

$$A(\mathbf{q}) \propto \exp(i\mathbf{q}\cdot\mathbf{r}). \tag{1.2}$$

[1]Even though this description is already heavily simplified, this description is sufficient for further considerations. Neglected are effects such as polarization, inelastic scattering, dipole radiation and so on.

In this context, $\mathbf{q} = \mathbf{k}_f - \mathbf{k}_i$ is called wave vector transfer. From this, it can easily been shown that

$$q = 2k \sin\frac{2\theta}{2} \tag{1.3}$$

holds where 2θ is the scattering angle (see Fig. 1.1). This is the master equation to transform scattering angles into q-space and vice versa. To understand the meaning of Eq. (1.2), we need the δ function:

$$\delta(x - x_0) = \left\langle \begin{array}{ccc} \infty & : & x = x_0 \\ 0 & : & x \neq x_0 \end{array} \right. \quad \text{with} \quad \int_{-\infty}^{\infty} \delta(x - x_0) dx = 1 \tag{1.4}$$

which directly implies that for an arbitrary smooth function $f(x)$

$$\int_{-\infty}^{\infty} f(x)\delta(x - x_0)dx = \lim_{\Delta \to 0} \int_{x_0 - \Delta}^{x_0 + \Delta} f(x)\delta(x - x_0)dx = f(x_0)$$

holds. Using the δ function, Eq. (1.2) can be rewritten as follows:

$$A(\mathbf{q}) \propto \exp(i\mathbf{q}\cdot\mathbf{r}) = \int_{-\infty}^{\infty} \delta(\mathbf{r} - \mathbf{r}_0)\exp(i\mathbf{q}\cdot\mathbf{r})d^3r\ .$$

This form is equivalent to the complex three-dimensional Fourier transformation of a δ function, which can be identified as the point-like charge of the particle at position $\mathbf{r}_0$.

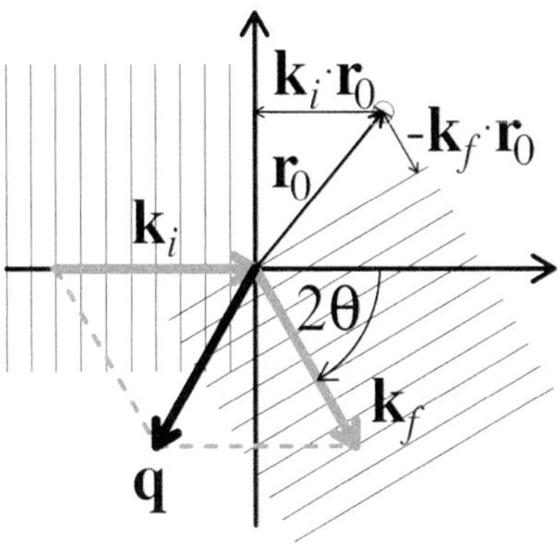

Figure 1.1 Sketch of a scattered wave. The scattering object is not located in the origin but at $\mathbf{r}_0$. The scattered wave from the object has phase difference, which is behind the wave scattered from the origin. The scattering angle is denoted by 2θ.

1.2 Scattering in Bulk Matter

Atoms are not distributed arbitrarily in condensed matter. In liquids and amorphous glasses a local order on an atomic scale is present. In perfect single crystals the order extends to macroscopic scales. Therefore, also the electrons are not arbitrarily positioned in space but exhibit order.

The scattering amplitude of a single electron has been derived in the earlier sections. As has been shown in Born approximation, the scattering amplitude of a multi electron system can be calculated by simply summing up the single scattering amplitudes of each electron in the far-field regime. By using Eq. (1.2), the scattered amplitude $A(\mathbf{q})$ of a many electron system with N electrons is

$$A(\mathbf{q}) \propto \sum_{j=1}^{N} A_j(\mathbf{q}) \propto \sum_j \exp[i(\mathbf{k}_f - \mathbf{k}_i)\cdot \mathbf{r}_j] = \sum_j \exp(i\mathbf{q}\cdot \mathbf{r}_j). \quad (1.5)$$

Such as for a single electron, this equation can be slightly rewritten by artificially introducing the δ function $\delta(\mathbf{r} - \mathbf{r}_i)$ at the position of the charges by

$$\sum_j \exp(i\mathbf{q}\cdot \mathbf{r}_j) = \sum_j \int \delta(\mathbf{r} - \mathbf{r}_j)\exp(i\mathbf{q}\cdot \mathbf{r})d^3r$$

$$= \int \left(\sum_j \delta(\mathbf{r} - \mathbf{r}_j) \right) \exp(i\mathbf{q}\cdot \mathbf{r})d^3r.$$

The sum over all δ functions is identical to the electron density distribution $\rho(\mathbf{r})$ of the illuminated matter. Consequently, the scattered amplitude is given by the following equation:

$$A(\mathbf{q}) \propto \sum_{j=1}^{N} A_j(\mathbf{q}) \propto \int \rho(\mathbf{r})\exp(i\mathbf{q}\cdot \mathbf{r})d^3r. \quad (1.6)$$

Equation (1.6) is the master equation for X-ray scattering in Born approximation. It means that the scattered amplitude is proportional to the Fourier transformation of the electron density. Usually, $A(\mathbf{q})$ is a complex function and may be written as $A(\mathrm{q}) = |A(\mathbf{q})|\exp[i\varphi(\mathbf{q})]$ with phase $\varphi(\mathbf{q})$. The measured intensity is given by the absolute square of the scattered amplitude $I(\mathbf{q}) = |A(\mathbf{q})|^2$. Due to this operation, the intensity lacks some information of the electron density $\rho(\mathbf{r})$ due to the missing phase. Consequently, together with $I(\mathbf{q})$, additional pre-knowledge or an independent measurement of $\rho(\mathbf{r})$ is required for an unambiguous interpretation of data analysis.

It is very important to understand that the approach of simply summing up the scattering amplitudes of single electrons neglects multiple scattering effects. This is the main drawback of the Born approximation against the exact dynamical scattering theory, which is also applicable at scattered intensities comparable to the incident intensity. Further, any kind of magnetic contribution to the scattering amplitude is fully neglected. Still, Eq. (1.6) is valid for a large number of scattering problems, and it is also very descriptive and easy to use.

1.2.1 Scattering in Disordered Matter

One important application of Eq. (1.6) is to calculate the X-ray scattering of disordered amorphous or liquid matter. In this case, the atoms (and as such the electrons) are not long-ranged ordered over the whole sample. However, due to repulsive and attractive interactions of atoms on very short length scales, short-range order is expected as a function of the distance R between two atoms. This short-range order is visible in the scattered X-ray intensity $I(\mathbf{q})$. Using $I(\mathbf{q}) = |A(\mathbf{q})|^2$ and Eq. (1.6), the intensity can be written as follows:

$$\begin{aligned} I(\mathbf{q}) &\propto \int \rho(\mathbf{r})\exp(i\mathbf{q}\cdot\mathbf{r})d^3r \cdot \int \rho(\mathbf{u})\exp(-i\mathbf{q}\cdot\mathbf{u})d^3u \\ &= \iint \rho(\mathbf{r})\rho(\mathbf{u})\exp(i\mathbf{q}\cdot[\mathbf{r}-\mathbf{u}])d^3r\,d^3u \\ &\overset{\mathbf{R}=\mathbf{r}-\mathbf{u}}{=} \iint \rho(\mathbf{r})\rho(\mathbf{r}-\mathbf{R})d^3r \cdot \exp(i\mathbf{q}\cdot\mathbf{R})d^3R \end{aligned}$$

$$\Rightarrow \quad I(\mathbf{q}) \propto \int g(\mathbf{R})\exp(i\mathbf{q}\cdot\mathbf{R})d^3R \tag{1.7}$$

Thus, the scattered intensity is a Fourier transformation of a function

$$g(\mathbf{R}) = \int \rho(\mathbf{r})\rho(\mathbf{r}-\mathbf{R})d^3r\,. \tag{1.8}$$

Bearing in mind that in matter the electron density $\rho(\mathbf{r})$ is actually a sum over distinct charges, the function $g(\mathbf{R})$ is identical to the well-known pair correlation function or pair distribution function (PDF) (see [10]). In the context of this book, the PDF is a measure of favorable distance vectors $\mathbf{R}$ between atoms. In the case of disordered matter, the atoms are distributed isotropically in space such that the PDF becomes a function of the distance $R = |\mathbf{R}|$ only. Using spherical coordinates with $d^3r = R^2\sin\theta \cdot d\varphi \cdot d\theta \cdot dR$ and with $q = |\mathbf{q}|$, the scattered intensity can be written as

$$I(\mathbf{q}) \propto \int g(R)\exp(i\mathbf{q}\cdot\mathbf{r})d^3r = 2\pi\int R^2 g(R)\exp(iqR\cos\theta)\sin\theta\cdot d\theta\cdot dR$$

$$= 4\pi\int R^2\frac{g(R)}{2iqR}\left[\exp(iqR\cos\theta)\right]_\pi^0 dR$$

$$\Rightarrow \quad I(q) \propto \int R^2 g(R)\frac{\sin(qR)}{qR}dR .$$

This result is the integral representation of the Debye formula. It states that for isotropically distributed charges, the intensity is given by a sin-Fourier transformation in real space. Consequently, the value of $g(R)$ can be determined by a simple Fourier back transformation. This approach is used for the so-called PDF analysis in which the measured intensity yields the PDF of the sample, which can then be correlated with physical properties.

1.2.2 Scattering in Crystalline Matter

In the case of crystalline matter, the symmetry of the crystal lattice determines the scattered X-ray intensity. In a perfect single crystal a parallelepiped, the so-called unit cell, is repeating in all three dimensions. In this book, the edges of the unit cell are vectors named **a**, **b**, **c**, respectively, and the cell contains a number of M atoms at well-defined position $\mathbf{u}_k$ (see Fig. 1.2) [3].

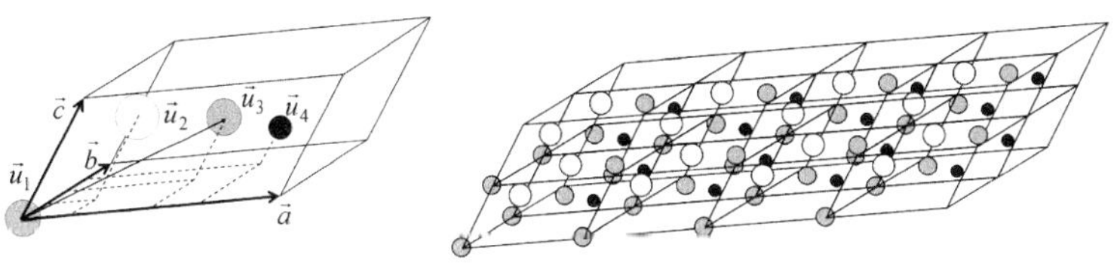

Figure 1.2 A trigonal unit cell with the unit vectors **a**, **b**, **c**, and four atoms at different positions $\mathbf{u}_k$ (left) and an assembly of some unit cells forming a small crystal (right).

By introducing the electron density $\rho_k(\mathbf{r})$ of an atom k in the unit cell, the electron density of the whole crystal can be written as

$$\rho(\mathbf{r}) = \sum_{n_a}^{N_a}\sum_{n_b}^{N_b}\sum_{n_c}^{N_c}\sum_{k}^{M}\rho_k(\mathbf{r}+n_a\mathbf{a}+n_b\mathbf{b}+n_c\mathbf{c}+\mathbf{u}_k)$$
$$= \sum_{n_a n_b n_c}\sum_{k}\int \rho_k(\mathbf{w})\delta(\mathbf{w}-\mathbf{r}-n_a\mathbf{a}-n_b\mathbf{b}-n_c\mathbf{c}-\mathbf{u}_k)d^3w$$

In this equation of the electron density, the values $N_{a,b,c}$ represent the finite character of the single crystal, whereas M is the number of atoms in the unit cell. Again, the δ function is introduced as a helper function, which makes the calculation of the scattered intensity easier. Following Eq. (1.6), the scattered amplitude for a single crystal becomes

$$A(\mathbf{q}) \propto \iint \sum_{n_a n_b n_c}\sum_{k}\rho_k(\mathbf{w})\delta(\mathbf{w}-\mathbf{r}-n_a\mathbf{a}-n_b\mathbf{b}-n_c\mathbf{c}-\mathbf{u}_k)\exp(i\mathbf{q}\cdot\mathbf{r})d^3w\cdot d^3r$$
$$= \int \sum_{n_a n_b n_c}\sum_{k}\rho_k(\mathbf{w})\int \delta(\mathbf{w}-\mathbf{r}-n_a\mathbf{a}-n_b\mathbf{b}-n_c\mathbf{c}-\mathbf{u}_k)\exp(i\mathbf{q}\cdot\mathbf{r})d^3r\cdot d^3w$$
$$= \sum_{n_a n_b n_c}\sum_{k}\int \rho_k(\mathbf{w})\exp[i\mathbf{q}\cdot(-\mathbf{w}+n_a\mathbf{a}+n_b\mathbf{b}+n_c\mathbf{c}+\mathbf{u}_k)]d^3w$$
$$= \sum_{n_a n_b n_c}\exp[i\mathbf{q}\cdot(n_a\mathbf{a}+n_b\mathbf{b}+n_c\mathbf{c})]\cdot \underbrace{\sum_{k}\exp(i\mathbf{q}\cdot\mathbf{u}_k)\cdot \underbrace{\int \rho_k(\mathbf{w})\exp(-i\mathbf{q}\cdot\mathbf{w})d^3w}_{f_k(\mathbf{q})}}_{S(\mathbf{q})}$$

$$\Rightarrow \quad A(\mathbf{q}) \propto S(\mathbf{q})\sum_{n_a n_b n_c}\exp[i\mathbf{q}\cdot(n_a\mathbf{a}+n_b\mathbf{b}+n_c\mathbf{c})]. \qquad (1.9)$$

Here, the scattering process has been separated mathematically into three parts. The contribution from a single atom of the unit cell $f_k(\mathbf{q})$ is called form factor. It determines the ability of an atom to scatter X-rays. Usually, but not necessarily, the form factor depends only on the modulus $|\mathbf{q}|$ and not on $\mathbf{q}$ as an atom has spherical symmetry. The functions $f_k(\mathbf{q})$ are tabled for almost every atom or ion [11].

The scattering contribution from one unit cell taking into account the position of the single atoms is called structure factor $S(\mathbf{q})$. The structure factor contains the phase information of the unit cell. Symmetry properties may cause amplification or extinction of the scattered amplitude for distinct wave vector transfers $\mathbf{q}$. The remaining triple sum is of the form $\Sigma_{n=0}^{N}x^n$ and can easily be calculated using the following expression:

$$\sum_{n=0}^{N} x^n = \frac{x^{N/2} - x^{-N/2}}{x^{1/2} - x^{-1/2}} \cdot x^{\frac{N-1}{2}}. \tag{1.10}$$

For a single crystal the size is given by the number of repetitions N_a, N_b, N_c in the directions of the three-lattice vectors **a**, **b**, **c**, and the scattered intensity is as follows:

$$I(\mathbf{q}) \propto \frac{\sin^2\left(\frac{1}{2}N_a\mathbf{q}\cdot\mathbf{a}\right)}{\sin^2\left(\frac{1}{2}\mathbf{q}\cdot\mathbf{a}\right)} \cdot \frac{\sin^2\left(\frac{1}{2}N_b\mathbf{q}\cdot\mathbf{b}\right)}{\sin^2\left(\frac{1}{2}\mathbf{q}\cdot\mathbf{b}\right)} \cdot \frac{\sin^2\left(\frac{1}{2}N_c\mathbf{q}\cdot\mathbf{c}\right)}{\sin^2\left(\frac{1}{2}\mathbf{q}\cdot\mathbf{c}\right)} \cdot |S(\mathbf{q})|^2 \tag{1.11}$$

Each of the three leading terms corresponds to a Laue function, which is displayed for one dimension in Fig. 1.3. The Laue functions exhibit pronounced maxima, which are called Bragg peaks or Bragg reflections and which are located at the Bragg condition:

$$\mathbf{q}_{\text{Bragg}} = \frac{2\pi}{\mathbf{a}\cdot(\mathbf{b}\times\mathbf{c})}\left(h[\mathbf{b}\times\mathbf{c}], k[\mathbf{c}\times\mathbf{a}], l[\mathbf{a}\times\mathbf{b}]\right)^T \tag{1.12}$$

with integer numbers *h*, *k*, *l*, which are called Miller indices. The according scattering angle θ_{Bragg} given by $q_{\text{Bragg}} = 2k \sin\theta_{\text{Bragg}}$ is called the Bragg angle.

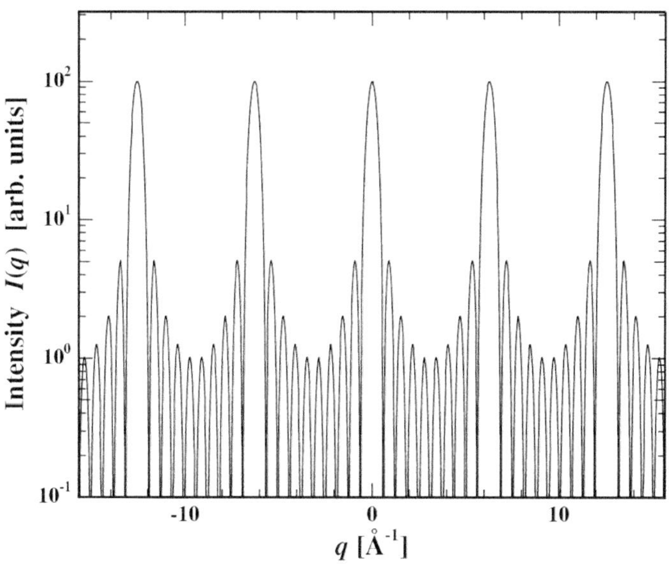

Figure 1.3 A one-dimensional Laue function for $N = 10$ calculated for a lattice constant $a = 1$ Å in log-representation.

The rapid oscillations between the Bragg reflections are called Laue oscillations. They are caused by an interference of the X-ray waves due to the presence of surfaces at $n_{a,b,c} = 0$ and $n_{a,b,c} = N_{a,b,c}$ and therefore contain information about the dimensions of the crystal in each direction. For large $N_{a,b,c}$ (for large crystals), the Laue oscillations vary rapidly, but they are also strongly damped as absorption effects reduce the interfering X-ray amplitudes. In this case, the crystal size, for example, $N_a a$ in a-direction, can be determined by a detailed peak shape analysis and analyzing the full at width half maximum $\Delta q_{\text{FWILM}} \approx 2\pi/(N_a a)$ of the Bragg reflections or, in better approximation, using the so-called Scherrer formula, $N_a a \approx K\lambda/(\Delta(2\theta)\cos\theta)$ with $K \approx 0.9$ and $\Delta(2\theta)$ as the FWHM of the reflection given in radian. For this, it is very important to compensate for the experimental resolution function, which can strongly modify the shape of the measured Bragg reflections.

For perfect and infinitely large crystals ($N_{abc} \to \infty$), δ functions are generated by Eq. (1.11) as Bragg reflections at the positions of the Bragg condition. This marks the breakdown of the Born approximation as the scattered intensity can never be larger than the incident intensity. The correct shape of the Bragg peaks for perfect crystals can be calculated by the dynamical scattering theory, which has been presented in other works [see 3,8]. In the last section of this chapter, some information about the exact shape of Bragg reflections is provided.

1.2.3 Scattering at Powders of Crystalline Matter

As explained earlier, single crystals scatter Bragg reflections when the sample is oriented in the X-ray beam to fulfill the scattering condition from Eq. (1.12). However, for a given Bragg reflection, this equation allows some freedom in orienting the sample. The reason is that the vectors of the incident beam, the exit beam, and the wave vector transfer, which in Bragg case, are identical to the reciprocal lattice vector, always form a two-dimensional plane, leaving the third dimension undefined. This is why a sample can be arbitrarily rotated around the direction of the incident beam with matching Bragg condition at all rotation angles (see Fig. 1.4).

This has consequences for powder samples, which consist of a huge amount of arbitrarily oriented single grains. From all grains, a large number fulfill the different valid Bragg conditions. Further, as the grains are randomly rotated around the incident beam, the so-called powder rings are observed, which represent the Bragg conditions. Of course, the characteristics of the ring depend strongly on the number of illuminated grains: For a small number, the ring appears spotted; only very large numbers result in a smooth intensity distribution around the ring (see Fig. 1.5a).

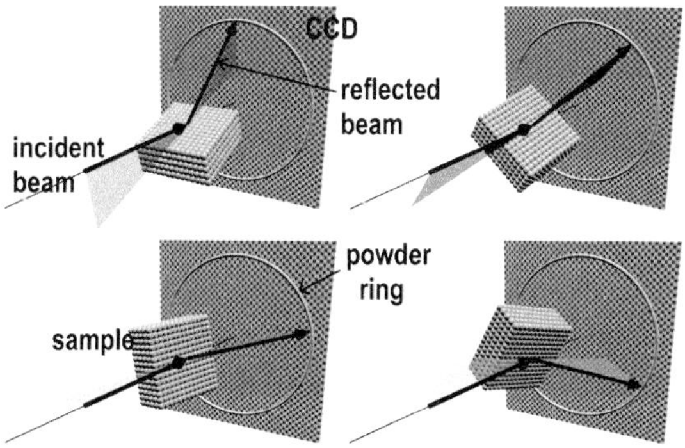

Figure 1.4 At this crystalline sample, the Bragg-condition is matched. On rotation of the sample around the incident beam, the Bragg condition remains valid. With arbitrary rotation (but fixed incident angle) a powder ring results.

In general, one can estimate that the illuminated volume of the powder sample should be roughly 10^9 times larger the grain size and that the grains should be arbitrarily distributed to achieve good powder rings. However, this may not be possible for different reasons. Mostly, restrictions apply to the illuminated sample volume; especially, if only small amounts are available. Concerning the angular distribution, the grains in the sample may be oriented in a preferred direction which may result from elongated grain shapes. Consequently, modulated powder rings are expected. To overcome this problem, powders are continuously spun in the X-ray beam to increase the angular acceptance to higher values (see Fig. 1.5b).

To analyze the data, the center of the powder rings has to be determined carefully and then the intensity of all the rings is integrated over ϕ at constant offset r from the center as shown in Fig. 1.5c. The resulting function $I(r)$ has to be converted to $I(q)$ by means of the photon wavelength, the distance sample-to-detector and r. This integration is especially mandatory if a slight granular structure in the two-dimensional powder pattern is still present (see Fig. 1.5c,d). The intensity of powder rings is not identical with the structure factor but has to be corrected by the multiplicity of the Bragg peaks. For example, for a cubic crystal, the reflections of the type $n00$ have a multiplicity of 6 because six permutations exist with identical Bragg reflections, namely, $n00$, $0n0$, $00n$, $-n00$, $0-n0$, $00-n$. Similarly, reflections of type $nn0$ have multiplicity of 12 in cubic crystals.

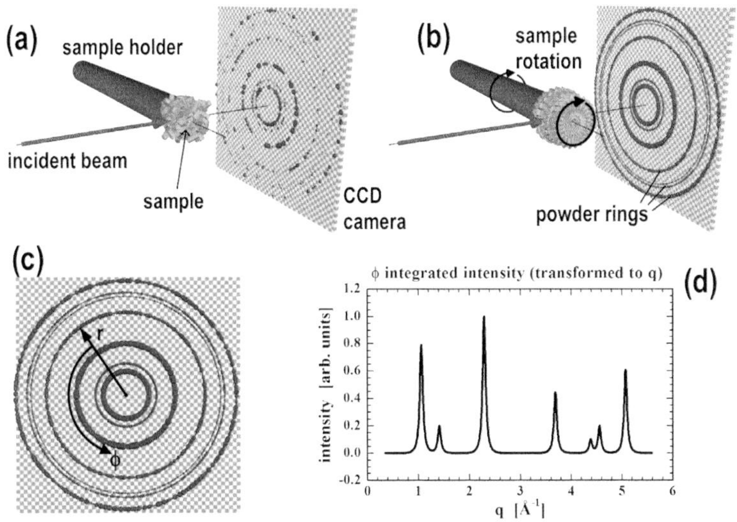

Figure 1.5 The powder sample is too small for homogeneous rings (a). Rotating the sample makes the rings continuous. By integrating (c) and converting to reciprocal space, (d) the typical powder pattern results.

For a detailed analysis of a powder diffraction data, many other considerations have to be taken into account, such as polarization of the photons. For this, we refer to the "*Fundamentals of Powder Diffraction*" by Pecharsky and Zavalij [12].

1.3 Scattering at Surfaces

Every type of condensed matter sample has a surface, even if the sample is accounted as "infinite" for the scattering experiment.

Thus, X-ray scattering in bulk samples always means that additional scattering contributions from the surfaces of the samples are present. The surface scattering may contaminate the bulk signal and has to be removed carefully before analyzing bulk data. However, surface scattering can also be extremely useful if surfaces and interfaces are the subject of the investigations [13]. In the following, the theoretical principles of X-ray surface scattering are presented.

It has already been shown previously that the scattered amplitude $A(\mathbf{q})$ is given by the complex Fourier transformation of the electron density $\rho(\mathbf{r})$ of the sample (see Eq. (1.6)). The function $\rho(\mathbf{r})$ is not only defined inside the sample but can be extended by $\rho(\mathbf{r}) = 0$ for all $\mathbf{r}$ outside the sample. Mathematically, bulk and surface properties can be separated by introducing $\rho_\infty(\mathbf{r})$ of a virtually infinite sample and a boundary function $B(\mathbf{r})$ to generate the surface with

$$\rho(\mathbf{r}) = \rho_\infty(\mathbf{r})\cdot B(\mathbf{r}). \tag{1.13}$$

Figure 1.6 depicts a simple example: A small crystal is extended to infinity. The boundary function of the crystal is given by a function with 1 inside the sample and 0 for the outside.

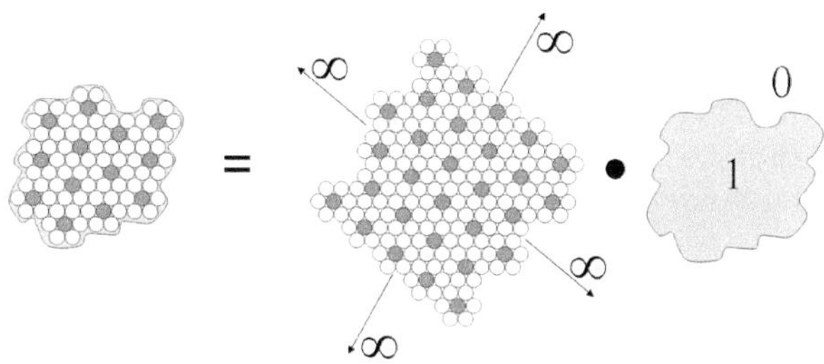

Figure 1.6 Schematic separation of the density of a finite sample into an infinite sample and a boundary function.

Equation (1.13) is inserted into the equation of the scattered amplitude. By using the convolution theorem [14], which states that a Fourier transformation of a product can be rewritten as a convolution of two Fourier-transformed functions, the form

$$A(\mathbf{q}) \propto \int \rho(\mathbf{r})\exp(i\mathbf{q}\cdot\mathbf{r})d^3r = \int \rho_\infty(\mathbf{r})B(\mathbf{r})\exp(i\mathbf{q}\cdot\mathbf{r})d^3r$$

$$\Rightarrow \quad A(\mathbf{q}) \propto \left[\int \rho_\infty(\mathbf{r})\exp(i\mathbf{q}\cdot\mathbf{r})d^3r\right] \otimes \left[\int B(\mathbf{r})\exp(i\mathbf{q}\cdot\mathbf{r})d^3r\right] \tag{1.14}$$

is achieved. In Eq. (1.14), the symbol ⊗ denotes the convolution. Consequently, a surface is mathematically similar to a resolution function. It strongly determines the shape and intensity of the X-ray scattering. It is important to mention that the function $B(\mathbf{r})$ does not necessarily represent a sharp bounding shape. It can and usually does include smooth boundaries, for example, caused by roughness.

1.3.1 Scattering at Crystal Surfaces

In the case of a crystalline substrate and a flat surface with a normal vector along z, the intensity of the Bragg reflections spreads along the q_z-axis. The surface scattering appears as so-called crystal truncation rods (CTRs), which fill the q_z-space between the Bragg peaks [15]. The shape of the CTRs is determined by the shape of the surface. Roughness, film properties, and surface reconstruction of atoms are visible with a resolution in sub-Angstroms (see Fig. 1.7 as an example with reconstructed Si-001 surface).

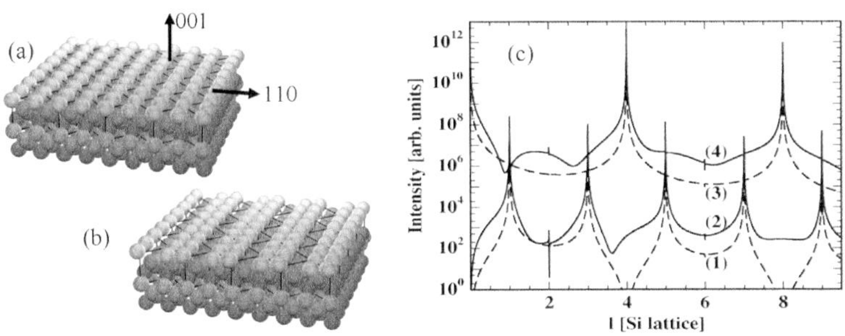

Figure 1.7 (a) Unrelaxed 001-surface of a silicon crystal. The unit cell is shown as black lines. (b) Energetic favorable 2 × 1-reconstruction of the Si-001 surface: Here, only the top monolayer is affected. (c) Comparison of two CTRs calculated for these samples with (1) 1l-scan (unrelaxed), (2) 1l-scan (relaxed), (3) 0l-scan (unrelaxed), and (4) 0l-scan (relaxed).

For the calculation of CTRs, it is usually easier to compute Eq. (1.9) with adequate values for n_c rather than using Eq. (1.14). The CTRs in Fig. 1.7 have been generated accordingly. In the case that a crystalline (epitaxial) layer is present on top of a crystalline substrate, the scattered amplitudes of the bottom and the top interface of the layer causes interference fringes depending on the layer thickness.

These fringes are equivalent to the Laue oscillations, which have been mentioned above.

The lattice constant of the layer material and the substrate are usually different so that at least two different Bragg reflections are expected. If two materials have similar lattice constants, it may be energetically favorable for the first few monolayers of the layer to grow with the same in-plane lattice constant as the substrate. In this case, the unit cells of the layer are more or less heavily strained.

At some distance from the substrate, the layer usually relaxes to the native, unstrained lattice constant. Model calculations of a typical example of this kind are shown in Fig. 1.8. For this, 8 layers of germanium (lattice constant 5.658 Å) are put onto bulk silicon (lattice constant 5.431 Å). The *l*-axes of both materials point along the surface normal. Four cases have been calculated, which differ in the amount of strained Ge-layers and relaxed Ge-layers. From the calculation, it is evident that for these kinds of measurements high resolution may be critical, especially if the layer is very thick or if the Bragg reflections are in close vicinity.

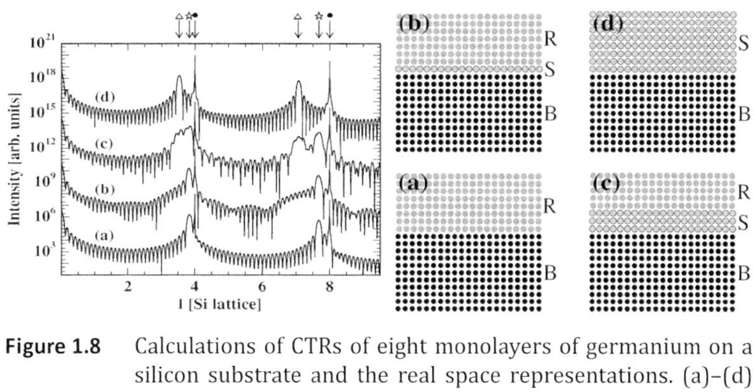

Figure 1.8 Calculations of CTRs of eight monolayers of germanium on a silicon substrate and the real space representations. (a)–(d) mean relaxed, 1 monolayer strained, 3 monolayers strained and fully strained Ge. The arrows on the left point to the Bragg-positions of Si (bullet) relaxed Ge (star) and strained Ge (triangle).

Not discussed in this context is grazing incidence diffraction. Here, the wave vector transfer is oriented in the surfaces. Accordingly, the scattering planes are perpendicular to the surface. The penetration depth can be tuned in a wide range by choosing adequate incident angle, which are very close to 0. This method is frequently used to determine the depth resolved in-plane structure of crystals, for example, in the case of surface phase transitions [16].

1.3.2 Scattering at Surfaces with Density Profile

A special case is the scattering in forward direction at flat surfaces. Forward direction means small exit angles close to $\mathbf{q} = 0$. This is mathematically equivalent to the $(h, k, l = 0,0,0)$-reflection. The 0,0,0-reflection always exists even in amorphous samples. It is the "Bragg peak" of homogeneous matter with the averaged electron density $\rho = <\rho(\mathbf{r})>$. The concepts of surface scattering are equally valid, and Eq. (1.14) can be used to calculate the intensity rod in q_z-direction. The surface scattering rod in forward direction is called reflectivity or specular scattering [7,17,18].

In the case of reflectivity, Eq. (1.14) can be simplified. The scattering rod points in the direction of q_z-axis with $(q_x, q_y) = (0, 0)$. Therefore, only components of $A(\mathbf{q})$ along q_z are considered. As a consequence, the shape function $B(\mathbf{r})$ only depends on z. In this case, $B(\mathbf{r}) = B(z)$ is proportional to the averaged electron density profile $\rho(z) = <\rho(\mathbf{r})>_{(x,y)}$, being averaged over the (x, y)-plane. The in-plane averaging has very important consequences: reflectivity measurements are by not sensitive to in-plane structures. Even surface islands or periodic gratings would appear as one complete layer but of reduced density. This situation is depicted in Fig. 1.9.

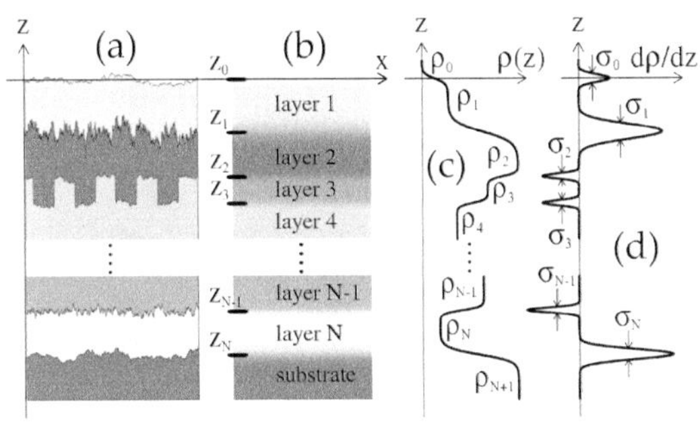

Figure 1.9 (a) Example of a complex layer system. (b) Same as (a) but averaged in the (x, y)-plane. Note that the grating in (a) has transformed to a homogeneous layer. (c) Resulting density profile $\rho(z)$. (d) Derivative of the density profile with the widths σ_j.

The function $B(z)$ can be split into two contributions by convolution of two other functions. One is the sudden step $\Theta(z)$ of the bulk density ρ down to 0 at the surface, and the other is the modification of the surface due to roughness and/or presence of thin films. With Eq. (1.14) and

$$B(z)=\int_{-\infty}^{z}\frac{dB}{d\zeta}d\zeta=\int_{-\infty}^{\infty}\frac{1}{2}(1-\Theta[\zeta-z])\cdot\frac{dB}{d\zeta}d\zeta=\frac{1}{2}(1-\Theta[z])\otimes\frac{dB}{dz}$$

and once again using the convolution theorem in the inverse representation, the scattered amplitude of a reflectivity can be written as follows:

$$\begin{aligned}A(q_z)&\propto\int\rho\cdot\exp(iq_zz)dz\otimes\int\left[\frac{1}{2}(1-\Theta[z])\otimes\frac{dB}{dz}\right]\exp(iq_zz)dz\\&\propto\rho\cdot\delta(q_z)\otimes\left[\int\frac{1}{2}(1-\Theta[z])\exp(iq_zz)dz\cdot\int\frac{dB}{dz}\exp(iq_zz)dz\right]\\&\propto\rho\int\frac{1}{2}(1-\Theta[z])\exp(iq_zz)dz\cdot\int\frac{dB}{dz}\exp(iq_zz)dz\\&\propto S_{\text{surf}}(q_z)\cdot\int\frac{d\rho(z)}{dz}\exp(iq_zz)dz\ .\end{aligned}$$

The correct calculation of the surface structure factor $S_{\text{surf}}(q_z)$ of a smooth and flat interface in Born approximation can be done in different ways. Easiest is to apply the exact Fresnel theory for optical media. The complex reflection coefficient $R_F(q_z)$ of a smooth surface of a homogeneous sample can be derived easily [9] and R_F would replace S_{surf} in the equation above. It can further be shown that in the frame of the Born approximation, $R_F \propto 1/q_z^2$ holds so that the final result is the following:

$$I(q_z)\propto\frac{1}{q_z^4}\left|\int\frac{d\rho(z)}{dz}\exp(iq_zz)dz\right|^2 . \tag{1.15}$$

The reflected intensity is the absolute square of a complex Fourier transformation and depends on the derivative of the electron density profile and thus on the film thickness, the roughness and the density contrast between the layers and the substrate. As mentioned earlier, and shown in Fig. 1.9, the roughness of interface j reduces to one parameter σ_j due to the average in the (x, y)-plane; σ_j is referred as the Gaussian width of the interface. With this knowledge, the electron density profile at an interface can be written as an Error function erf(z), which is the integral of the Gaussian distribution function [19].

For an N-layer system, the complete density profile is the sum of $N + 1$ Error-functions (see Fig. 1.9):

$$\rho(z) = \sum_{j=0}^{N} \Delta\rho_j \frac{1}{2}\left(1 - \operatorname{erf}\left[\frac{x - x_j}{\sqrt{2}\sigma_j}\right]\right) \quad \text{with} \quad \Delta\rho_j = \rho_{j+1} - \rho_j .$$

Knowing that the derivative of an Error function is of Gaussian type and that the Fourier transformation of a Gaussian is of the same form and using Eq. (1.15) the reflected intensity of a multilayer system is given by

$$I(q_z) \propto \frac{1}{q_z^4}\left[\sum_{j=0}^{N} \Delta\rho_j^2 \exp\left(-q_z^2\sigma_j^2\right) + 2\sum_{j=0}^{N-1}\sum_{k=j+1}^{N} \Delta\rho_j \Delta\rho_k \exp\left(-q_z^2 \frac{\sigma_j^2 + \sigma_k^2}{2}\right)\cos\left(q_z[z_j - z_k]\right)\right]. \tag{1.16}$$

This has two parts. The single sum contains the incoherent scattering from each interface. Each of the terms is multiplied with a damping exponential, which is determined by the roughness of the respective interface. The interference of the scattering from the interfaces is represented by the double sum. It can be shown that for N layers with $N + 1$ interfaces $N(N + 1)/2$ interference terms exist, which appear as fringes in the reflectivity data. The amplitude of the fringes is determined by the electron density step (the contrast) at the corresponding interfaces. The damping of each of the interference terms depends on the average roughness of the two corresponding interfaces [20].

Figure 1.10 displays a two-layer sample and some calculated reflectivity data for different parameters. The effect of the contrast and the roughness can be seen by the damping of the oscillations. The layer thickness determines the period of the oscillation. It is evident that the scattering pattern can be very complex depending on the number layers. On the contrary, if the contrast is small, layers can be more or less invisible in the reflectivity.

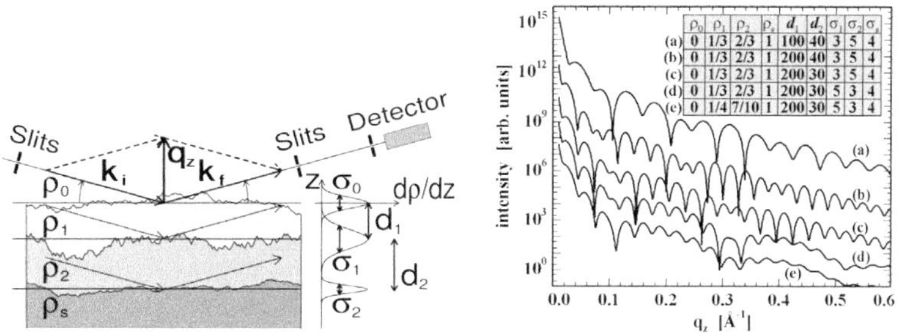

Figure 1.10 Schematic view of a two-layer sample grown on a substrate (left) and right some reflectivity curves calculated for different parameters of a two-layer system using Born approximation (right).

A special case is the scattering at multilayer systems. For multilayers, a characteristic sequence of monolayers is repeated many times. In this case, the sequence can be accounted as a unit cell and the repetition creates a crystal-like periodicity perpendicular to the surface. Consequently, Bragg reflections appear in q_z-direction where the intensity is determined by the structure factor of the layer sequence [see 21].

1.3.3 Scattering at Rough Surfaces

Surfaces with height functions depending on the in-plane coordinate $\mathbf{R} = (x, y)^T$ scatter X-rays not only in q_z-direction (reflectivity) but also off-specular [22]. In the following, the off-specular scattering of a surface with height function $h(x, y)$ following the Gaussian statistics is calculated. The surface is localized in z-direction at 0. Furthermore, the sample is considered to have a uniform density, ρ_0. In this case, with a semi-infinite sample and following Eq. (1.1), the scattered amplitude in Born approximation can be written as follows:

$$A(\mathbf{q}) \propto \rho_0 \int_{-\infty}^{\infty}\int_{-\infty}^{\infty}\int_{-\infty}^{h(x,y)} \exp(i[q_x x + q_y y + q_z z])dzdxdy$$

$$\propto \frac{1}{iq_z}\iint_S \exp(i[q_x x + q_y y])\exp(iq_z h(x,y))dxdy - 0\,.$$

For the second line, it has been assumed that the scattering from the "bottom" of the semi-infinite sample is negligible. Also, the area in the *xy*-plane is not infinity but limited by the illuminated area, *S*. The intensity can now be written as

$$I(\mathbf{q}) \propto \frac{1}{q_z^2} \iint_S \iint_{S'} e^{i\mathbf{Q}(\mathbf{R}-\mathbf{R}')} \cdot e^{iq_z[h(\mathbf{R})-h(\mathbf{R}')]} d^2R \cdot d^2R'$$

where the coordinates in the q_x, q_y-plane are identified by the vector **Q**. As $h(\mathbf{R})$ is a statistical function, appropriate averaging has to be applied. Here, we discuss the incoherent superposition of coherent scattering from small parts of the sample due to beams with very small coherence lengths. In the case of fully or largely coherent beams, further calculations must be done differently (see Chapters 12 and 13 in this book). The intensity can be rewritten as follows:

$$I(\mathbf{q}) \propto \frac{1}{q_z^2} \iint_S \iint_{S'} e^{i\mathbf{Q}(\mathbf{R}-\mathbf{R}')} . < e^{iq_z[h(0)-h(\mathbf{R}-\mathbf{R}')]} > d^2R \cdot d^2R'.$$

Here, it has been used that the averaging only depends on the distance of two points on the surface. Introducing the distance vector $\boldsymbol{\rho} = \mathbf{R}-\mathbf{R}'$, one gets

$$I(\mathbf{q}) \propto \frac{1}{q_z^2} \iint_S \iint_{S'} e^{i\mathbf{Q}\cdot\boldsymbol{\rho}} . < e^{iq_z[h(0)-h(\boldsymbol{\rho})]} > d^2\rho \cdot d^2R'$$

$$= \frac{S}{q_z^2} \iint_S e^{i\mathbf{Q}\cdot\boldsymbol{\rho}} . < e^{iq_z[h(0)-h(\boldsymbol{\rho})]} > d^2\rho.$$

with the illuminated surface *S*. For Gaussian distributed $h(\boldsymbol{\rho})$, this equation can be rewritten as follows:

$$I(\mathbf{q}) \propto \frac{S}{q_z^2} \iint_S e^{i\mathbf{Q}\cdot\boldsymbol{\rho}} e^{<\{iq_z[h(0)-h(\boldsymbol{\rho})]\}^2/2>} d^2\rho$$

$$= \frac{S}{q_z^2} \iint_S e^{i\mathbf{Q}\cdot\boldsymbol{\rho}} \cdot e^{-q_z^2/2[2<h(0)^2>-2<h(0)h(\boldsymbol{\rho})>]} d^2\rho.$$

By introducing the auto-correlation function (ACF)

$$C_{hh}(\mathbf{R}) = < h(0)h(\mathbf{R}) > = \iint h(\boldsymbol{\rho})h(\boldsymbol{\rho}-\mathbf{R}) d^2\rho \quad \text{and} \quad C_{hh}(0) = \sigma^2 \quad (1.17)$$

which is virtually identical to the PDF $g(\mathbf{r})$ of the surface, the scattered intensity from a rough surfaces is given by the following expression:

$$I(\mathbf{q}) \propto \frac{S}{q_z^2} \exp(-q_z^2\sigma^2) \iint_S \exp[q_z^2 C_{hh}(\mathbf{R})] \cdot \exp(i\mathbf{Q}\cdot\mathbf{R}) d^2R\,.$$

This equation contains not only the off-specular scattering but also the specularly scattered part, which can be eliminated by subtracting the reflectivity Eq. (1.15).

In the case of Gaussian distributed $h(\mathbf{R})$, the ACF is a symmetric function and the complex Fourier transformation becomes real. Thus, for the off-specular scattering of a single rough surface, the so-called diffusely scattered intensity [22] is

$$I_{\text{diff}}(\mathbf{q}) \propto \frac{S}{q_z^2} \exp(-q_z^2\sigma^2) \iint \left(\exp[q_z^2 C_{hh}(\mathbf{R})] - 1\right) \cos(\mathbf{Q}\cdot\mathbf{R}) d^2R. \tag{1.18}$$

To understand the meaning of this equation, in the following small $q_z\sigma$ are discussed. Using Eq. (1.18), we gain the following approximation:

$$I_{\text{diff}}(\mathbf{q}) \overset{q_z\sigma<<1}{\propto} S\exp(-q_z^2\sigma^2) \iint C_{hh}(\mathbf{R}) \cos(\mathbf{Q}\cdot\mathbf{R}) d^2R = S\exp(-q_z^2\sigma^2) P(\mathbf{Q}).$$

The Fourier transformation of the ACF $C_{hh}(\mathbf{R})$ is $P(\mathbf{Q})$ and called power spectral density (PSD). Thus, in a first-order approximation, the diffuse scattered intensity is proportional to the PSD of the surface. This is a very important result: The surface diffuse X-ray scattering is directly linked with the surface physics of a sample via the PSD [see for example 23].

Taking the Fourier transformation of statistically rough surfaces, the PSD is a measure of the number of modes in a given frequency interval. For a given surface, the frequencies and amplitudes of the modes follow from the PSD, but the phases are random. This is shown in Fig. 1.11. For two different PSDs, surfaces have been created using the same random phases, φ. The Gaussian PSD creates a smooth-looking surface, whereas the Lorentzian PSD makes a jagged surface as the high frequencies in the PSD are not so much suppressed. The respective diffuse scattering signals are also shown.

Depending on the low-frequency spectrum of the PSD, the intensity can exhibit a very narrow peak at q = 0. This is seen especially for liquid surfaces as discussed in Chapter 9. The roughness of a liquid surface is caused by capillary waves with a low-frequency

$\alpha_c \propto \lambda$. This dependence is used in X-ray optics to generate a low-pass filter. The difference between the reflectivity in Born approximation and the exact dynamical scattering theory is displayed in Fig. 1.12.

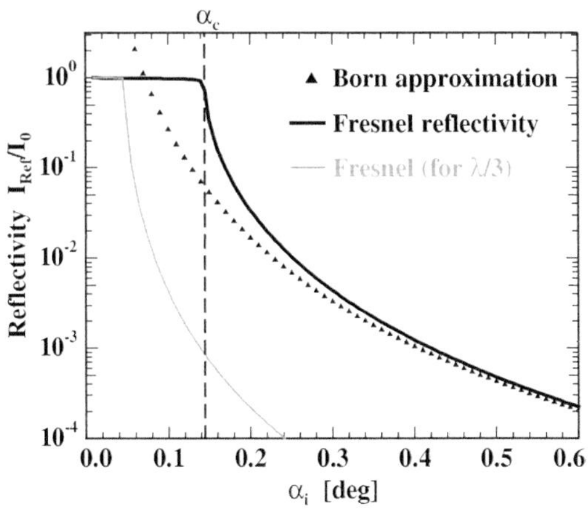

Figure 1.12 Comparison between a calculated reflectivity of flat silicon in Born-approximation and in the exact dynamical theory (Fresnel reflectivity). The black curves are calculated using a photon energy of 12.4 keV, the gray for 37.2 keV. For the Fresnel reflectivity the critical angle α_c is visible as the large angle limit of the plateau of total external reflection.

Considering Bragg reflections with $(h, k, l) \neq (0, 0, 0)$, the dynamical theory yields different results for the so-called Laue case and the Bragg case. In a simplified picture, for the Laue case, the **q**-vector of the excited Bragg peak is perpendicular to the surface normal **n**. For the Bragg case, this **q**-vector is parallel to **n** (see Fig. 1.13) [8].

Qualitatively, these two different geometries have the following consequences for the scattering: In the Laue case, the X-ray beam is reflected in transmission and in the Bragg case in reflection. Therefore, in the Laue case, the thickness of the crystal must be close to some absorption lengths $\Lambda = 1/\mu$ to allow significantly high reflection and transmission values. If the thickness of the crystal would be too large, absorption effects would reduce the reflected intensity; small thicknesses would result into low reflectivity due to small scattering volumes.

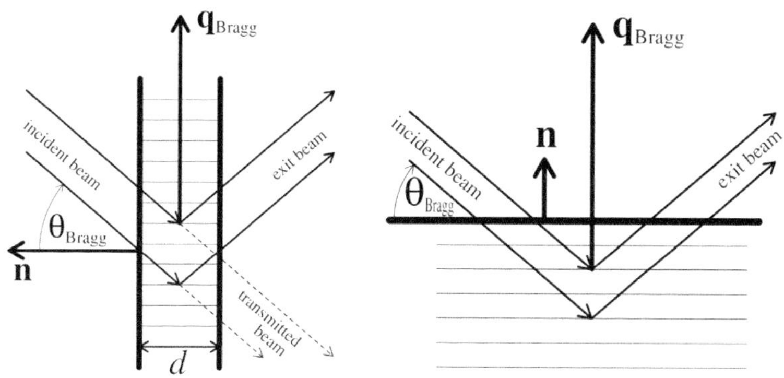

Figure 1.13 Left a sketch of the Laue-geometry at a sample with thickness *d*, right of the Bragg-geometry at a semi-infinite sample. In thin parallel lines, the lattice planes are drawn.

In the Bragg case, the crystal can be semi-infinitely thick. In both cases, interference patterns are expected for finite crystal thickness (as predicted by the so-called Pendellösung ansatz). However, in the Bragg case, the crystal thickness (when used for X-ray optics) is usually some millimeters such that the interference beating does not appear and only the envelope of the curve is visible. In the Laue case, the beating is suppressed in many cases, e.g. if the surfaces are not well polished interferences are suppressed.

Further, it turns out that the shape of the reflections and the incident angle θ depend on the reflection geometry, the wave length, and also on the crystal type. This is depicted in Fig. 1.14. It shows the Bragg reflections of a Silicon 111-reflection at different photon energies different crystal thickness and for the two geometries (Laue and Bragg).

The scattered intensity (in Fig. 1.14 given as the reflectivity of the sample) is <1 in the exact dynamical theory, and thus the intensity does not diverge as in the case using the Born approximation. Ignoring the Pendellösung effects, for perfect Laue crystals, the maximum reflectivity of 0.5 can be reached by sufficiently thick crystals and neglecting the absorption. The shape of the Bragg reflection in the Laue case is essentially Lorentzian and well centered to the Bragg angle θ_{Bragg}. For a cubic crystal and scattering as indicated in Fig. 1.13, the width of the reflection $\Delta\theta$ is given by the following equation:

$$\Delta\theta = \frac{r_e \lambda^2}{\pi V_c} \cdot \frac{2\left|S(\mathbf{q}_{\text{Bragg}})\right|}{\sin 2\theta_{\text{Bragg}}} \tag{1.20}$$

where $S(\mathbf{q}_{\text{Bragg}})$ is the structure factor from Eq. (1.9) at the Bragg position $\mathbf{q}_{\text{Bragg}}$. The crystal's unit cell volume is denoted by V_c.

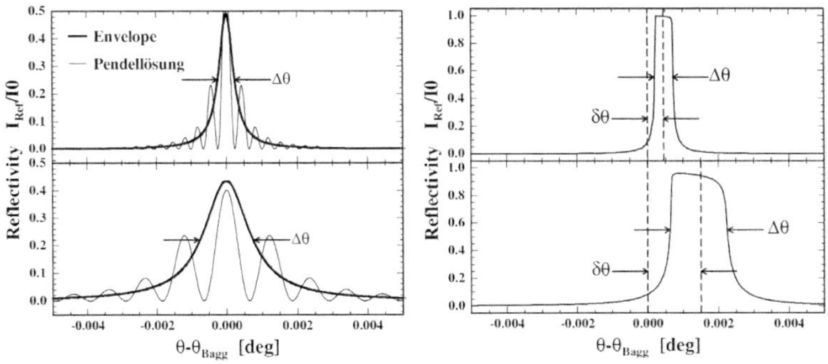

Figure 1.14 Left: a sketch of a silicon 111 Bragg reflection in Laue geometry. On the top, the crystal thickness d is 52 μm and the photon energy E = 30 keV. At the bottom, d = 18 μm and E = 0 keV. Right: a sketch of a silicon 111 Bragg reflection in Bragg geometry using a semi-infinite crystal. On the top, the photon energy is E = 30 keV and at the bottom E = 10 keV. The symbols are explained in the text.

For Bragg crystals, the reflectivity can be very close to one depending on the absorption effects. The shape differs significantly from the reflection in Laue geometry. It forms a plateau with almost constant reflectivity and steep tails and, furthermore, the position is shifted to larger values due to refraction effects. The width of the reflection $\Delta\theta$ is given by Eq. (1.20). The shift $\delta\theta$ from the Bragg position is given by

$$\delta\theta = \frac{r_e \lambda^2}{\pi V_c} \cdot \frac{\left|S(0)\right|}{\sin 2\theta_{\text{Bragg}}}\,. \tag{1.21}$$

References

1. Kroll NM, Ron A, Rostoker N (1964) Optical mixing as a plasma density probe, *Phys Rev Lett,* **13,** 83–86.
2. DuBois DF, Goldman MV (1967) Parametrically excited plasma fluctuations, *Phys Rev,* **164,** 207–222.

3. Warren BE (1990) *X-Ray Diffraction,* Dover, New York.
4. Stetsko YP, Chang S-L (1997) An algorithm for solving multiple-wave dynamical X-ray diffraction equations, *Acta Cryst A,* **53,** 28–34.
5. Batterman BW, Cole H (1964) Dynamical diffraction of X-rays by perfect crystals, *Rev Mod Phys,* **36,** 681–717.
6. Stepanov SA, Ulyanenkov AP (1994) A new algorithm for computation of X-ray multiple Bragg diffraction, *Acta Cryst A,* **50,** 579–585.
7. Als-Nielsen J, McMorrow D (2001) *Elements of Modern X-Ray Physics,* Wiley, New York.
8. Zachariasen WH (1967) *Theory of X-Ray Diffraction in Crystals,* Dover, New York.
9. Born M, Wolf E (1980) *Principles of Optics,* Cambridge University Press, Cambridge.
10. Brückel Th (2002) *Soft Matter: Complex Materials on Mesoscopic Scale* (Dhont JKG, Gompper G, Richter D, eds) (Schriften des Forschungszentrums Jülich, Vol 10, Jülich).
11. Shmueli U (2008) *International Tables for Crystallography, Volume B,* Wiley, New York.
12. Pecharsky VK, Zavalij PY (2003) *Fundamentals of Powder Diffraction and Structural Characterization of Materials,* Springer, New York.
13. Fuoss PH, Brennan S (1990) Surface sensitive X-ray scattering, *Annu Rev Mater Sci,* **20,** 365–390.
14. Bracewell RN (2000) *The Fourier Transform and its Applications,* McGraw-Hill, Boston.
15. Robinson IK (1986) Crystal truncation rods and surface roughness, *Phys Rev B,* **33,** 3830–3836.
16. Dosch H (1992) *Critical Phenomena at Surfaces and Interfaces,* Springer, Berlin, Heidelberg; *Springer Tracts in Modern Physics,* vol. 126.
17. Tolan M (1999) *X-Ray Scattering from Soft-Matter Thin Films,* Springer, Berlin, Heidelberg; *Springer Tracts in Modern Physics,* vol. 148.
18. Daillant J, Gibaud A (2009) *X-Ray and Neutron Reflectivity,* Springer, Berlin, Heidelberg.
19. Abramowitz M, Stegun I (1972) *Handbook of Mathematical Functions,* Dover, New York.
20. Parratt LG. (1954) Surface studies of solids by total reflection of X-rays, *Phys Rev,* **95,** 359–369.

21. Nitz V, Tolan M, Schlomka J-P, et al. (1996) Correlations in the interface structure of Langmuir-Blodgett films observed by X-ray scattering, *Phys Rev B,* **54,** 5038–5050.

22. Sinha SK, Sirota EB, Garoff S, Stanley HB (1988) X-ray and neutron scattering from rough surfaces, *Phys Rev B,* **38,** 2297–2311.

23. Harden JL, Pleiner H, Pincus PA (1991) Hydrodynamic surface modes on concentrated polymer solutions and gels, *J Chem Phys,* **94,** 5208–5221.

24. Braslau A, Deutsch M, Pershan PS, Weiss AH, Als-Nielsen J, Bohr J (1985) Surface roughness of water measured by X-ray reflectivity, *Phys Rev Lett,* **54,** 144–147.

25. Prince E. (2008) *International Tables for Crystallography, Volume C,* Wiley, New York.

Chapter 2

Scattering and Diffraction Beamlines at Synchrotron Radiation Sources

Oliver H. Seeck
Deutsches Elektronen Synchrotron (DESY), Notkestraße 85, 22607 Hamburg, Germany
oliver.seeck@desy.de

As has been shown in Chapter 1, X-ray scattering and diffraction methods can be used to investigate microscopic properties of samples from micrometers size down to atomic distances. Therefore, X-ray scattering is frequently used by scientists and by industry and is very often the only available method.

For good and reliable results an X-ray experiment must be performed very carefully. Most importantly, the X-ray beam has to be prepared in compliance with the needs of the experiment, which means one needs to tune the beam parameters by adequate choices of optics and X-ray sources. The instrument that is capable of tuning the X-ray beam parameters is called the beamline. The beam parameters of interest are the following:

- F [cts/s] is the flux (or intensity).

X-Ray Diffraction: Modern Experimental Techniques
Edited by Oliver H. Seeck and Bridget M. Murphy

ISBN 978-981-4303-59-0 (Hardcover), 978-981-4303-60-6 (eBook)
www.panstanford.com

- $\sigma_{x,y}$ [mm] is the size in horizontal (x) and vertical (y) direction.
- $\sigma'_{x,y}$ [mrad] is the divergence in horizontal and vertical direction.
- E [eV] is the average photon energy.
- $\Delta E/E$ [%] is the energy bandwidth.
- τ_p[fs] is the pulse length.

Further parameters are the degree of polarization and the transverse and longitudinal coherence length. They are not discussed in this chapter. At X-ray beamlines, all parameters can be more or less tuned but usually not independently. For example, maximum flux can only be achieved at a particular energy, or reducing the bandwidth means reducing the flux. To find the optimum beam parameter set, it is important to know that, depending on the experiment, all the parameters are not equally relevant. For instance for high-resolution diffraction the energy bandwidth and the divergence must be very small whereas the beam size may not be critical.

The staff members of X-ray beamlines are able to determine the optimum settings for a specific experiment and tune the beamline accordingly. However, for external users it is also very helpful to be able to estimate the beam parameters from the beamline settings. For this a basic understanding of X-ray sources and beamline components is required. Figure 2.1 displays a sketch of a beamline with the major components.

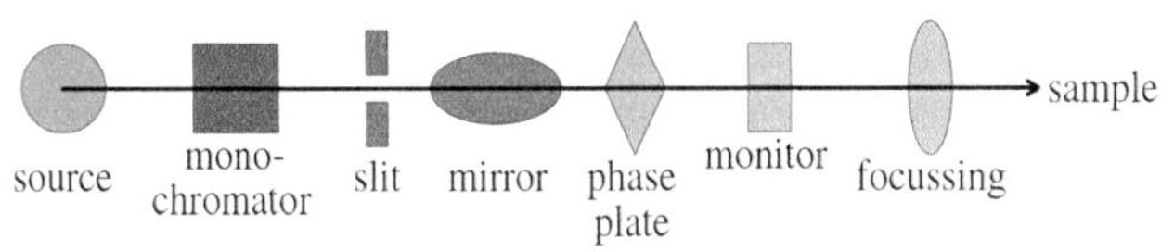

Figure 2.1 Prototype of an X-ray scattering beamline. The order of the devices is artificial. Also, some devices may not be used and others may be duplicated.

In the following sections, a selection of beamline components is explained with respect to their function in scattering experiments. In particular, it will be demonstrated how the beam parameters at the sample can be estimated. This chapter provides an overview, and therefore the theory is kept simple and sometimes qualitative.

2.1 Synchrotron Radiation Sources

The most important part of an X-ray scattering beamline is the X-ray source. In small laboratories, X-ray tubes are used. However, the beam parameters can only be varied within a very limited range. The photon energy and bandwidth are more or less given by the anode material. Also, the tube radiates with relatively large divergence and source size. Despite recent developments in optics and X-ray tube technology leading to small beamsizes or high resolution, for many challenging experiments the only feasible option is a synchrotron radiation source. They deliver intense and tunable X-ray beams that are already comparably small and well collimated at the source position and therefore well suited for high resolution measurements.

Dedicated synchrotron radiation sources are specially designed storage rings that are loaded from pre-accelerators with a regular bunch pattern of charged particles: electrons or positrons.[1] The kinetic energy of the particles is constant at a few gigaelectron volts. At this energy the electrons move with almost the speed of light. At well-defined positions in the storage ring insertion devices are installed. They force the electrons to radiate. As the electrons form bunches, the radiation is concentrated in flashes.

The ring-like structure of a synchrotron radiation source has the advantage that bunches radiate periodically as they repeatedly enter the insertion devices. The ring structure also has disadvantages: as the electron bunches are formed by particles of the same charge, the bunch size increases with time due to electrostatic repulsion. This can be partially but not fully compensated by adequate focusing. If a storage ring can maintain stable bunches at a size of several micrometers, it is called "third- generation synchrotron radiation source."

If the effects of the storage ring on the bunch shape are not acceptable, the bunch is directly guided from the accelerator to the insertion device on a linear path. After radiating, the bunch is dumped. This concept is used at "fourth- generation synchrotron radiation sources," the X-ray lasers. In the following section, different

[1]There is no theoretical difference between storage rings filled with electrons or positrons. Therefore, positrons will not be mentioned any more.

types of X-ray sources are explained in more detail. For a detailed introduction to synchrotron radiation sources, we would like to refer to the book of H. Wiedemann [1].

2.1.1 Bending Magnets

Without applying any external force, electrons travel in a straight line. To constrain them to the storage ring, magnetic fields are used. For this, every few meters along the ring bending magnets or dipole magnets are installed. With the magnetic field vector pointing down, electrons experience a Lorentz force directed to the center of the storage ring (see Fig. 2.2). This means that the path of the particles in the storage ring is actually approximated by a polygon with the bending magnets positioned at the corners.

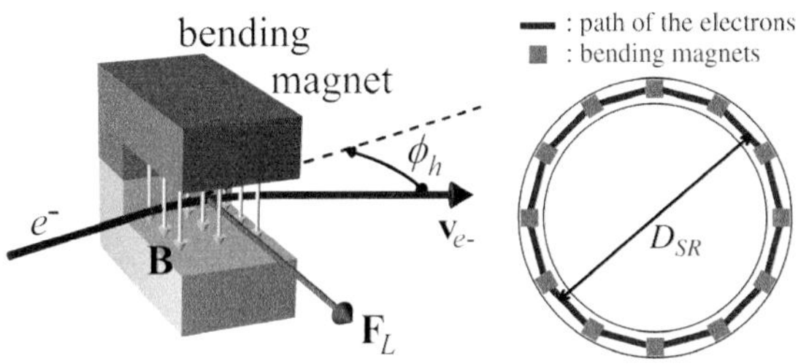

Figure 2.2 On the left is a sketch of a bending magnet with the speed of the electrons $\mathbf{v}_{e-}$, the field **B**, the resulting force $\mathbf{F}_L$ and the horizontal deflection angle ϕ_h. On the right is a sketch of a circular storage ring with the polygonal path of the electrons and the bending magnets. The diameter of the ring is denoted by D_{SR}.

At the bending magnets, the electrons are kicked by the Lorentz force to follow the storage ring curvature. This kick means acceleration. From electrodynamics, it is well known that an accelerated charged particle radiates dipole radiation [see 2]. Therefore, this radiation appears at each of the bending magnets. In the non-relativistic case, the field of the dipole radiation is concentrated on a ring around the center with the ring radius perpendicular to the acceleration direction, which is parallel to the force (see Fig. 2.3). However, at storage rings, the kinetic particle energy E_k^{e-} is some gigaelectron volts and the speed v_{e-} of the electrons, given in units of the speed of light c, can be calculated using the following equation:

$$\frac{v_{e-}}{c}=\sqrt{1-\left(\frac{E_0^{e-}}{E_0^{e-}+E_k^{e-}}\right)^2}, \tag{2.1}$$

where the rest energy of an electron $E_0^{e-} = m_0^{e-}c^2 \approx 0.51$ MeV, and the rest mass of an electron $m_0^{e-} \approx 9.1 \cdot 10^{-31}$ kg. From this follows that for, for example, 1 GeV electrons, the speed is $v_{e-} = 0.9999999 \cdot c$, which consequently means that they move in a highly relativistic manner. At this speed, the relativistic aberration of light deforms the dipole radiation to a fan which is extremely collimated toward the forward direction [3]. The opening angle ϕ of the radiation is

$$\phi=\sqrt{1-\left(\frac{v_{e-}}{c}\right)^2}=\sqrt{1-\left[1-\left(\frac{E_0^{e-}}{E_0^{e-}+E_k^{e-}}\right)^2\right]}=\frac{E_0^{e-}}{E_0^{e-}+E_k^{e-}}\overset{E_k^{e-}>>E_0^{e-}}{\approx}\frac{E_0^{e-}}{E_k^{e-}} \tag{2.2}$$

with ϕ given in radians (see Fig. 2.3) [1,3]. In the example with a particle energy of 1 GeV, the vertical opening angle is $\phi = 0.5$ mrad, and the radiation is emitted as a narrow beam tangential to the storage ring. Looking at the whole bunch rather than the single electron, the vertical opening angle of the bunch ϕ_v is basically identical to ϕ. The horizontal opening angle of the bunch ϕ_h is much wider than ϕ and is given by the deflection angle of the bending magnet, which is determined by the magnet's length and magnetic field **B**. For X-ray beamline purposes, ϕ_h is usually accounted as "infinite."

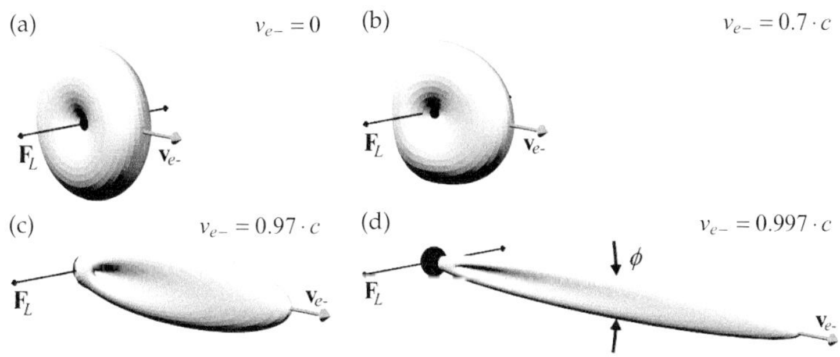

Figure 2.3 Isosurfaces of dipole radiation calculated at different $\mathbf{v}_{e-}$. The effect of the relativistic aberration creates a beam radiating in (longitudinal) forward direction. See the text for explanation of the symbols.

To estimate the pulse length τ_p of the radiation flash two contributions have to be added: the time τ_{pb} which is needed for the bunch to enter the bending magnet (defined by the bunch length) and the time τ_{pe-} of one electron traveling through the magnetic field. The time τ_{pb} is determined by the ability of the storage ring to compress the bunch in longitudinal direction. The bunch length is usually some $l_b = 30$ mm in full width half maximum (FWHM), which corresponds to a time $\tau_{pb} = l_b/c \approx 100$ ps (FWHM) [1]. An estimate of the time τ_{pe-} is more difficult. Due to its high speed, the bunch runs after the emitted radiation, which is therefore highly compressed. From this, $\tau_{pe} \approx l_d/v_{e-} - l_d/c$ follows. If one particular observation direction in the horizontal is considered, the length l_d is not identical to the length of the dipole. In fact, $l_d = R_{BM} \cdot \sin\phi$ is the length at which the two radiated cones with opening angle ϕ and bending radius R_{BM} can still overlap in observation direction (see Fig. 2.4). Finally, the time τ_{pe-} can be estimated by

$$\tau_{pe-} = \frac{m_{e-}}{2eB} \cdot \left(\frac{E_0^{e-}}{E_k^{e-}} \right)^2 \tag{2.3}$$

with the elementary charge of $e \approx 1.6 \cdot 10^{-19}$ C. Following our example with the kinetic energy $E_k^{e-} = 1$ GeV and the magnetic field 1 T, the time τ_{pe-} is of the order of 10^{-18} s. This number is extremely small and does not contribute to the pulse length $\tau_p = \tau_{pb} + \tau_{pe-}$ of one bunch.

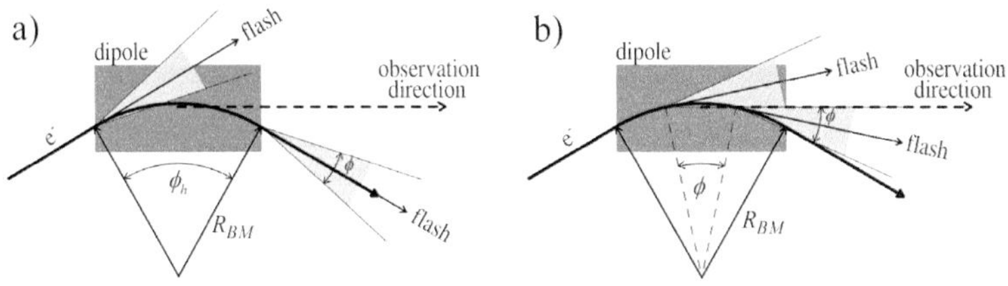

Figure 2.4 Situation at a bending magnet with an electron traveling through it. Both flashes in (a) do not coincide in observation direction and do not contribute to the time τ_{pe-}. In (b) the situation is displayed which is used to calculate this time.

The short time scale of τ_{pe-} has immediate consequences for the energy distribution of the emitted radiation. From Heisenberg's uncertainty principle, $\Delta E \cdot \Delta t > 2\hbar$, with $\hbar$ being the Planck-constant, follows that the photon energy is uncertain within a FWHM of at least 1 keV, using again that $\Delta t = \tau_{pe-} \approx 10^{-18}$ s at $E_k^{e-} = 1$ GeV. After more careful calculations, the so-called critical energy for bending magnet radiation can be derived

$$E_c = \frac{3e\hbar B}{2m_{e-}} \cdot \left(\frac{E_k^{e-}}{E_0^{e-}} \right)^2 \quad \text{reducing to} \quad E_c[\text{keV}] = 0.6650 \cdot (E_k^{e-}[\text{GeV}])^2 \cdot B[\text{T}] \tag{2.4}$$

in convenient units[4]. The critical energy splits the broad energy spectrum of the photons into two parts of equal area. E_c is in the range of some keV for standard storage rings with particle energies of some GeV. Having in mind that the radiation is very well collimated in vertical direction, bending magnets are very good X-ray sources for a wide energy range, from some keV to some tens of keV, and are used at dedicated diffraction beamlines at many synchrotron radiation facilities.

For a high resolution X-ray experiment, the angular flux density Φ_{BM} is important. Most beamlines are built along the central radiation cone. In this case, the radiation of a bending magnet is given by

$$\begin{aligned} \Phi_{BM}\left[\frac{\text{cts/s}}{\text{mrad}^2 \cdot (0.1\%\text{BW})}\right] &= \frac{d^3 F_{BM}}{d\phi_h \cdot d\phi_v \cdot (\Delta E/E)} \\ &= 1.33 \cdot 10^{13} \cdot (E_k^{e-}[\text{GeV}])^2 \cdot I[\text{A}] \cdot H_2\left(\frac{E}{E_c}\right) \end{aligned} \tag{2.5}$$

where I is the storage ring current and 0.1% BW is the unit of the energy bandwidth $\Delta E/E$ at photon energy E. The total flux is denoted as F_{BM}. The function $H_2(x)$ is plotted in Fig. 2.5 and is basically the flux distribution function of a bending magnet. The divergence of the beam in vertical direction depends on the photon energy and is therefore not simply $\phi = E_0^{e-}/E_k^{e-}$. For the vertical divergence σ'_y

$$\sigma'_y[\text{mrad}] = 0.408 \cdot \frac{C(E/E_c)}{E_k^{e-}[\text{GeV}]} \tag{2.6}$$

can be derived where the function $C(x)$ is again plotted in Fig. 2.5, and σ'_y is a root mean square (rms) value. It follows that for high photon energies, the vertical beam divergence is smaller than for low energies.

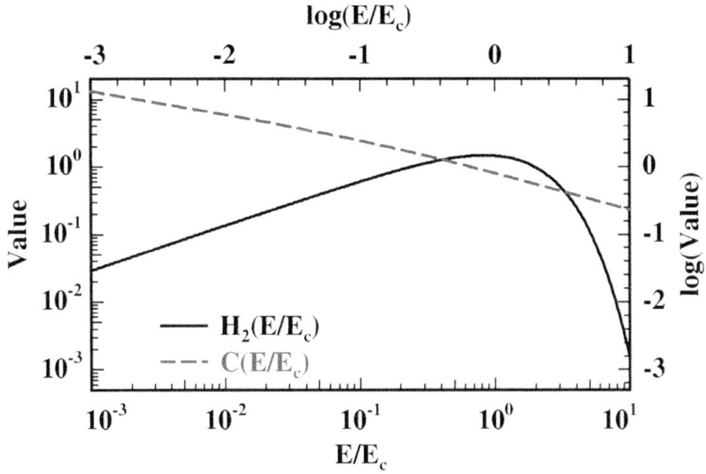

Figure 2.5 The functions H_2 und C plotted versus E/E_c.

As an example, the beam parameters will be calculated for a beamline at a bending magnet of B = 0.4 T at a storage ring with E_k^{e-} = 6 GeV particles and loaded with I = 100 mA = 0.1 A current. The beam defining slit in front of the sample, at D_{sl} = 30 m distance from the source, should have the vertical size $d_{sl,y}$ = 0.2 mm and $d_{sl,h}$ = 6 mm horizontally. The photon energy in the experiment should be E = 12 keV with a band width of $\Delta E/E = 10^{-4} = 0.01\%$ which would be typical for a diffraction beamline. Focusing is not implemented.

The first step is to calculate the critical energy. Equation (2.4) yields E_c = 9.58 keV. This means a value of $\log(E/E_c) = \log(1.25) \approx 0.1$ and therefore from Fig. 2.5 $\log H_2(E/E_c) = 0.15 \Rightarrow H_2(E/E_c) = 10^{0.15}$ = 1.4. From this, the angular flux density $\Phi_{BM} \approx 6.7 \cdot 10^{13}$ (cts/s)/(mrad^2 0.1% BW) can be calculated.

The flux density does not include the information about the beam divergence, which is defined as the acceptance of the last slit in front of the sample. From the values of the slit given above, the vertical rms-divergence is $\sigma'_{y,sl} = \arctan[(d_{ls,y}/2)/D_{sl}] \approx \arctan(10^{-4}\text{m}/30\text{m}) \approx 0.003$ mrad and the rms of the horizontal divergence $\sigma'_{y,sl} = \arctan[(d_{ls,y}/2)/D_{sl}] \approx 0.1$ mrad. Now, the vertical divergence, which is defined by the slit, must be compared with the limiting natural divergence σ'_y. For the above bending magnet, the value of $\sigma'_y \approx 0.043$ mrad results with the estimated value of $C(E/E_c) \approx 10^{-0.2}$ = 0.63 using Fig. 2.5. To estimate the flux after the slit, the smaller value of $\{\sigma'_y, \sigma'_{y,\,sl}\}$ has to be used here: $\sigma'_{y,\,sl}$. Finally, taking into account the desired energy bandwidth of $10^{-4} = 0.1 \cdot (0.1\%\text{BW})$ the flux after the slit can now be estimated by correcting the flux density with the according numbers and one gets $F_{BM} \approx \Phi_{BM} \times 0.1$ mrad $\times$ 0.003 mrad $\times$ 0.1(0.1%BW) = $2 \cdot 10^9$ cts/s.

Not taken into account in this example is the vertical size of the bunch which causes an extended X-ray source. With a vertical source size (rms) of σ_y = 30 μm (as a typical source size at bending magnets at the European Synchrotron Radian Facility ESRF [5]) the natural beamsize in front of the slit would be $\sigma_{y,\,sl} = D_{SL} \cdot \tan(\sigma'_y) + \sigma_y$ = (210+30)μm. This means that the beam is 15 percent larger than calculated for a point-like source. Therefore, the beam intensity is additionally reduced by this factor.

In summary, the flux is larger by a factor 10 ... 20 than compared to modern lab sources, but the bending magnet radiates at a wide energy range. The flux can only be increased by higher ring current or appropriate focusing. Higher magnetic field would not necessarily mean higher intensity, as the field only shifts the critical energy E_c.

2.1.2 Wigglers

The limited flux is not the only problem of bending magnet beamlines. Also, the bending radius of the magnet and therefore the magnetic field is determined by the storage ring curvature. This means that at large storage rings, due to the small curvature, only small magnetic fields can be used at each bend and that the critical energy E_c is small (in the range of some keV). Quite some applications, especially, in materials science require high photon energies, which can only be achieved at sources with large E_c. To get around this problem, since the late 1970s, bending magnets with alternating field direction have been constructed (see Fig. 2.6). These insertion devices are called wigglers [6,7]. As one wiggler period consists of two bends, the intensity of the wiggler radiation in forward direction is a factor of $2N_W$ larger than at a single bending magnet, where N_W is the number of wiggler periods. Usually, the effective field of a wiggler is zero due to the alternating fields so that the path of the electrons after the device is identical to the initial path. In contrast, the local field can be very strong.

As an example, a wiggler with a local field B = 2 T and N_W = 35 periods is considered. Otherwise, the parameters of the bending magnet example hold. The critical energy is E_c = 40 keV. From this, at E = 12 keV, $\log(E/E_c) \approx -0.52$ and $\log(H_2(E/E_c)) \approx 0 \Rightarrow H_2(E/E_c) \approx 1$ follows. The flux density of the wiggler beam is $\Phi_W = 2N_W \cdot \Phi_{BM} \approx 3.6 \cdot 10^{15}$ (cts/s)/(mrad^2 0.1% BW) using Eq. (2.5) for the calculation of Φ_{BM}. This would finally lead to a flux of the unfocused beam of $F_W \approx 1 \cdot 10^{11}$ cts/s after the slit. Accordingly, the flux is a factor of 50, larger than compared to that of the radiation from single bending magnet. The reason that this factor is smaller than the expected value of $2N_W = 70$ is due to mismatch between photon energy (E = 12 keV) and critical energy (E_c = 40 keV). For 40 keV photons, the flux after the slit would be $F_W \approx 1.4 \cdot 10^{11}$ cts/s which is exactly within expectations.

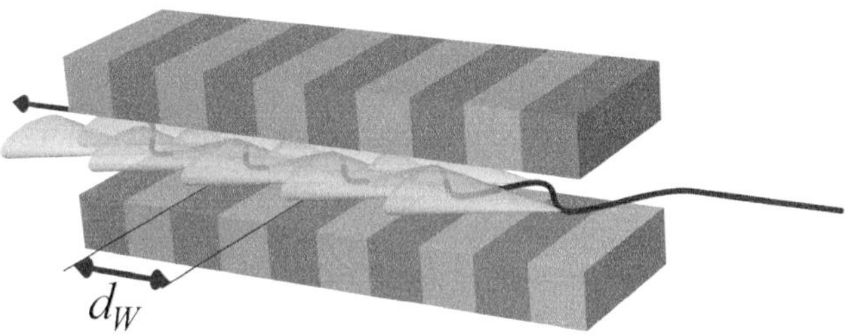

Figure 2.6 Principle of a wiggler. Alternating bending magnets create a wiggling path of the electrons with each kink creating an X-ray fan. The period is denoted by d_W. The number of periods is $N_W = 5$.

2.1.3 Undulators

As has been mentioned in the previous section, the single X-ray fans from each of the wiggler dipoles strongly overlap. For this reason, the radiated intensities of each of the bends have to be summed up. However, this means that the single emitted waves are added incoherently without taking into account interference effects. Especially at old wiggler devices, the magnetic structure is rather imperfect such that incoherent superposition of the waves may be valid.

At perfectly periodic magnetic structures, such as undulators [8], some of the emitted waves are fully in phase and therefore interfere constructively, whereas all other emitted waves are incoherent and form the wriggler background (see Fig. 2.7). In the case for coherent waves with constructive interference, the amplitudes rather than the intensities have to be summed up. This has great consequences for the flux F_U of the undulator. In the case of constructive interference, the flux is given by $F_U = |\sum A_U|^2 = |2N_W A_U|^2$ for the coherent part of the amplitude A_U. For incoherent waves the wiggler result holds, thus $F_U = \sum|A_W|^2 = 2N_W|A_U|^2$. Thus, at an undulator the amplification factor for the constructively interfering waves is $4N_W^2$, which is an enormous increase compared to wiggler radiation. It is important to note that only the waves which are emitted from one and the same electron at the bends of the undulator can interfere [1]. The waves radiated from different electrons are not in phase and cannot interfere, as in storage rings the electrons in a bunch are uncorrelated. This is in contrast to X-ray lasers, as will be shown in the next section.

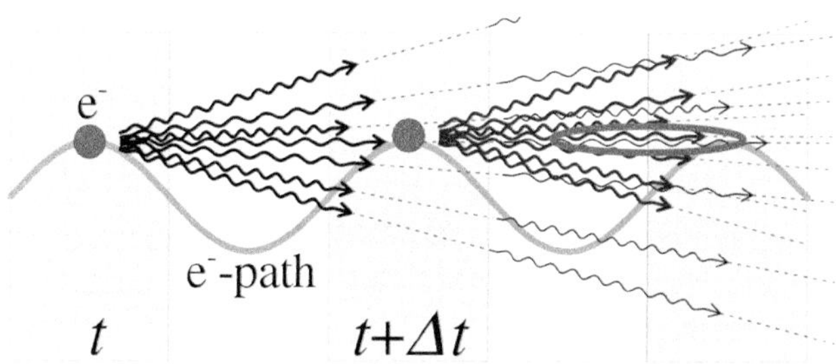

Figure 2.7 Principle of an undulator. One electron radiates at different times and the radiation overlaps. Only in forward direction and for a particular wavelength (see ellipsis on the right side), the two waves interfere constructively.

It is clear that constructive interference appears only if all waves have the same wavelength with a very small bandwidth. Furthermore, the directions of the photons have to be virtually identical for constructive interference. This has immediate implications on the angular divergence of the harmonics.

At the wavelengths of constructive interference, the emitted flux peaks and the so-called undulator harmonics appear in the undulator spectrum. As can be seen in Fig. 2.8, the harmonics raise at integer multiples of the fundamental energy. Additionally, the odd harmonics are more narrow and intense than the even ones. The energy of the fundamental and, therefore, also of all other harmonics depends on the **B**-field of the undulator, which is usually adjusted by the gap between the bottom and the top part of the magnetic structure.

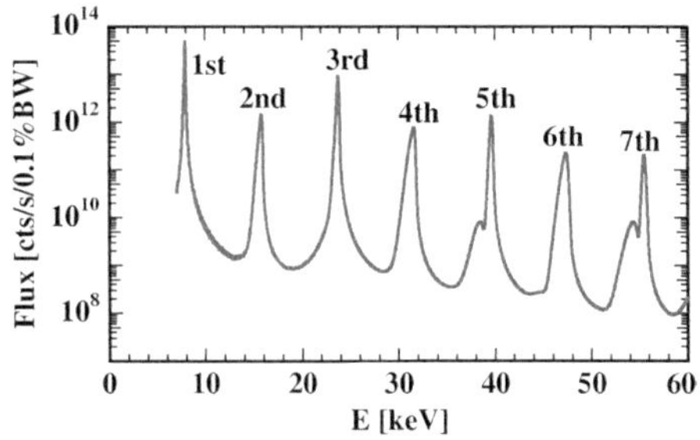

Figure 2.8 Typical undulator spectrum measured at a fixed gap. The odd harmonics result in sharp and intensive peaks. The even harmonics are much smaller and the bandwidth is larger.

The calculation of the flux, divergence and energy bandwidth of the undulator harmonics can be carried out using a set of equations. For this, the so-called deflection parameter $K \approx 0.934 \cdot \lambda_U\,[\text{cm}] \cdot B[\text{T}]$ (also called K-parameter) is introduced, which is determined by the **B**-field and the period length λ_U of the undulator [9]. The upper limit of the K-parameter usually has a value between 1 and 5 and is given by the maximum field and, therefore, by the minimum gap distance.

From the K-parameter, the photon energy E_n of the nth odd harmonics can be calculated via

$$E_n[\text{keV}] = \frac{0.95 \cdot n \cdot (E_k^{e-}[\text{GeV}])^2}{\lambda_u[\text{cm}](1+\text{K}^2/2)} \quad \Rightarrow \quad K = \sqrt{\frac{1.9 \cdot n \cdot (E_k^{e-}[\text{GeV}])^2}{\lambda_u[\text{cm}] \cdot E_n[\text{keV}]} - 2} \tag{2.7}$$

for beams in forward direction [4]. For beams which are not in forward direction the photon energy generally becomes softer. The bandwidth in forward direction can be estimated by

$$\frac{\Delta E_n}{E_n} \approx \frac{1}{nN_U} \tag{2.8}$$

with an undulator of N_U magnetic periods. The angular flux density Φ_U in forward direction is then given by [4]

$$\Phi_U\left[\frac{\text{cts/s}}{\text{mrad}^2 \cdot (0.1\%\text{BW})}\right] = \frac{d^3F_U}{d\phi_h \cdot d\phi_v \cdot (dE/E)}$$
$$= 1.744 \cdot 10^{14} \cdot N_U^2 \cdot (E_k^{e-}[\text{GeV}])^2 \cdot I[\text{A}] \cdot f_n(K). \tag{2.9}$$

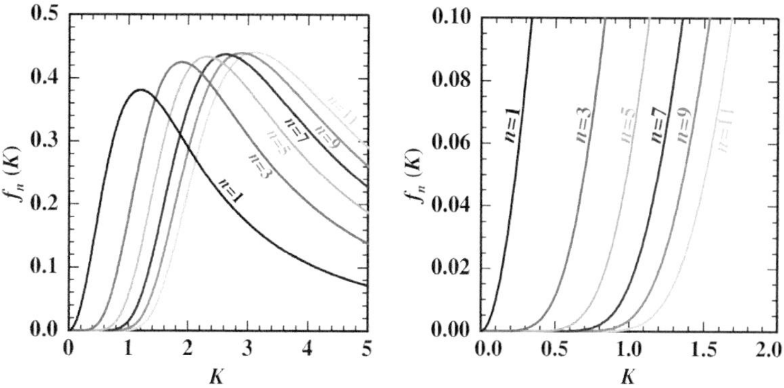

Figure 2.9 The functions $f_n(K)$. The left graph is a magnification of the low K range.

The functions $f_n(K)$ are displayed in Fig. 2.9 for odd integer numbers n which denote the undulator harmonics. In most cases the X-ray beam from the odd harmonics is superior to that of the even harmonics in terms of energy bandwidth, flux and divergence. Therefore, even harmonics are not discussed further.

As stated earlier, to allow interference the undulator radiation must be basically parallel and should be, therefore, much narrower when compared to the beam of wigglers or bending magnets. The rms of the natural divergence of an undulator harmonics can be estimated by [4]

$$\sigma'_r[\text{rad}] \approx \frac{E_0^{e-}}{E_k^{e-}}\sqrt{\frac{1+K^2/2}{4nN_U}} \approx \sqrt{\frac{\lambda}{2N_U\lambda_U}} \tag{2.10}$$

with the wavelength λ. From Eq. (2.11), the value of σ'_r is identical in horizontal or vertical direction, which is, however, idealized. In fact, the electron bunch parameters have strong influence on the beam divergence. The real divergences of the X-ray beams σ'_x and σ'_y are calculated by a convolution of the natural divergence σ'_r with parameters from the electron bunch. One finally gets

$$\begin{aligned}\sigma'_x[\text{rad}] &\approx \sqrt{\frac{\varepsilon_x}{\beta_x}+\left(\frac{E_0^{e-}}{E_k^{e-}}\right)^2\cdot\frac{1+K^2/2}{4nN_U}} \approx \sqrt{\frac{\varepsilon_x}{\beta_x}+\frac{\lambda}{2N_U\lambda_U}}\\ \sigma'_y[\text{rad}] &\approx \sqrt{\frac{\varepsilon_y}{\beta_y}+\left(\frac{E_0^{e-}}{E_k^{e-}}\right)^2\cdot\frac{1+K^2/2}{4nN_U}} \approx \sqrt{\frac{\varepsilon_y}{\beta_y}+\frac{\lambda}{2N_U\lambda_U}}\end{aligned} \tag{2.11}$$

where σ'_x and σ'_y are virtually identical to $d\phi_h$ and $d\phi_v$ from Eq. (2.9). Here, $\varepsilon_{x,y}$ is the emittance of the storage ring, and $\beta_{x,y}$ is the betatron function, which is a property of the storage ring at the position of the undulator [1].

As an example, the beam parameters will be calculated for a beamline at an undulator with a maximum K-parameter of $K_{\max} = 1.8$ and a period length of $\lambda_U = 3$ cm with $N_U = 100$ periods. A model storage ring with an emittance of $\varepsilon_x = 4\text{nm} \cdot \text{rad}$ and $\varepsilon_y = 0.04\ \text{nm} \cdot \text{rad}$ is filled with electrons at $E_k^{e-} = 6$ GeV and loaded with $I = 100$ mA current. The betatron functions at the undulator are $\beta_x = 16$ m/rad and $\beta_y = 4$ m/rad. The beam defining slit in front of the sample, at $D_{sl} = 30$ m distance from the source, should have the vertical size of $d_{sl,v} = 0.2$ mm and horizontal size of $d_{sl,h} = 6$ mm. The photon energy should be $E = 30$ keV with a band width of $\Delta E/E = 10^{-4} = 0.01\%$. Again, focusing is not implemented.

For this example, we start to calculate the possible K-parameters. Considering the fundamental harmonics $n = 1$, Eq. (2.7) results in an invalid K-parameter. Therefore, the first harmonics of the undulator radiation cannot be used. The 3rd harmonic yields $K_{n=3} = 0.53$ and the fifth harmonic yields $K_{n=5} = 1.34$. Further, harmonics are not allowed as the K-parameter would exceed $K_{\max} = 1.8$. The corresponding functions f_n are $f_3(K_{n=3}) = 0.015$ and $f_5(K_{n=5}) = 0.20$ (see Fig. 2.9). From this follows that the angular flux densities are $\Phi_{U,n=3} = 9.4 \cdot 10^{16}$ (cts/s)/(mrad^2 0.1% BW) and $\Phi_{U,n=5} = 1.3 \cdot 10^{18}$ (cts/s)/(mrad^2 0.1% BW) using Eq. (2.9). The next step is to calculate the divergences using Eq. (2.11): $\sigma'_{x,\,n=3} = 0.016$ mrad, $\sigma'_{x,\,n=5} = 0.016$ mrad, $\sigma'_{y,\,n=3} = 0.004$ mrad and $\sigma'_{y,\,n=5} = 0.0037$ mrad. The horizontal acceptance of the slit (see Section 2.1.1) is $\sigma'_{x,\,sl} = 0.1$ mrad and therefore much larger than the horizontal beam divergence. In vertical direction the value of $\sigma'_{y,\,sl} = 0.003$ mrad is smaller than the natural vertical divergence of the beam. Finally, knowing from Eq. (2.8) that the undulator bandwidths of $\Delta E/E_{n=3} = 3 \cdot 10^{-3}$ and $\Delta E/E_{n=5} = 2 \cdot 10^{-3}$ are larger than the required value of 10^{-4}, the results for the flux after the slit are

$$F_{U,n=3} \approx \Phi_{U,n=3} \cdot 0.016\ \text{mrad} \cdot 0.003\ \text{mrad} \cdot 0.1(0.1\%\text{BW}) = 4.5 \cdot 10^{11}\ \text{cts/s}.$$

$$F_{U,n=5} \approx \Phi_{U,n=5} \cdot 0.016\ \text{mrad} \cdot 0.003\ \text{mrad} \cdot 0.1(0.1\%\text{BW}) = 6.2 \cdot 10^{12}\ \text{cts/s}.$$

For the fifth harmonic, the flux is 1.5 orders of magnitude higher as compared to a wiggler and 3.5 orders of magnitude higher as compared to the flux at a bending magnet. Furthermore, in horizontal direction the slit is only partially illuminated. With the natural horizontal divergence of $\sigma'_{x,n} \approx 0.016$ mrad, the undulator beam at D_{sl} is only 1 mm wide. This means that a slit size of $d_{sl,h} = 1$ mm would be perfect to shape the X-ray beam without cutting much intensity.

Similar to the case for the bending magnet, the non-zero source size means larger beams at the experiment which reduces the flux after the slit accordingly. This will be discussed in more detail in the section "brilliance" later in this chapter.

2.1.4 Undulators at X-Ray Free Electron Lasers

As explained in the previous section, at a storage ring the undulator radiates intensity which is amplified at particular wavelengths. This is due to constructive interference of the wave fields emitted by one electron at different bends in the undulator. The electrons in the bunch are uncorrelated, which means that the radiation of different electrons adds up incoherently. A further gain in radiated intensity can be achieved by creating structured bunches such that the emitted radiation of all electrons at every bend is in phase. In this case, the radiation at the experiment would be not only extremely intense but also perfectly coherent and laser-like.

Fourth-generation synchrotron radiation sources such as X-ray free-electron lasers (XFELs) are linear accelerators [e.g., 10]. This means that the electron bunches cannot be reused after each turn such as at storage rings. However, there are also advantages in this concept. Each bunch is "fresh" with optimum parameters concerning the size and energy distribution, and the undulator length is not limited by the curvature of the storage ring.

On entering the undulator, the electrons in the bunch radiate according to the equations in the previous section. At the undulator harmonics with the energy bandwidth $\Delta E/E_n$, strong amplification of the waves occurs. As the bunch travels further, the radiation field $\mathbf{E}_{ph}$ becomes very strong. The electrons interact with $\mathbf{E}_{ph}$ by aligning along this field. However, as $\mathbf{E}_{ph}$ consists from many wavelengths with the distribution $\Delta E/E_n$, the aligning process would be more or less suppressed at some point. However, there is a certain probability that some modes are stronger than others. The stronger modes have high capability to modulate the bunch which, as a consequence, radiates stronger at this particular wavelength. If the undulator is very long, finally one mode dominates the process and laser-like radiation is achieved (see Fig. 2.10). This whole process is called self-amplified spontaneous emission (SASE) [11]. As the SASE principle as a statistical process and the radiation develop from noise, the emitted X-rays differ from bunch to bunch in intensity and in wavelength. This problem can be overcome by seeding the XFEL with a predefined wave field [12].

As the radiation from an XFEL is made up of a huge number of coherent electrons, the intensity of each flash is expected to be extremely high and again many orders of magnitude larger as compared to undulator radiation. Typical beam parameters for a high energy XFEL are E = 10 keV, $\Delta E/E$ = 0.1%, pulse duration τ_p = 100 fs, flux per pulse I_p = 5.10^{12} cts, beamsize (rms) at source $\sigma_x = \sigma_y$ = 50 μm and beam divergence (rms) $\sigma'_x = \sigma'_y$ = 0.5 μrad. With this already in one XFEL pulse the flux of a 3rd generation source undulator integrated over one second is achieved. Also, the divergence is smaller and, even more importantly, isotropic in x and y. In summary, XFELs deliver extremely short X-ray pulses with very high flux and can be used for time resolved diffraction and imaging measurements. However, radiation damage is an issue.

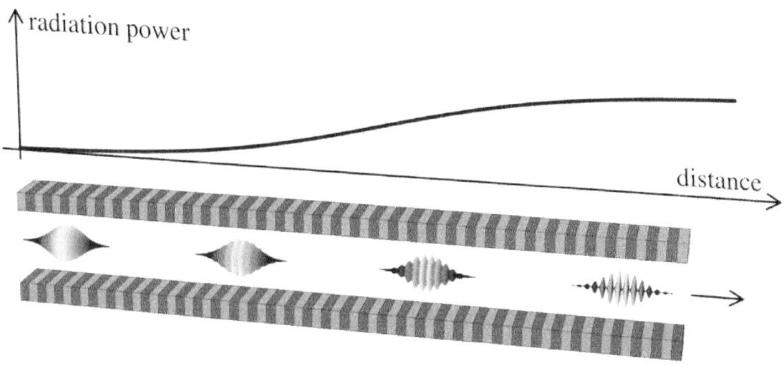

Figure 2.10 Principle of an XFEL. In a very long undulator, the bunch is structured due to the radiation field. This process is self-amplifying.

2.2 Brilliance

It has been shown that all sources can be described by a few set of parameters. The flux F is a measure of the number of photons and the energy E of the average energy of the photons which are distributed with the bandwidth $\Delta E/E$. For the focusing capability and the resolving power the divergence $\sigma'_{x,y}$ and the source size $\sigma_{x,y}$ is important.

From Liouville's theorem [13] follows that an ensemble Ξ of particles takes a constant phase space. The phase space is spanned by the three-dimensional position and momentum. For photons emitted from synchrotron radiation sources, the momentum is related with the angle and the position with the source size. Thus, the combination of source divergence and size can be considered as constant. As a consequence, the *size* of the phase space can only be tuned by cutting an appropriate smaller ensemble out of the original Ξ. In contrast, the *shape* of the phase space can be modified without changing the number of particles.

To account for the Liouville Theorem and for the conservation of energy, the so-called brilliance or spectral brightness B_{ph} has been introduced [4]. The brilliance

$$B_{ph} = \frac{\text{Number of photons}}{\text{Divergence}\cdot\text{Source Size}\cdot\text{Bandwidth}\cdot\text{Time}} = \frac{\Phi}{\sigma_x[\text{mm}]\sigma_y[\text{mm}]} \tag{2.12}$$

stands for the number of photons in phase space, energy bandwidth and time, emitted from the X-ray source. X-ray optics can tune the beam but cannot increase the brilliance.

In the following section, the source size will be discussed in more detail as it determines B_{ph}. The photon source size is not only given by the electron bunch size, but it is convoluted with optical parameters. For an undulator source, the rms values of the photon beam is determined by (see [14])

$$\begin{aligned}
\sigma_x &\approx \sqrt{\varepsilon_x\beta_x + \left(\frac{E_0^{e-}}{E_k^{e-}}\right)^2 \cdot \frac{N_U\lambda_U^2(1+K^2/2)}{16\pi^2 n}} \approx \sqrt{\varepsilon_x\beta_x + \frac{2\lambda N_U\lambda_U}{16\pi^2}} \\
\sigma_y &\approx \sqrt{\varepsilon_y\beta_y + \left(\frac{E_0^{e-}}{E_k^{e-}}\right)^2 \cdot \frac{N_U\lambda_U^2(1+K^2/2)}{16\pi^2 n}} \approx \sqrt{\varepsilon_y\beta_y + \frac{2\lambda N_U\lambda_U}{16\pi^2}}
\end{aligned} \tag{2.13}$$

Following the example from the undulator section, the rms source size would be $\sigma_{x,n=5}$ = 250 µm horizontally and $\sigma_{y,n=5}$ = 13 µm vertically. With an angular flux density of $\Phi_{U,n=5} = 1.3 \cdot 10^{18}$ (cts/s)/(mrad^2 0.1% BW), and using Eq. (2.12) and the rms source sizes, the brilliance of the undulator beam at the fifth harmonics would be $B_{ph,\,n=5} = 1.3 \cdot 10^{18}/0.25/0.013 = 4 \cdot 10^{20}$ (cts/s)/(mm^2 mrad^2 0.1% BW).

This value of the brilliance is averaged over 1 second. However, due to the pulsed structure of the radiation, the so-called peak brilliance B_{peak} of the radiation differs from B_{ph}. For the peak brilliance, the photon-less time between the pulses is not taken into account. To estimate B_{peak} let us assume a storage ring with circumference of 900 m and with a filling of 300 bunches. This corresponds to X-ray flashes with a repetition rate of 10^8 flashes per second for bunches traveling with the speed of light. As mentioned earlier in this chapter, one flash has the duration of $\tau_{ph} \approx 100\text{ps} = 10^{-10}$ s. Summing over all 10^8 flashes means an effective time of 10^{-2} s, which is just 1 percent of 1 second. Dividing B_{ph} by 10^{-2} we get $B_{peak} = 4 \cdot 10^2$ (cts/s)/(mm^2 mrad^2 0.1% BW) for the peak brilliance.

The number of photons N_{bunc} emitted by one bunch can be estimated using the peak brilliance B_{peak}, and the beamsize $\sigma_{x,\,y}$, the divergence $\sigma'_{x,y}$ and the pulse duration τ_{ph} by $N_{peak} = B_{peak} \cdot \sigma_x \cdot \sigma_y \cdot \sigma'_x \sigma'_y \cdot \tau_{pb}$. Following the example (using $\sigma'_{x,\,n=5} = 0.016$ mrad and $\sigma'_{y,\,n=5} = 0.0037$ mrad from the previous sections), the number of photons expected in one flash and 0.1 percent bandwidth is $N_{bunch} \approx 7.5 \cdot 10^5$ photons. This is a very low number and for this reason single pulse experiments at third generation source are very difficult to conduct. This is different at XFELs, where peak brilliance is approximately 10^{12} (cts/s)/(mm^2 mrad^2 0.1% BW) [see Ref. 15]. Taking the divergence and source size from the XFEL section the number of photons in one pulse and 0.1 percent bandwidth is $N_{\text{peak,XFEL}} \approx 5 \cdot 10^{12}$ photons. Consequently, XFELs are the only available X-ray source which can be used for single pulse diffraction measurements.

The brilliance is also useful to estimate the focusing or collimation capability of X-ray optics. For example, on focusing, the beamsize is reduced but unfortunately this goes along with a larger divergence, since the brilliance remains constant. Similar arguments hold for collimation[2] of the beam, which can never be perfect in the sense that the beam becomes exactly parallel.

[2]Collimation means reducing the beam divergence to virtually zero angle.

2.3 Beamline Optics

The source is the device which determines the brilliance of the X-ray beam. By further optics, the beam shape, bandwidth and divergence can be tuned. If necessary the brilliance has to be reduced (usually by reducing the flux).

For high-resolution X-ray scattering applications, it is mandatory to use a very well-defined bandwidth. For this, the undulator spectrum is tuned with the gap such that the most intensive harmonics matches the desired photon energy. For the diffraction experiment, all other harmonics have to be suppressed. Also, the natural bandwidth of the chosen harmonics given by Eq. (2.8) is not sufficiently small and has to be decreased. Additionally, it may be necessary to reduce the divergence or beamsize.

The most import device that is installed at every diffraction beamline is the monochromator. It is usually built as a fixed exit double crystal monochromator (DCM), using certain Bragg reflections to pick photon energies with small bandwidth (see Fig. 2.11). One disadvantage of the crystal monochromator is that it does not only select a single energy but also selects the integer harmonics which match the Bragg condition. To estimate the flux throughput and the resolving power of the DCM, the so-called DuMond [16] diagrams can be used. They are useful to present the characteristics of an X-ray device in energy-momentum space where the momentum is equivalent to the divergence of the beam. For each of the optical devices, a rhomb-like shape can be drawn with the area symbolizing the reflecting part. In a DuMond diagram several different devices can be overlaid. The intersection of all rhombuses is the throughput of the beamline. In the following section, an example will be used to demonstrate this.

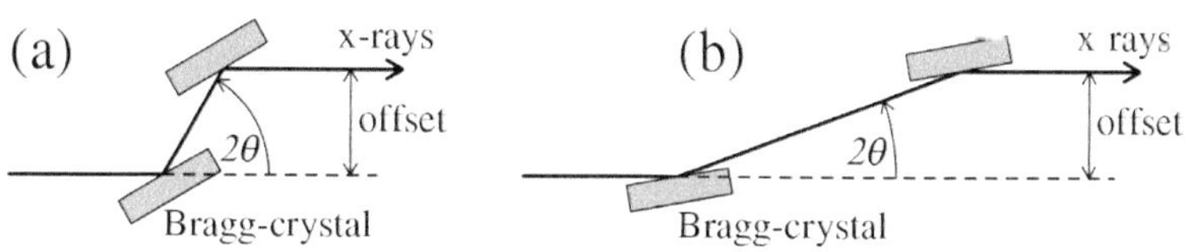

Figure 2.11 Principle of a double crystal monochromator with fixed offset. (a) at low photon energy and (b) at high energy.

Let us consider the example of the undulator section with a beam of $E_{n=5}$ = 30 keV and using the fifth harmonic of the undulator spectrum. We will use a quite complicated schematic optical setup to explain the single devices in the DuMond diagrams. The setup is as follows (see Figure 2.12 undulator, silicon double crystal monochromator (111 Bragg reflection, vertical scattering geometry, offset pointing up), vertical collimation mirror, silicon double crystal monochromator (311 Bragg reflection, vertical scattering geometry, offset pointing down), and silicon crystal monochromator (111 Bragg reflection, horizontal scattering geometry).

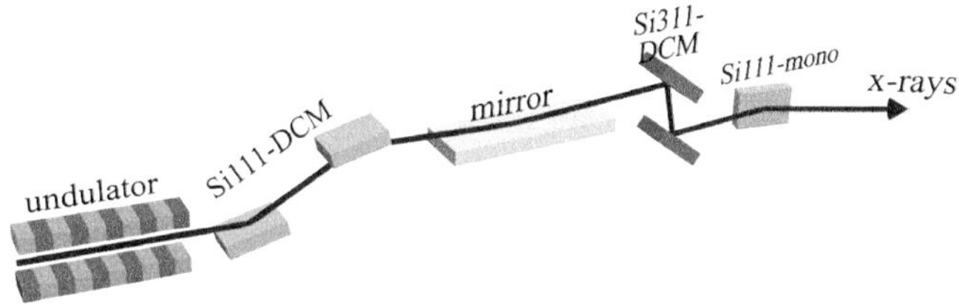

Figure 2.12 Example beamline as described in the text.

The first step is to create the DuMond diagram for the undulator. For this, the energy axis is centered around $E_{n=5}$ = 30 keV and the divergence axis around 0. The area of photon flux is a rectangle (see Fig. 2.13). The half-height is given by $\Delta E/E_{n=5} = 2 \cdot 10^{-3}$, which corresponds to ΔE = 60 eV. The vertical half-width is given by $\sigma_{y,n=5}$ = 0.0037 mrad = 3.7 μrad. This rectangle contains an integrated flux $F_{U,n=5} = \Phi_{U,n=5} \cdot \sigma'_{x,n=5} \cdot \sigma'_{y,n=5} \cdot \Delta E/E_{n=5}(0.1\%BW) \approx 10^{14}$ cts/s from the whole 5th harmonic. In the DuMond diagram, it has an area of $A_{ph} = 7.4 \cdot 120\mu\text{rad} \cdot \text{eV} = 888\ \mu\text{rad} \cdot \text{eV}$.

The second step is to include the scattering characteristics of the first Si111 monochromator. In a DuMond diagram with a relative small energy and angle range, the position of the Bragg condition follows a straight line with the slope given by Bragg's law, as has been shown in the chapter about scattering theory (see Chapter 1). The maximum of the Bragg reflection is not positioned at the Bragg condition but slightly off toward larger incident angles and with a non-zero width $\Delta\theta_{111}$ (the so-called Darwin width, see Chapter 1 and references). For this reason, a perfect crystal oriented with incident angle equal to the Bragg condition actually reflects photons at energies of a few eV larger than expected. In the DuMond-diagram, the area of the monochromator which would reflect with almost 100 percent is drawn as a striped rhombus in Fig. 2.13. The intersection between the undulator radiation and the monochromator rhombus represents the reflected photons. The area of the intersection can be estimated easily by $2\sigma y \cdot 2\Delta E_{111}$ = 31 µrad·eV with $2\Delta E_{111}$ = 4.2 eV. This is only 3.5 percent of the original area and of the same amount the flux is reduced.

However, the photons now have a much narrower bandwidth of $2\Delta E$ = 7.4 eV. For maximum throughput, the second Si111 crystal of the double crystal monochromator has to be exactly oriented as the first one. For this reason, the reflection rhombus of the second crystal is exactly overlapping the rhombus of the first crystal.

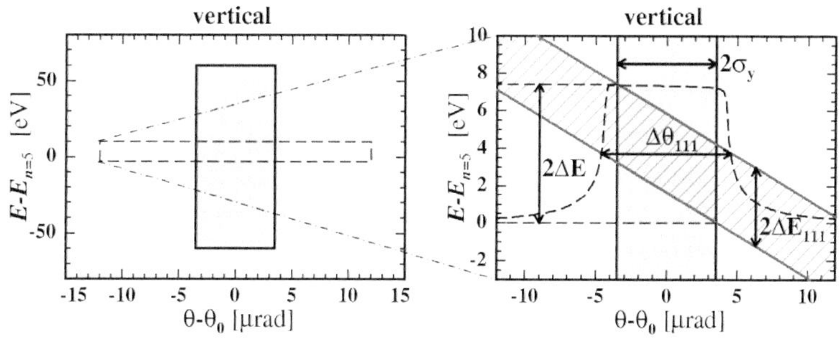

Figure 2.13 Left: DuMond diagram of the fifth undulator harmonics at 30 keV. The shaded area contains the flux. The dashed rectangle symbolizes the cut, which is displayed on the right. Right: The canted striped polygon is the area at which the Si111 crystal reflects (Darwin profile is shown as a dashed line). The shaded and striped rhombus marks the reflected flux.

The next optical device is the vertically collimating mirror. If the incident angle of this mirror is set to an angle smaller than the critical angle, the reflectivity is almost 100 percent. To account for the collimation effect, the angular width σ_y of the X-rays shrinks accordingly without effecting flux. Basically, the mirror compresses the area of reflection, in our example to half of the original value (σ_y = 1.85 µrad), thus $2\sigma_y \cdot 2\Delta E_{111}$ = 15.5 µrad·eV. Further, it suppresses the higher harmonics from the undulator spectrum due to the wavelength dependence of the critical angle α_c (see Chapter 1 and references).

The following device is a Si311 double monochromator which scatters downwards. For this reason and taking into account a fixed direction of the rotation for all crystals, the slope of the Si311 is reverse compared to the Si111 crystal. This is displayed in Fig. 2.14 (left) as a dashed polygon. Further, the dynamical scattering theory yields a specific Darwin width and Darwin shift that are different from the results using the Si111 reflection. To match the reflection condition of the Si111, the Si311 crystal has to be slightly detuned from the original Bragg condition, visible as a shift of the reflection polygon towards the center. Again, the reflected signal from the first Si311 crystal is the intersection of the flux reflected by the mirror and the Si311 polygon. This new shape is marked in Fig. 2.14 (left) as a shaded checked rhombus. Using that $2\Delta E_{311}$ = 0.85 eV, this area is now approximately $2\sigma_y \cdot 2\Delta E_{311}$ = 3.1 µrad·eV, which is 0.7 percent of the original area. The effective energy resolution is now $2\Delta E$ =1.5 eV, and the vertical angular FWHM $2\sigma_y$ = 3.7 µrad. As for the Si111 case the second Si311 monochromator is optimized for maximum flux, and therefore overlaps the first crystal in the DuMond diagram.

The last device is a Si111 crystal that scatters in horizontal direction. For this reason, the shaded checked rhombus has to be transformed from the vertical direction. This can be done by keeping the energy bandwidth of $2\Delta E$ =1.5 eV and using the horizontal divergence instead of the vertical divergence. This yields an area of 48 µrad·eV in the horizontal divergence, which is displayed as the shaded checked area on the right of Fig. 2.14 (right). The flux in this area is the same as that calculated before for the vertical case after the last Si311 monochromator. As shown earlier, the reflection area of a Si111 crystal has to be drawn and the intersection, shown here as a shaded densely checked rhombus, is the final area of reflection with an area of $\Delta\theta_{111} \cdot 2\Delta E$ =9.2·1.5 = 13.8 µrad·eV. This corresponds to 29 percent of the original area in horizontal geometry, and the whole flux has to be reduced by this factor.

Finally, the flux after all optical devices is only 0.2 percent of the original flux from the whole 5th harmonic and corresponds to $2 \cdot 10^{11}$ ph/s. The energy bandwidth of this beam is $2\Delta E$ =1.5 eV, and the divergences are $2\sigma_x$ =12.7 µrad and $2\sigma_y$ =3.7 µrad. It has not been taken into account that the reflectivity of the single devices is typically only 80–90 percent, which for six devices would reduce the flux by another factor of 3. Additionally, polarization effects have been neglected. The last monochromator scatters horizontally with an angle of $2\theta_h$. For a vertically polarized X-ray beam, the flux is reduced by the Lorentz polarization factor $P \propto \cos^2 2\theta_h$. In our example $2\theta_h = 7.56°$, and the polarization factor is very close to one [17].

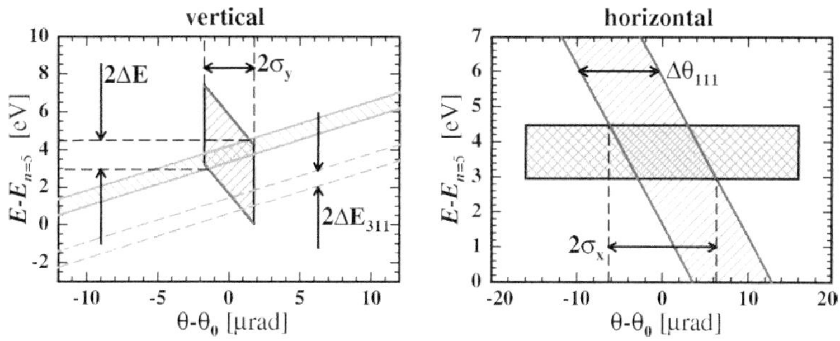

Figure 2.14 Left: DuMond diagram containing the compressed area from the Si111 monochromator. Also drawn is the shifted Si311 reflection area. Right: After transformation to the horizontal, the area is again cut by the horizontally scattering Si111 crystal.

At beamlines with very high **q**-resolution other effects such as multiple scattering effects may appear which may reduce the flux further but only at selected photon energies [18]. Depending on the in-plane orientation, at perfect single crystals such as monochromators, the Bragg condition may be fulfilled for more than one reflection at the same time. For example, a silicon crystal with a 111 surface and oriented with the incident beam along 1–10 shows strong multiple reflections at exactly 8309.9 eV. By tuning the incident angle to the reflection condition of the 111 reflection, additionally, the following reflections appear: 1–10, 11–2, 133, 044, 06–2, -264. They decrease the flux of the 111 reflection significantly and, further, give a characteristic fingerprint to the shape of this Bragg reflection. Figure 2.15 displays the 111 Bragg reflection at this position.

To get around this problem, the monochromators may be rotated around the surface normal (azimuth rotation). For crystals with symmetric Bragg condition of the main reflection, rotation is possible without affecting the scattered intensity. However, the parasitic reflections can be detuned and suppressed accordingly.

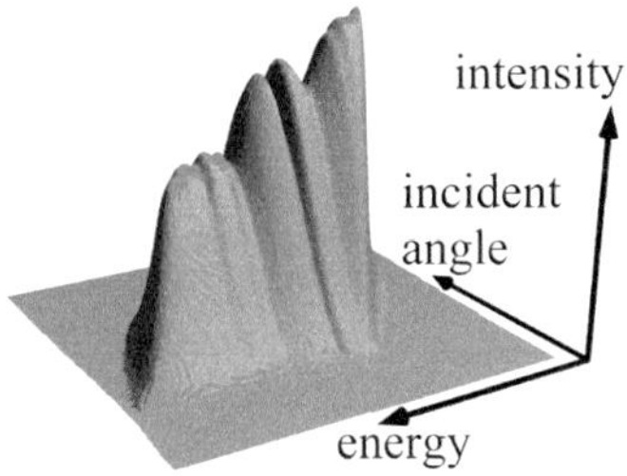

Figure 2.15 Intensity distribution of the Si111 Bragg-reflection at 8309.9 eV. The shape is strongly influenced by multiple scattering effects.

References

1. Wiedemann H (2003) *Synchrotron Radiation*, Springer, Berlin, Heidelberg.
2. Landau LD, Lifshitz EM (1994) *The Classical Theory of Fields*, 4th ed, Butterworth-Heinemann, Amsterdam.
3. Jackson JD (1999) *Classical Electrodynamics*, Wiley, New York.
4. Thompson AC, Vaughan D (2001) *X-Ray Data Booklet*, Lawrence Berkeley National Laboratory, Berkley.
5. Hettel RO (2002) Beam stability at light sources, *Rev Sci Instrum*, **73**, 1396–1401.
6. Berndt M, Brunk W, Cronin R, et al. (1979) *Initial Operation of SSRL Wiggler in SPEAR, IEEE Trans Nuc Sci*, **26**, 3812–3815.
7. Gürtler P, Jackson A (1983) A multiplole wiggler magnet for DORIS, *Nuc Instrum Meth*, **208**, 163–166.
8. Motz H (1951) Applications of the radiation from fast electron beams, *J Appl Phys*, **22**, 527–535.
9. Hofmann A (1978) Quasi-monochromatic synchrotron radiation from undulators, *Nucl Instrum Meth*, **152**, 17–21.
10. Deacon DAG, Elias LR, Madey JMJ, Ramian GJ, Schwettman HA, Smith TI (1977) First operation of a free-electron laser, *Phys Rev Lett*, **38**, 892–894.
11. Kondratenko AM, Saldin EL (1980) Generation of coherent radiation by a relativistic electron beam in an ondulator, *Part Accel*, **10**, 207–216.
12. Feldhaus J, Saldin EL, Schneider JR, Schneidmiller EA, Yurkov MV (1997) Possible application of X-ray optical elements for reducing

the spectral bandwidth of an X-ray SASE FEL, *Opt Commun*, **140**, 341–352.

13. Landau LD, Lifshitz EM (1980) *Statistical Physics*, Pergamon Press, Oxford.
14. Balewski K, Brefeld W, Decking W, Franz H, Röhlsberger R, Weckert E (2004) *PETRA III: A Low Emittance Synchtrotron Radiation Source*, Deutsches Elektronen-Synchrotron DESY, Hamburg.
15. Boutet S, Williams GJ (2010) The Coherent X-ray Imaging (CXI) instrument at the Linac Coherent Light Source (LCLS), *New J Phys*, **12**, 035024.
16. DuMond JWM (1937) Theory of the use of more than two successive X-ray crystal reflections to obtain increased resolving power, *Phys Rev*, **52**, 872–883.
17. Warren BE (1990) *X-Ray Diffraction*, Dover, New York.
18. Stetsko YP, Chang Sh-L. (1997) An algorithm for solving multiple-wave dynamical X-ray diffraction equations, *Acta Cryst A*, **53**, 28–34.

Chapter 3

Micro- and Nanodiffraction

Christina Krywka and Martin Müller
Helmholtz-Zentrum Geesthacht, Max-Planck-Str 1,
21502 Geesthacht, Germany
christina.krywka@hzg.de

3.1 Introduction

During the past decade, there has been a considerable success in concentrating low-divergence, hard X-rays available from third-generation sources onto ever smaller beams. Due to their short wavelength, it is physically possible to confine X-rays in spot sizes of only a few nanometers [1]. The recent advance in microfabrication and high-precision surface processing technologies has reduced the achievable X-ray focal spot size by a factor of 10^{-3} compared with the first attempts of hard X-ray microdiffraction (the earliest undertaken as early as the 1940s [2]). As the first microdiffraction experiments relied on collimation only, the main driving force behind X-ray focusing was the desire to increase flux density. The availability of micrometer-sized focal spots provided an unprecedented spatial resolution in X-ray diffraction experiments. Both wide and small

X-Ray Diffraction: Modern Experimental Techniques
Edited by Oliver H. Seeck and Bridget M. Murphy

ISBN 978-981-4303-59-0 (Hardcover), 978-981-4303-60-6 (eBook)
www.panstanford.com

angle X-ray scattering and X-ray crystallography experiments provide huge additional information content when combined with the micrometer or submicrometer spatial resolution provided by a micro- or nanofocused X-ray beam in a scanning experiment. Many interesting systems found in life science (e.g., single cells), high-performance biological materials (e.g., silk fibers), hierarchically structured systems (e.g., wood, muscle tissue) as well as micro- or nanostructured synthetic materials can be investigated with X-ray diffraction methods—provided that a spatial resolution corresponding to the structural dimensions of these systems can be achieved. Details of particular structures only become visible when using such a local probe.

The underlying conceptually simple idea of X-ray micro- and nanofocusing is—as is its application in X-ray microdiffraction experiments—very challenging from the technical point of view. Not only is the quality of the recorded data highly sensitive to any sort of mechanical instabilities (e.g., vibrations, temperature drifts) and source instabilities (e.g., beam drifts), but it is also crucial for the precision of the mechanical sample alignment to match the targeted spatial resolution of the experiment, that is, down to a few tens of nanometers. Micro- and nanodiffraction experiments (for simplicity referred to as "microdiffraction experiments" in the following) require a very low background signal, in particular because the sample volume illuminated by a small beam is very small as well.

In this chapter, we first present an overview of the most extensively used techniques to generate micro- and nanofocused hard X-ray beams. Experimental aspects of microdiffraction experiments are first discussed generally. The aspect of beam damage due to the high flux density in focused X-ray beams is addressed as well. Examples of experiments were chosen from the systems mentioned earlier, also taking into account that X-ray scattering can be carried out in situ , whereas external parameters such as temperature, mechanical stress, or water content are being changed.

3.2 X-Ray Focusing Optics

The aim of an optical element to be used as an X-ray focusing device in a microdiffraction experiment is to demagnify the X-ray source, which in general produces a more or less divergent beam, and to project it onto the sample. The demagnification is limited by the imperfections of the optical elements being used, by the geometrical optics, and most fundamentally by the diffraction limit. The diffraction limit is the best possible resolution $\Delta_{diff} = W\lambda/N_A$ where W is an aperture shape dependent constant, λ is the wavelength of the focused radiation, and N_A is the numerical aperture of the optical system given by $N_A = n\sin\theta$. Here, n is the index of refraction of the optical element and 02θ is its opening angle. For a given optical element, the demagnification ratio equals (and therefore limits) the ratio of the distances, that is, $G/B = g/b$. Here, G is the size of the object, g is the object-lens distance, b is the lens-image distance, and B is the size of the image, that is, the focal spot size.

Thus, the price to pay for a large demagnification is a short working distance for the experiment behind the optical element and an increased divergence of the X-ray wave field. This divergence in turn limits the minimum accessible scattering angle in a diffraction experiment and its angular resolution. These limits cannot be annulled; however, they can be overcome by trading in other performance features. A small spot can still be achieved with a low divergence if the source size is reduced by an aperture. In this case, photon flux is sacrificed for a low divergence. Using diffractive rather than refractive optical elements will remove the limits of geometrical optics, but will impose new limitations such as a strong dependence of the focal length on the energy E of the incoming radiation (where $E = hc/\lambda$ with c and h being the light speed and the Planck constant, respectively). Consequently, optical elements for focusing of X-rays do not only exploit refraction, but there is also a wide range of commonly used diffractive, reflective, and beam-concentrating elements. The choice of the optical element strongly depends on the main concern of the experiment as for any of the mentioned types of focusing devices one has to cope with some kind of trade-off, mainly between conserving incoming flux, spatial and angular resolution, or coherence.

What follows is a general overview of the most extensively used optical elements along with their main assets and drawbacks. This chapter intends to present the basic principles of these optical devices only, that is, no explicit mathematical considerations are performed as these can be easily found in the relevant references provided.

3.2.1 Refractive Optics

The interaction of X-rays with matter is very weak. The real part of the index of refraction can be written as $n = 1-\delta$, where δ is the refractive index decrement. For X-rays with a photon energy of several keV (λ of the order of 0.1 nm), which penetrate an air–solid or vacuum–solid interface, it will be $\delta \approx 10^{-6}$, i.e. n is only slightly less than unity [3]. For visible light, this value is considerably larger than unity, that is, $n \approx 1.6$ for an air–glass interface. The effect of the small deviation from unity is that a refractive X-ray optical element (e.g., a single lens) will provide only a very small deflection angle, leading to an extremely long focal distance, in strong contrast to lenses for visible light. As $n < 1$, an X-ray focusing lens will be concave, rather than the convex shape of focusing lenses for visible light.

Although the concept of using refractive optics for X-ray focusing was already addressed over 60 years ago [4], it was put into practice for the first time only almost 50 years later [5]. As a consequence of the inherent weak refractive power, many single lenses need to be stacked along their optical axes. Consequently, to maximize the transmission of the lens stack, it is crucial to use a material with a low electron number (Z) as for low-energy X-rays the photoelectric absorption is dominant and proportional to Z^4 while δ is proportional to Z. In the high-energy X-ray range, however, Compton scattering is dominant also being proportional to Z [6].

3.2.1.1 Metal compound refractive lenses

In the thin-lens approximation, the focal length f_0 of a stack of N biconcave, parabolic lenses with an apex curvature radius R is $f_0 = R/2N\delta$. If the lens thickness is nonnegligible (i.e., high N), then the deviations can be corrected by applying the thick lens formula (see [7]).

The first compound refractive lens (CRL) was realized by drilling a row of 30 holes each with a radius of 300 μm into a block of aluminum, as shown schematically in Fig. 3.1. This simple, low-cost approach led to a one-dimensionally focusing CRL with a focal length of 1.8 m for 14 keV radiation, which is sufficient to produce a focal line of 8 μm width (FWHM) [5].

The CRLs available today are much more sophisticated as they have a rotational parabolic shape (see Fig. 3.1), thereby providing focusing in both directions free of spherical aberrations, just as their glass counterparts for visible light. Being manufactured by embossing in low-Z metals such as aluminum ($Z = 13$) or beryllium ($Z = 4$), they can readily handle even the high heat load of an undulator beam. CRLs are easy to align and robust, and no deterioration of beryllium or aluminum has been observed even after years of operation. They can be used over a wide energy range of 5 keV to about 120 keV and for the apex radius of curvature R a range of 50–1000 μm is available. The geometrical aperture A of these lenses is typically about 1 mm; however, it has to be borne in mind that the effective aperture A_{eff} is smaller than the geometrical aperture due to the increased absorption in the outermost parts of the lenses. The focal length is a function of the number of single lenses comprising the lens stack (N typically 20–200) and typically lies within the range of 30–0.5 m.

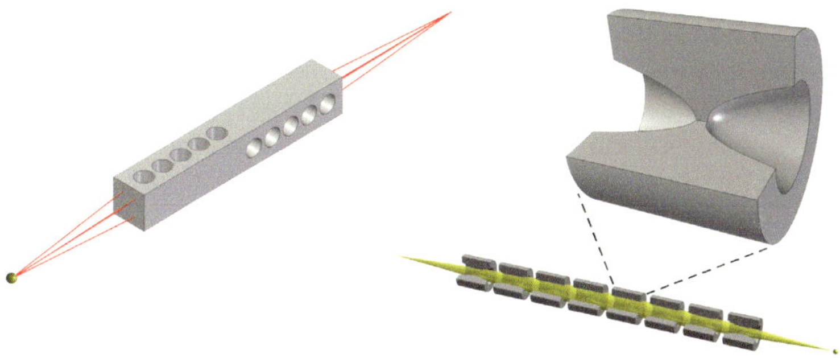

Figure 3.1 Left: The first hard X-ray refractive lens was manufactured by drilling a series of holes into an aluminum block. Two series of holes tilted by 90° provided focusing in both directions. Right: Today, refractive lenses are made by indenting the required parabolic shape into a thin metal plate or foil and arranging them into a stack (compound refractive lens, CRL) to increase the refractive power. Note the concave shape of each single focusing lens. Figure from Ref. [8].

Unlike diffractive optical elements, they are almost achromatic, that is, the focal length varies only slightly with the energy of the incident radiation. The smallest spot sizes achieved with Be-CRLs are of the order of ~1 μm (for a high-β ESRF undulator source at E = 12 keV) and even submicrometer sizes are potentially feasible along with focal lengths of significantly <1 m [7].

3.2.1.2 Silicon nanofocusing refractive lenses

In order to achieve focal lengths in the range of centimeters to millimeters, that is, to increase the demagnification ratio, either the number of lenses in a CRL lens stack must be further increased, or the apex curvature radius, R, has to be reduced. Since a high number of lenses will also reduce the transmission of the lens system, smaller lenses with a curvature radius of $R = 1.5$ μm are much more efficient. A rotational parabolic shape with these dimensions, however, is very difficult to manufacture at present. Lithography-based fabrication methods allow to create a composite nanofocusing refractive lens (NFL), comprising N multiple one-dimensionally focusing parabolic lenses [9] with $N \sim 100$ being a typical value.

As shown in Fig. 3.2, two such planar lenses, each generating a line focus with a slightly different focal length, can be arranged in a crossed geometry along a common optical axis to generate a point focus. In principle, the resulting image is far from being free of geometrical distortions, but due to their high refractive power NFLs allow large demagnifications into the 100 nm range even at short source-to-lens distances of 40–70 m and relatively large source sizes. One of the smallest spot sizes achieved so far with crossed NFLs is close to 50 nm (at a low-β ESRF undulator source at $E = 21$ keV) [9].

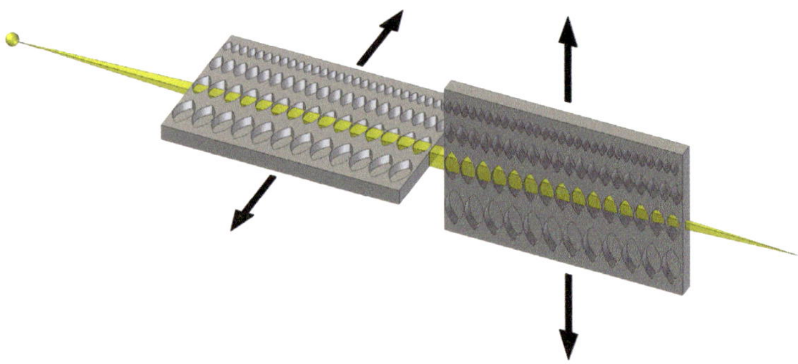

Figure 3.2 A composite nanofocusing refractive lens (NFL) consists typically of two planar lens arrays arranged in a crossed geometry for two-dimensional focusing. To select a specific focal length from an array, each of them can be moved independently. From Ref. [8].

Being typically manufactured of silicon ($Z = 14$) and having very low thicknesses of the material between two individual lenses

(~5 μm), NFLs provide a very low attenuation. However, due to their small apex radii of curvature, the geometrical aperture values are relatively small (as compared to CRLs), being typically <50 μm. Just as in CRLs, the effective aperture A_{eff} is smaller than the geometrical aperture due to the increased absorption in the outermost parts of the lenses, although the effect is not as pronounced because of the smaller amount of material. NFLs can be used over a wide energy range of 8 keV to about 100 keV [7]. The alignment is more complicated than for a CRL as two separate lenses must be aligned into the beam and must have coinciding optical axes.

Usually, many NFLs with slightly different focal lengths are manufactured onto a silicon wafer and placed parallel to each other. In this way, one can easily switch between NFLs by simply moving the wafer to change the focal length or to keep it fixed at a different energy. A main factor limiting the efficiency and focusing power of NFLs is—just as in CRLs—the high absorption in the outer parts of each single lens. This reduces the effective aperture of a lens system with a large number of lenses. One approach to overcome this limit is an arrangement of lenses with a gradually increasing refractive power (adiabatic lens) [10]. In this concept, the aperture of each single lens segment is adapted to the decreased size of the beam coming from the preceding lens segment. This adiabatic focusing is expected to be able to dramatically decrease the achievable focal spot size to <10 nm, hence close to the theoretical limit predicted in Ref. [1].

3.2.1.3 Diamond lenses

Current research aims at manufacturing NFLs of diamond. Due to its higher density (as compared to silicon), diamond allows for a higher refraction power, thus less material is needed to achieve the same focal length. Therefore, a higher demagnification at a comparable degree of absorption is feasible and because of diamond's superior heat conductivity, these lenses will be able to better cope with the high heat load of an undulator source [11]. This is especially important considering the advent of new high-brilliance synchrotron sources such as PETRA III or free-electron laser (FEL) sources.

The exploitation of diamond's high refracting power is also regarded as vital for the fabrication of adiabatically focusing lens system. However, technological obstacles of manufacturing the required high-aspect ratio, free-standing diamond structures need to be overcome first.

3.2.1.4 Polymer lenses

Unlike diamond, materials used in lithography-based fabrication techniques can be processed with much fewer geometrical restraints. Thin, free-standing high-aspect ratio structures with heights of several millimeters can be manufactured of the polymer SU-8, a commonly used epoxy-based negative photoresist. As shown in Fig. 3.3, a crossed geometry of two one-dimensionally focusing sets of SU-8 lenses can even be fabricated in one manufacturing process onto one substrate, and the polymer is reported to be stable against radiation damage [12]. For higher X-ray energies, (>40 keV), where the refractive power of the polymer is no longer sufficient, the same crossed-lens geometry can also be fabricated from nickel [13].

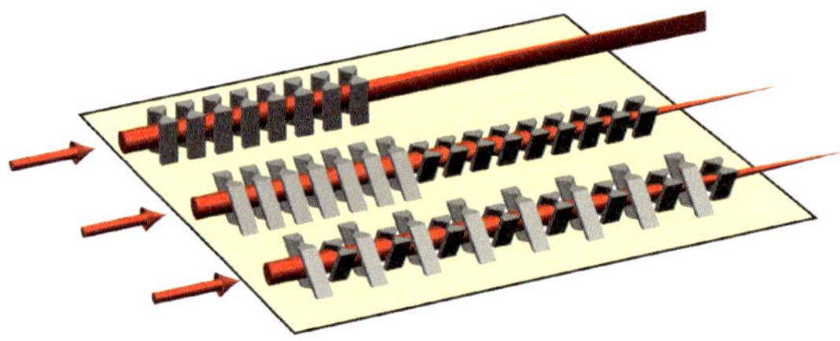

Figure 3.3 Compound polymer lenses can be used either for one-dimensional focusing (top) or can be arranged in a crossed geometry (middle, bottom) to provide two-dimensional focusing. Note that for a crossed geometry the lens elements are tilted by 45° against the substrate onto which they are fabricated. Figure from Ref. [8].

3.2.2 Diffractive Optics

3.2.2.1 Fresnel zone plates

As mentioned earlier, the absorption in the outer parts of a regular refractive lens is one of its main limiting factors. Therefore, it is convenient that most of the material in a classically shaped lens only provides a phase shift of only $2\pi N$, and therefore can simply be removed [3]. This applies for lenses in both the optical and X-ray regime, hence the well-known concept of Fresnel lenses used for visible light can be transferred for X-ray optics (see Fig. 3.4). However, it is very difficult to manufacture the Fresnel lens shape correctly on the microscopic length scale required for an X-ray lens, that is, as an arrangement of parabola pieces. Rather than a composition of parabolic segments, Fresnel X-ray optics are approximated by an arrangement of flat, concentric rings (zones) of alternating refractive indices. In such a circular grating called Fresnel zone plate (FZP), the zone width gradually decreases as the zone radius is getting larger.

The focusing power of an FZP is limited by the Rayleigh resolution, being $\Delta s \approx 1.2\ \Delta r_m$ where Δs denotes the smallest first order focal spot size that can be achieved with a given FZP whose outermost zone width is Δr_m. Despite the technological challenges in fabricating zone widths on the nm scale, a 15 nm outer zone width FZP was recently demonstrated [14]. Theoretical approaches even predict the feasibility of an FZP with a resolution of <10 nm—provided the technological advance required to fabricate zone widths close to the limit given by the atomic structure of matter [15].

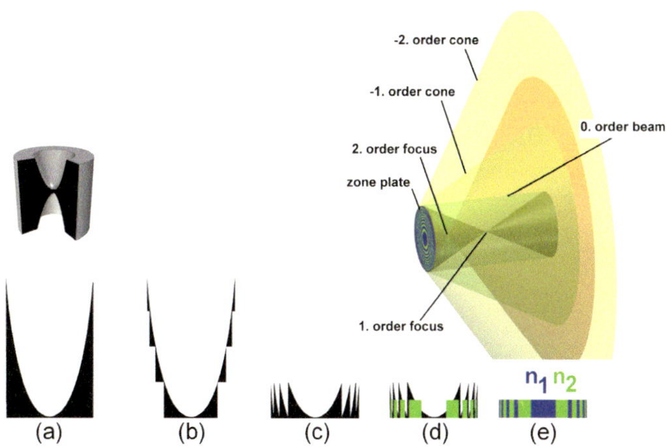

Figure 3.4 From a refractive lens to a Fresnel zone plate (FZP). As most parts of the refractive lens body (cross-section (a)) only contribute by a phase shift of $2\pi N$, they can be omitted (b). The remaining structure of parabola sections can be arranged onto a common baseline (c). As this geometry turns out to be too difficult to microfabricate, an approximation is used (d). The approximation consists of flat, concentric rings (zones) of alternating refractive indices (n_1, n_2) whose widths gradually decrease as the zone radius increases (e). Their thickness, however, remains constant at $\lambda/2$. Partially from Ref. [8].

However, compared to a true refractive lens, the imaging quality of an FZP is worse, and most prominently the FZP generates not only a single focal spot but also multiple orders of foci—a consequence of its diffractive nature. Still, FZPs are routinely used in X-ray microscopy setups. The main advantage of FZP lies within their easy handling and specifically for microsocopy techniques within the short focal length of typically some centimeters.

Also the efficiency is relatively high, that is, the ratio of the incident flux against the flux concentrated into the focal spot can be as high as 40 percent at 8 keV [16]. However, both the focusing efficiency and the focal spot sizes worsen with increasing X-ray energies. As energy increases, in order to retain the phase-shift required for a certain focal length, the thickness of the FZP needs to increase as well, leading to a zone width-to-thickness aspect ratio that is very difficult to manufacture. Hence, for focusing hard X-rays, refractive or reflective optical elements are better suited.

There is a wide range of other diffractive optical elements optimized for higher energies or easier production, most of which being either Fresnel lens variations (e.g., flat prism arrays and CLESSIDRA lenses [17], multilayer Laue lens [18], photon sieves [19]), or combinations of the Fresnel's lens concept with other methods (e.g., Bragg–Fresnel lens [20]). Figure 3.5 displays some examples of this kind.

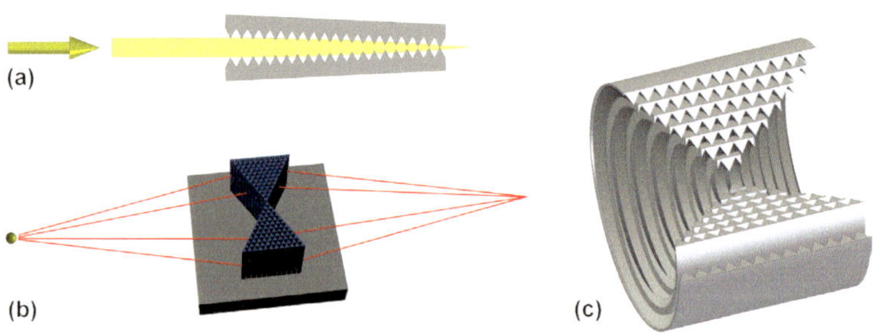

Figure 3.5 The concept of a prism array for X-rays was first put into practice by cutting two radial sections of a vinyl record and arranging them grove-to-grove with a slight tilt angle (a), also known as alligator lens. A different approach is the microfabricated CLESSIDRA lens, where the prism array resembles an hourglass (b). A rolled X-ray prism provides two-dimensional focusing (c). Figure from Ref. [8].

3.2.3 Reflective Optics

As the interaction of X-rays with matter is very weak, conventional reflection happens only at a negligibly small rate. However, as the real part of the index of refraction for X-rays at an air–solid (or vacuum-solid) interface is slightly smaller than unity, total external reflection occurs in the regime below the critical angle θ_c given by $\sin\theta_c = \sqrt{2\delta}$, where δ is the refractive index decrement. Although only very small reflection angles are possible (θ_c being typically <0.3°) mirrors for X-rays exist and are being widely used.

3.2.3.1 Kirkpatrick–Baez mirror

Named after its proposers a Kirkpatrick–Baez mirror (KB mirror) is a combination of two curved mirrors arranged in a crossed configuration [4], as shown in Fig. 3.6. However, the originally proposed spherically or cylindrically curved mirrors suffered from spherical aberrations. Only elliptically curved mirrors arranged in KB geometry provide aberration-free focusing in two dimensions, limited only by the quality of the curvature and mirror surface roughness. Hence, the demagnifying rate varies with the rate of the image distance and focal length. The required elliptical shape can be achieved by either elastically bending an initially flat mirror or by machining and polishing it to fit the desired curvature.

Either way, due to the extremely small incidence angles, the substantial technical challenge is to obtain the elliptical shape with the required accuracy and low surface roughness. A roughness too high creates diffuse scattering while an excessive figure error will cause the image to be blurred. For a mirror system used at a typical hard X-ray synchrotron beamline in order to generate a demagnification required for a sub-100 nm focal spot, the surface roughness must be significantly <0.5 nm and the figure height error well <1 nm [21].

Single-layer mirror-bending techniques have been used for many years, however, appear to have reached a technological limit at a focal spot size of ~200 nm. The main drawback of the bending technique is the time-consuming alignment procedure required to be performed repeatedly as the bent optics will relax over time due to the bending mechanics involved.

Within the past five years, a highly optimized polishing technique has been established and has made possible the fabrication of elliptical-shaped mirrors with a figure accuracy of only 0.15 nm (rms) [22]. After a conventional rough figuring, a polishing technique called elastic emission machining (EEM) is used whose removal mechanism relies on chemical reactions rather than mechanical removal. The so-far best mirror produced using this technique is reported to provide a line focus with a beam width of 25 nm at a photon energy of 20 keV at the 1 km-long beamline BL29XUL of Spring8. Combined with additional adaptive, deformable mirrors used to counteract the wavefront distortions caused by remaining figure errors, the authors report on a 20 keV X-ray focused to 7 nm in one direction [23].

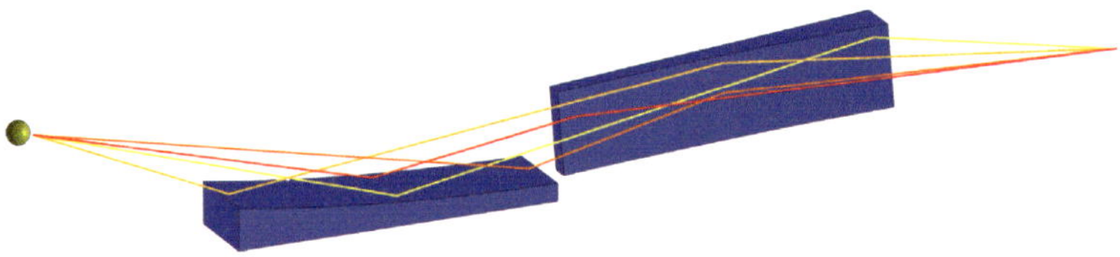

Figure 3.6 Two mirrors, each with a specific elliptical curvature, arranged in a crossed geometry provide a two-dimensional aberration-free focusing. The source position (left) and the focus position (right) match the two foci of the corresponding ellipse. Figure from Ref.[8].

Although reflective optical elements are conceptually achromatic, KB mirrors do have a limited energy range in which they can be used. The upper limit is given by the critical angle of incidence θ_c which at a given wavelength is only a function of the electron density of the mirror coating material. Above θ_c the reflectivity decreases significantly, consequently this value must not be exceeded at any region of the elliptically curved mirror surface. Hence, for a given elliptical mirror there is an energy cut-off below which its focal length is independent of the wavelength.

To be suitable as a mirror coating the material must have a high density to provide a high deflection angle. Typically, heavy noble metals are used (Au, Pt, Rh). As the material cannot be chosen arbitrarily and figure errors become more important at smaller incidence angles, there is an energy limit of $E \sim 30$ keV above which single-layer KB mirrors cannot be used efficiently.

3.2.3.2 Multilayer KB mirror

If an X-ray mirror needs to be operated at incident angles larger than the critical angle of a given coating material—for example, to increase the energy cut-off or to improve the mirror's acceptance—it can be equipped with a multilayer coating. A multilayer coating consists of typically many hundreds or thousands of alternating layers of high- and low-refractive index materials each with a nonlinear gradient in period along the mirror length [24]. Next to a high difference in *n*, the unit cell geometries of the materials must also fit to allow an epitaxial deposition of alternating layers. Typically used material combinations are W/C, Ni/C, W/B_4C, and Mo/Si.

Accordingly to the reflectivity of a flat multilayer, there are reflectivity maxima at incidence angles larger than θ_c [3], whose exact positions obey the Bragg criteria. A multilayer KB mirror, strictly speaking a diffractive element, therefore also has some monochromating capabilities and can be effectively used in a white X-ray beam [24].

Focal spot sizes of slightly <50 nm were reported [25]; however, due to the larger incidence angles involved, the focused X-ray wavefield exhibits a larger divergence as compared to single-layer KB mirrors.

3.2.4 Beam Concentrating and Collimating Elements

Next to the optical devices described earlier, all of which have in common that they create a more or less precise, demagnified image of the source in the focus at the sample position, there are devices that can concentrate or collimate the incoming wavefield in order to confine the radiation to a much smaller area, rather than create a focal spot. In other words, these elements condense the incoming wavefield, hence do not image the source.

3.2.4.1 Capillaries

Glass capillaries were used in the first microdiffraction experiments [2] and were also the first type of optics to achieve a beam confined to the order of 100 nm and below [26]. Similar to fiber optics for visible light, X-rays can be guided inside a hollow capillary due to the total reflection at the inner surface of the capillary, that is, provided the critical angle of reflection θ_c does not exceed the curvature of the fiber. Unlike fiber optics, where the core material needs to be solid to have a higher index of refraction than the outer material, total reflection of X-rays occurs within hollow capillaries due to the index of refraction of air being larger than that of glass. When the capillary is tapered, that is, the cross-section of the inner opening decreasing along the capillary, one will find a condensed beam at the exit of the capillary, however, with an equally increased divergence corresponding to θ_c (as shown in Fig. 3.7). If the relatively small acceptance of a single capillary (straight or tapered) poses a problem, a polycapillary array can be used, which consists of a high number of capillaries assembled in a bunch-like structure.

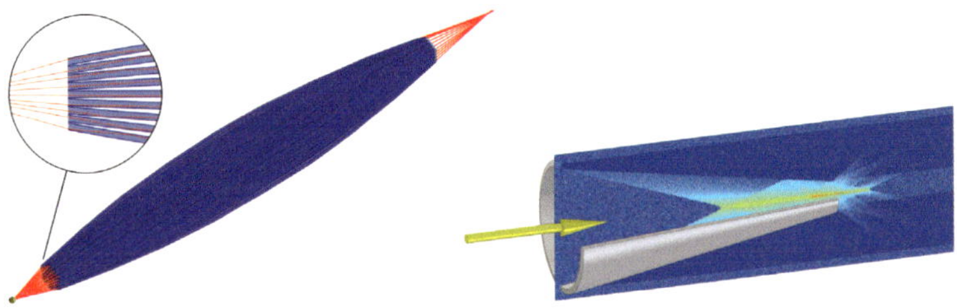

Figure 3.7 Left: A bunch of specifically tapered or curved capillaries (polycapillary array) provide a beam concentration at a short, specific distance behind the exit of the array with the advantage of having a relatively large input area. Only elliptical-curved capillaries provide true focusing, that is, an image of the source at the focal spot. Otherwise, only a nonimaging beam concentration is achieved. Right: Intensity distribution inside a single, nonelliptically tapered capillary. Note the intensity enhancement at the capillary exit, due to the multiple total reflection of the incoming wavefield on the inner capillary walls, but also the strongly divergent field outside the capillary. Figure from Ref. [8].

Although tapered capillaries provide an easy-to-handle method for beam concentrating, they are not ideally suited for diffraction methods. This is mainly due to the poorly controllable taper profile as it forms during the combined heating and pulling of the originally straight capillaries. Along with the inability to control the microroughness inside the capillary, this results in an unpredictable divergence distribution of the exiting wavefield, limiting the resolution in a microdiffraction experiment.

This drawback can be overcome by using an elliptical-shaped capillary profile (rather than tapered), leading to an optical element capable of truly imaging the source (rather than condensing the beam). When coupled to a short focal distance optical element (e.g., an FZP) to create a demagnified source in one focus of the ellipse, the capillary provides an efficient, inexpensive focusing device with an extended focal distance (as compared to the FZP alone). However, like all types of capillaries, the elliptical-shaped capillaries suffer from the nonpolishable inner surface and the demonstrated focal spot size of 250 nm is far above the theoretically achievable value [27]. Recent approaches to apply a metal coating onto the inner surface may, however, decrease the surface roughness and make elliptical capillaries a more attractive optical element.

3.2.4.2 Waveguides

The working principle of X-ray waveguides (further referred to only as "waveguides") is strictly analogous to hollow waveguides for radio frequency fields (e.g., microwave resonators). As shown in Fig. 3.8, a waveguide is a long, thin, straight channel,[1] open on both ends, either empty or filled with a low density material (called "core") and is manufactured into a substrate material, for example, silicon. The length of the waveguide is usually several millimeters, and the sides of its cross-section are of the order of several 100 nm down to currently ~20 nm [28].

Divergent or focused radiation is coupled into the waveguide by simply illuminating the entrance of the waveguide that, ideally, is aligned to coincide with the optical axis of the radiation source. The portion of radiation that overilluminates the entrance of the waveguide is simply absorbed in the substrate material. Unlike in capillaries, which are capable of confining even incoherent waves between reflecting surfaces and where the flux density enhancement is due to the continuously decreasing capillary cross-section, in a waveguide a standing wave is promoted; therefore, a coherent superposition of the electromagnetic field amplitudes of the incident wave occurs and enhances the field intensity.

[1]The first waveguides used for X-rays were thin, laterally extended layers, sandwiched in between substrate layers, hence confining the wavefield in only one direction. This type is continuously being used and to distinguish it from the channel-like waveguides, the latter are commonly referred to as 2D-waveguides.

At the waveguide exit, the dimensions of the exiting wavefield are constricted by the waveguide and are a function of the resonance order. The number of modes that can propagate in the waveguide is defined by the material properties and the geometric parameters. For the first-resonance mode, the size of the exiting wavefield can even be as small as half of the waveguide dimensions. The exiting beam is inherently coherent.

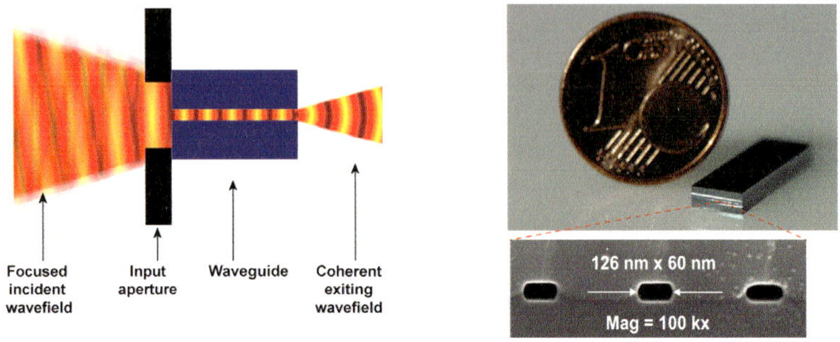

Figure 3.8 Left: Schematic principle of a waveguide based beam collimation and coherence filtering with a color indicated amplitude distribution [35]. Right: Actual waveguide, comprising a silicon wafer with multiple parallel channels, each of which can be used as a 2D-waveguide. The input openings of some channels are shown in the blow-up [29].

Although waveguides can be manufactured for a wide energy range, any specific waveguide has an optimum energy to be operated at. The efficiency of a waveguide is strongly dependent on the substrate flatness, the interfacial roughness, the absorption of the core material and the energy of the incident radiation. The optimum energy for a specific waveguide is influenced by two conflicting effects: higher energy photons are less absorbed in the substrate material, whereas the incidence angle for which resonance occurs is reduced with higher energy, and with this the portion of the beam profile and the beam divergence is accepted by the waveguide. The opening angle of the acceptance cone is in the range of the critical angle of the waveguide's inner interface, that is, few milliradians. The exiting wavefield is subject to diffraction at an aperture, the latter of course being the opening of the waveguide. In the first-resonance mode (TE_0), the exiting beam is Gaussian-shaped and virtually tail-less. For higher-order resonances, split beams are generated, separated by the internal reflection angle. The wavefield exiting the waveguide is very well approximated by a point source located at the waveguide exit [30]. Therefore, in a waveguide-based microdiffraction experiment, the resolution can be even slightly smaller than the waveguide beam size, due to the coherence properties.

Because of the inherently small entrance of the waveguide and its narrow acceptance cone an efficient pre-focusing (e.g., using a KB mirror) is crucial for obtaining a good efficiency of the waveguide-based optical system. The alignment is complex and requires navigational structures next to the waveguide as both a high-angular and positional precision is essential.

It has recently been shown [1] that there is a universal, wavelength-independent lower limit for the spot size that can be achieved with a waveguide, whose exact value depends only on the electron density of the confining material. Effectively, the numerical aperture of waveguides is limited by the critical angle $\theta_c \approx \sqrt{2\delta}$ of total reflection inside the confining material. This limit is of the order of slightly <10 nm, and the most recent advances in fabrication of waveguides have led to a waveguide with a cross-section of 30 nm by 40 nm [29], hence close to this theoretical boundary.

3.3 Experiments

Micro- or nanofocus X-ray beams can, in principle, be used to increase the spatial resolution of various X-ray imaging and scattering techniques. Here, we will focus on giving some examples of small- and wide-angle scattering experiments. With a spatial resolution corresponding to the beam size of the X-rays (thus, down to some tens of nanometers), parameters such as crystal structure, lattice constants and orientation distribution functions (texture) of crystalline or semicrystalline materials and density fluctuation or mesoscopic structures in noncrystalline material are determined, leading to high-resolution maps of such quantities and enabling a local structure analysis of hierarchically structured materials. Another diffraction application, the use of X-ray microbeams for the structure determination of small crystals, in particular of proteins, is not covered in this chapter (see [31,32] for an introduction).

3.3.1 X-Ray Micro- and Nanodiffraction Instrumentation

The various X-ray optical elements described earlier have to be integrated into an X-ray scattering instrument. However, there are certain requirements for an X-ray microbeam to be used in diffraction experiments. First of all, the total flux in the (small) beam cross-section has to be high enough to obtain a diffraction diagram within reasonable time (a few seconds would be desirable, ideally subsecond time resolution is achieved). This requirement means that only synchrotron radiation may be used to fully exploit microdiffraction. Typical flux values for microbeams at synchrotron radiation source microfocus beamlines are of the order of 10^{10} photons/s. Background elimination is another crucial issue for diffraction experiments. This applies to intensity "tails" of the beam profile (normally eliminated by slit systems or pinhole apertures) as well as to air scattering from the primary beam passing through air. The latter can be reduced by minimizing the beam path in air by a compact design of the space around the sample and by placing a very small lead beamstop immediately behind the sample.

One of the first operational synchrotron radiation beamlines for microdiffraction experiments was the Microfocus Beamline ID13 at the European Synchrotron Radiation Facility (ESRF, Grenoble, France) [32], which has been dedicated to X-ray microdiffraction and micro-small-angle scattering experiments on weakly scattering samples (soft matter, polymers, biological materials). Focusing elements used at ID13 include an ellipsoidal mirror, also in combination with glass capillaries, graded multilayer mirrors and a one-dimensional X-ray waveguide. Beam diameters are between 1 and 10 μm, with the waveguide setup being an exception [33]. Here, the beam was horizontally focused by a bent multilayer mirror (3 μm focus) and vertically compressed by a waveguide to 80 nm.

ID13 at ESRF now has a dedicated station for nanodiffraction. Such a setup is also currently being commissioned at PETRA III (DESY, Hamburg, Germany), the most brilliant synchrotron radiation source so far. The Nanofocus Endstation at beamline P03 (MiNaXS) [34] has already produced waveguide-generated nanobeams (50 nm × 50 μm waveguide) with sufficient flux for nanodiffraction experiments [35]. An upgrade with KB mirrors, which may also be used as prefocusing devices for waveguides, will significantly increase the flux density of beams with diameters <200 nm. Both experimental stations are operated at about 13 keV X-ray energy.

Such small beams require an excellent mechanical stability of the instrumentation. Similar to X-ray microscopes, massive granite blocks provide low-vibration levels and low-thermal expansion coefficients. The influence of the remaining vibrations is greatly reduced by the fact that focusing elements and sample positioning/scanning stages (see discussion below) are mounted on the same support and thus hardly move relative to each other, while a powerful air conditioning is used to ensure temperature constancy of the order of 0.1° C at the experimental endstation. Finally, the limited acceptance of basically all focusing devices is turned into an advantage as instabilities of the primary beam do not automatically lead to instabilities of the micro- or nanobeam.

Figure 3.9 shows the schematic set-up of an X-ray microdiffraction experiment at a synchrotron radiation source. Typical X-ray wavelengths for such experiments are around $\lambda = 1$ Å. The sample, for example, a wood cross-section, is positioned in the microbeam using highly precise translation stages with repeatability better than 1 μm. The stages allow for scanning the sample through the beam with microscopic position resolution. Diffraction patterns are collected with a two-dimensional detector. In order to correlate optical images with X-ray scans, a video microscope is situated at a calibrated position so that specific positions on an extended sample can be investigated. For nanodiffraction, submicron positioning is required. For calibration purposes, the use of markers on the samples, to be found with an additional X-ray fluorescence detector, may be useful.

Two-dimensional detectors for microdiffraction should have a good spatial resolution as generally the beam size at the detector position is (despite the increased beam divergence by focusing) still very small, that is, the resolution of the detector images is rather limited by the pixel size than by the beam parameters. In that respect, sensitive CCD detectors are still a very good choice for microfocus applications. By contrast, novel pixel detectors with still larger pixels have the advantage of negligible background noise, which can be of advantage if the scattering signal to be detected is very weak, at the expense of a non-optimized resolution of the recorded diffraction pattern.

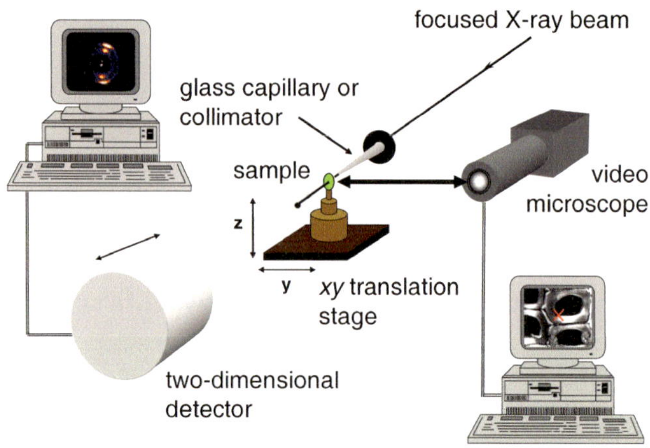

Figure 3.9 Schematic setup of an X-ray microdiffraction experiment at a synchrotron radiation source. The X-ray beam from the synchrotron is focused, for example, by an ellipsoidal mirror and a glass capillary to a diameter of 2 μm at the sample position. The sample can be moved in two orthogonal directions perpendicular with high precision and repeatability. The scattering pattern is collected using an area detector. A video microscope at a calibrated distance is used to select the beam position on the sample. After [36].

3.3.2 Examples of Micro- and Nanodiffraction Experiments

The following examples have been chosen from the authors' main research areas. Many are studies on biological materials (mainly based on polymers) as those are well adapted to a specific function by hierarchical structuring on various length scales. They can (as pointed out above) be well addressed by position-resolved X-ray scattering techniques (small-angle scattering and diffraction). Also engineering materials such as metallic alloys will exhibit structure gradients over microscopic distances if they have been processed, for example, by welding or by the growth and further treatment of thin films for coatings.

Not very many synchrotron radiation beamlines are suited for this kind of research, which is why with two exceptions (in Section 3.3.2.1, Nanofocus Endstation at P03) all experiments were performed at the Microfocus Beamline ID13 at ESRF. Remarkably, many setups used a combination of focusing elements. In detail, these were an ellipsoidal mirror and a glass capillary (see Sections 3.3.2.2 and 3.3.2.3), compound refractive lenses and a 2D waveguide (see Section 3.3.2.1). Only the nanobeam used for the high-resolution scan reported in Section 3.3.2.3 and the microbeam for the in situ experiment reported in Section 3.3.2.4 were produced by KB mirror setups alone.

3.3.2.1 Small beams and crystallographic parameters

A microdiffraction experiment with a 2D detector (such as sketched in Fig. 3.9) on a polycrystalline sample yields crystallographic parameters with a high spatial resolution. Most prominently, lattice constants, crystals size, and texture (orientation distribution) can be determined from positions, widths and azimuthal intensity distributions of Debye–Scherrer rings. The ratio between the beam cross-section and the crystal size has a major impact on those rings. Continuous "powder rings" only appear if the crystals are much smaller than the area illuminated by the beam. This is, however, unless for nanocrystalline materials very unlikely for micro- or even nanobeams. Using a quasi-1D waveguide beam of 100 nm × 3 μm, this was demonstrated by an experiment on a powder mixture present in ancient ceramics [33] with a large range of grain sizes present (and consequently more or less grainy diffraction rings). Similarly, grain fining in a friction stir weld of an aluminum alloy drastically changes the appearance of Debye–Scherrer rings (Fig. 3.10; nanobeam generated using a 50 nm × 50 μm waveguide).

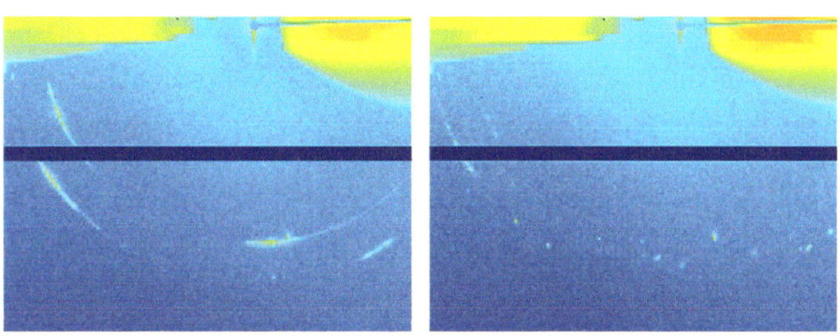

Figure 3.10 Al diffraction rings from a friction-stir welded Al alloy. Left: Fine grains and pronounced deviations from powder texture in the weld. The grains are significantly smaller than the beam size of 50 nm × 50 μm. Right: Grainy rings with individual reflections from the coarse-grain material.

Among the first experimental approaches to use the submicrometer spatial resolution provided by the waveguide-generated nanobeam in the P03 (PETRA III) Nanofocus Hutch was a diffraction experiment on a nanocrystalline TiN coating. Because of their high hardness and wear resistance, protective polycrystalline TiN hard coatings are used in various machining applications, especially for metal cutting or turning. The lifetime and the performance of the coated tools depend decisively on the local gradients of microstructure and residual stress across the hard coatings [37]. The aim of the experiment was to reveal the local gradients across a blasted TiN coating. A polycrystalline 6 μm-thick TiN coating was deposited on WC/Co substrates using a CVD process. The as-deposited coating was blasted using a pressure of $3{\cdot}10^5$ Pa applying Al_2O_3 particles with a diameter of approximately 50 μm. From the coating/substrate composite, a 10 mm-thick slice was prepared by cutting the composite perpendicular to the interface, by mechanical polishing using a diamond saw and finally by focused-ion beam milling.

The 10 μm thick slice was characterized in transmission WAXS geometry by vertically scanning the horizontally aligned coating and the substrate through a waveguide beam of 50 μm (h) × 50 nm (v) using a step size of 0.2 μm [35]. The setup is shown schematically in Fig. 3.11. Also shown in Fig. 3.11 is the distribution of the experimentally observed intensities (integrated Debye–Scherrer rings) as a function of the diffraction angle and of the vertical scanning position. The intensity variation along the TiN reflections indicates the presence of a strong crystallographic texture in the coating. The relative changes of the peak positions (marked by arrows) can be interpreted as the presence of compressive strains, which are obvious especially close to the blasted coating surface. At vertical positions of about 7.5 μm, a WC reflection can be observed at about $2\theta = 2°$.

The above examples from the field of engineering materials science demonstrate the potential of X-ray diffraction studies with small beams. Microstructure and strain (and even depth profiles of these properties) can be two-dimensionally quantitatively mapped with very high spatial resolution. By scaling down the technique of three-dimensional X-ray Diffraction (3D-XRD [38]) for nanobeams even 3D experiments on nanocrystalline materials will soon be enabled at the P03 Nanofocus Endstation.

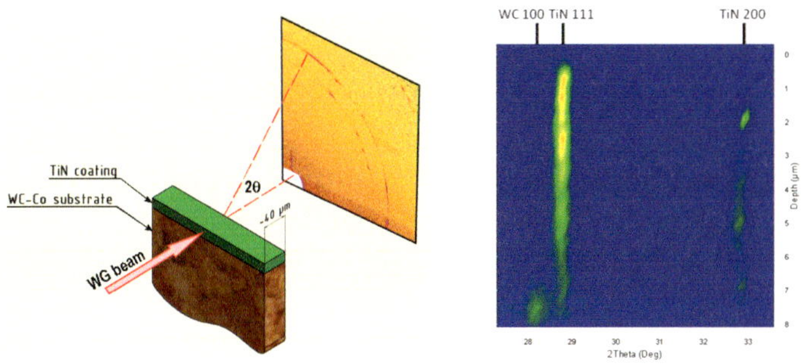

Figure 3.11 Left: schematic setup of the nanodiffraction experiment on a TiN coating deposited on a WC–Co substrate (inset shows a transmission electron microscope image of the actual sample). Right: composite image of the experimentally observed intensities recorded in the nanodiffraction experiment. The intensity modulations of the TiN 111/200 and substrate WC 100 reflections indicate the presence of a strain gradient within the microstructure of the coating. From Ref. [35].

3.3.2.2 µSAXS on single-cellulose fibers

When synchrotron radiation is used for microfocusing X-rays the resulting microbeam still has a small divergence so that small-angle scattering with medium resolution is possible. Even for nanobeams this is true [33]. The main experimental issue in small-angle scattering is the separation of scattering at low angles (arising from contrast on mesoscopic length scales in the sample) from the direct beam, which has to be blinded out by a beamstop in order to protect the detector.

For larger beams, the classical small-angle scattering camera consists of sets of slits (or round apertures) for defining beam size and divergence, a long (frequently several meters) evacuated flight tube behind the sample, with the beam stop just in front of the detector at the end of the tube. The camera length then results from the requirement that small scattering angles have to be well defined for an extended primary beam. For microbeams, much shorter setups provide an adequate resolution, which has been shown to be mainly limited by the detector resolution [39,40].

The first experiment of its kind, an investigation of cellulose nanocrystals (so-called microfibrils [41]) orientation in flax fibers [42], was still carried out with a "classical" SAXS set-up. The fiber with a diameter of 20 μm was scanned through a 2 μm beam. The 2D SAXS pattern is an equatorial streak due to the scattering contrast between the cellulose microfibrils and the embedding matrix (Fig. 3.12a). The smallest accessible wave vector transfer Q is 0.04 $Å^{-1}$ corresponding to a largest resolvable structure of $d = 2\pi/Q = 160$ Å. The experiment revealed the homogeneously high orientation of the cellulose nanocrystals within the fiber (Fig. 3.12b; 7° FWHM). Measurements of fiber bundles would always yield a too large deviation from perfect alignment as the individual fibers would never be aligned perfectly parallel (in the experiment with the same fibers 12° FWHM). Another advantage of a single fiber SAXS experiment is the reduction of pore and surface scattering at low Q [42].

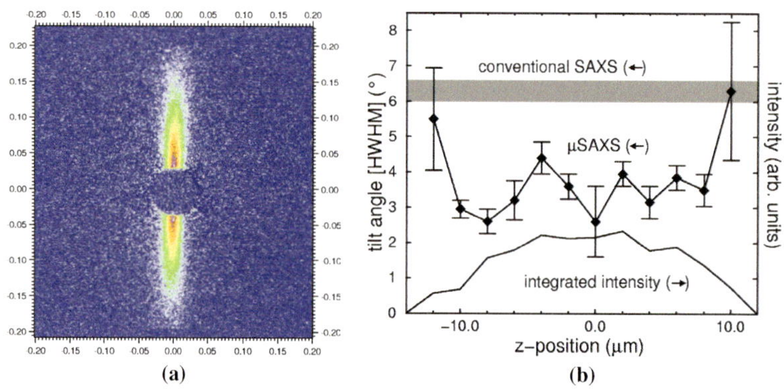

Figure 3.12 (a) Two-dimensional μSAXS pattern of a horizontally oriented single flax fiber in the water-swollen state. (b) Position (z) dependence of integrated scattering intensity (solid line in lower part) and microfibril misalignment determined from the streak widths (hwhm) in the two-dimensional μSAXS patterns (full circles). For comparison, the misalignment measured in a conventional SAXS experiment (averaged over a bundle of fibers) is also displayed (the shaded area visualizes the error bars). From Ref. [41].

3.3.2.3 2D microdiffraction scanning of the wood cell wall

The quite regular structure of softwood tracheids makes them an ideal object for systematic studies of the layered plant cell wall structure. Most tracheids have a rectangular shape (with rounded corners) and a typical diameter between 20 and 40 μm. The main structural material of wood cell walls is cellulose; its crystalline units are called microfibrils. In the case of softwood, typical microfibril sizes of ~30 Å laterally (i.e., perpendicular to the molecular chain direction) are found in the secondary walls, whereas the length of the crystals may be of the order of 100 nm [41].

The main difference between the individual wood cell wall layers concerns the arrangement of the microfibrils with respect to one another. The primary wall exhibits little order, the microfibrils form a disordered fibrillar network. In contrast, in the secondary walls the microfibrils are laid down in a parallel fashion: The c axes of all crystals are all well aligned in a particular direction. There is, however, no preferred orientation of the a and b directions. Such a fiber texture is observed in many nanocrystalline biological systems. Usually, a fiber texture is not perfect in the sense that there is some variation of the c axis direction around a mean orientation. X-ray diffraction is extremely sensitive to such minute details of microfibril orientation.

On the next hierarchical structure level, the parallel microfibrils in the secondary cell wall layers are wound in a helical fashion around the wood cell. The angle of the microfibril helix with respect to the longitudinal cell axis is called the microfibril angle (MFA) and is thought to vary between the individual secondary cell wall layers. The thickest layer, S2, dominates the cell wall's mechanical properties.

A direct imaging of the local orientation of the cellulose microfibrils in wood cell walls is possible by scanning a thin (e.g., 10 μm thick) cross section through a microbeam. Such an experiment was carried out for the first time by Lichtenegger et al. [36] on Norway spruce (*Picea abies*) latewood xylem tissue. A tracheid cell cross-section and parts of the neighboring tracheids, in total, an area of 52 × 42 μm^2 was scanned in a mesh scan with 2 μm step size. At every point, an asymmetric fiber diffraction pattern was recorded (Fig. 3.13a). The origin of this asymmetry is the tilt of the fiber-textured oriented cellulose microfibrils against the X-ray beam in the case of a non-zero MFA [43,44]. The complete trace of the microfibrils around the central tracheid's cell wall and the attached neighboring cell walls could be covered.

Every single asymmetric diffraction pattern was "translated" into an arrow [36,43]. The direction of the arrow corresponds to the meridian of the diffraction pattern and indicates the direction a microfibril would point upward when viewed along the microbeam. The arrow length is proportional to the projection of a unit length microfibril viewed in the same way. By evaluating all diffraction patterns of the composite image in Fig. 3.13a in this way, a complete map of microfibril orientation is obtained (Fig. 3.13b). It can clearly be seen that the sense of the helices in neighboring cells is the same: at the border between two cell walls, the direction of the arrows changes. Concerning the homogeneity of the MFA, there are small deviations from the mean MFA (20°), in particular in the cell corners.

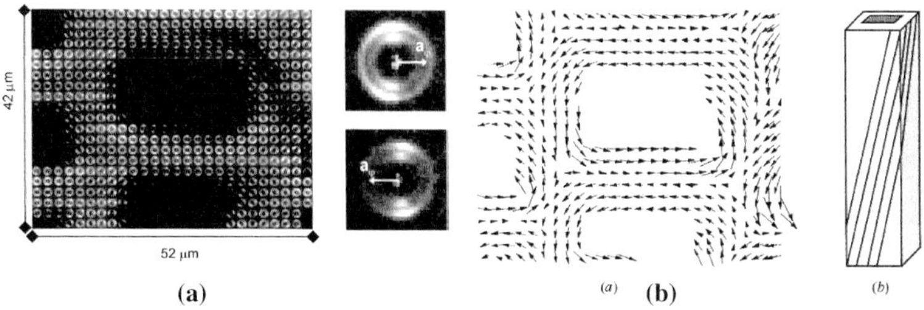

Figure 3.13 (a) Mesh scan of a spruce wood cross-section with an X-ray microbeam (pixel size: 2 × 2 µm). A complete tracheid and parts of neighbouring cells were investigated. Dark regions correspond to lumina, bright regions showing a scattering signal correspond to cell walls. Two typical diffraction patterns are shown at greater magnification. Note the asymmetric angular intensity distribution. The direction of asymmetry indicates the local orientation of the cellulose microfibrils and can be translated into the arrows shown. (b) Complete map of local cellulose microfibril orientations as calculated from Fig. 3.13a. Following the arrows readily yields the trace of the microfibrils around the cell. Longer arrows denote a more pronounced asymmetry of the diffraction patterns corresponding to smaller local MFA; shorter arrows denote larger MFA. The arrow map can be translated into a three-dimensional model: the cellulose microfibrils trace a *Z* helix around the tracheid. From Ref. [36].

The position resolution of 2 µm, given by the beam size, is insufficient to yield any information on MFA variation between the different secondary wall layers or within a layer. However, by reducing the beam size further this limitation can be overcome. In recent test experiments with a nanobeam at ID13 (ESRF) the individual cell wall layers could be visualized. In such experiments the sample preparation is of major importance as an obliquely cut cross section would make it difficult to orient the sample such that the X-ray beam only penetrates a single cell wall layer. Only then the details of the transition between two layers can be clarified. The high spatial resolution allows one to observe the local breakdown of fiber texture [44].

3.3.2.4 In situ deformation of single wood cells

The in situ combination of mechanical tests with structural investigations on a molecular level is clearly a domain of X-ray diffraction using synchrotron radiation. Single tracheid (see discussion earlier) deformation experiments combined with X-ray microbeam diffraction experiments help distinguishing between collective deformation mechanisms in the tissue and those of a single wood cell. An example is shown in Fig. 3.14 [45]. The experiment was carried out at ID13 with a beam size of about 5 μm in diameter. During the tensile test, the 30 μm thick cell was continuously scanned through the beam in order to minimize beam damage effects. The graph shows the macroscopic stress–strain curve of the mechanically extracted dry (3 percent RH in the sample cell) pine wood tracheid together with the crystal lattice strain of the cellulose microfibrils (from the 004 cellulose reflection) in fiber direction. It is possible to scale this curve such that it closely follows the macroscopic stress-strain curve, suggesting a direct mechanical stress transfer to the microfibrils. This serial arrangement of crystals and embedding matrix represents the Reuss model in polymer science [45]. It usually breaks down at higher strain in case of intact wood tissue.

Using special sample environments, such in situ experiments can be carried out under controlled humidity conditions. The simplest way to do so is to enclose the air volume around the sample by, for example, using weakly scattering Kapton foil (polyimide) and inserting a basin containing water (100 percent RH), silica gel (a few percent RH) or a saturated salt solution for intermediate relative humidity. Inside thin polar polymer fibers equilibrium conditions corresponding to specific moisture content are reached within a few seconds. The drastic changes of the mechanical properties of, for example, wood and silk are reflected in changes of the load transfer between the embedding matrix and the nanocrystals.

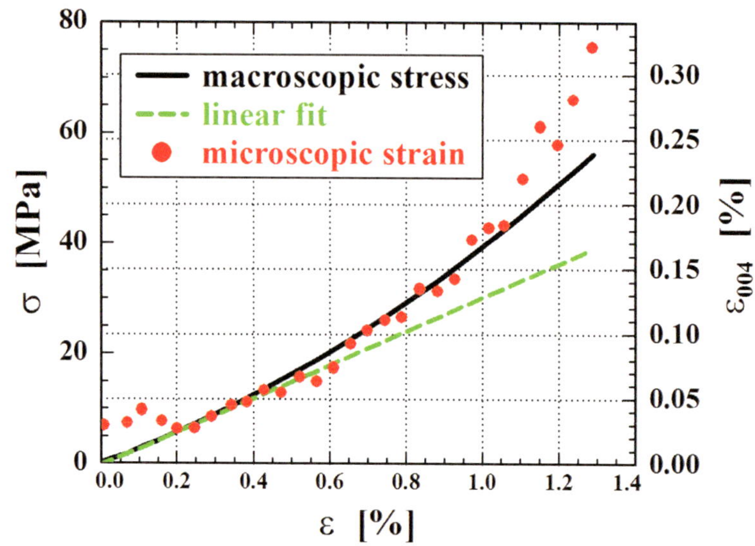

Figure 3.14 Stress-strain curves of a single pine wood tracheid cell (solid lines, right-hand side scale). The symbols give the strain ε_{004} (right-hand side scale) of the crystal lattice in axial direction, as determined from the meridional 004 reflection. From [45].

3.3.3 Beam Damage in Microdiffraction Experiments

The high X-ray flux density in a microfocus beam—10^{10} photons/s in a 5 μm diameter beam yield $5{\cdot}10^{16}$ photons/s/cm^2—means that the technique is not completely nondestructive as is normally assumed for X-ray scattering experiments. In particular, organic material is quickly destroyed by beam damage. Despite this obvious problem, no extensive studies of beam damage effects in high-flux microbeams have been published yet, mainly owing to the fact that true high-flux micro- and nanobeams have become available only very recently.

Beam damage limits to some extent the in situ capabilities of microdiffraction. For native fibers such as silk or cellulose data should always be taken at the same spot; otherwise the natural spatial variation of structural properties (see [46] for cellulose fibers) could be mistaken for changes under, for example, mechanical load. In the beam of the above properties, Bragg reflection intensities of silk, a protein fiber, start to decay significantly only after 2 min integral exposure (repeatedly 2 s exposures, 0.5 s delays). Tensile test experiments such as the one described in Section 3.3.2.4 (wood cellulose withstands the beam about 10 times longer than silk) thus have to be finished within that time, challenging the experimental setup, in particular the detector speed (efficiency and read-out).

The nature of the beam damage occurring in polymers is not completely clear. There is some evidence that secondary beam damage, that is, the diffusion of by primary beam damage created radicals, plays a major role. Diffusivity of water and other molecules can be reduced by cooling of the sample low temperatures. Another option for less secondary beam damage is a treatment of the sample with free-radical scavengers that will immobilize or bind the radicals that are inevitably created by the incident radiation (e.g., antioxidants) [47]. A different way to reduce secondary beam damage would require faster detectors. If a point on the samples can be measured before the diffusing radicals have reached the spot flux density could even be increased further. Micro- or nanofocus maps would then be destructively, however, date would still be obtained from always undamaged regions of the sample.

3.4 Summary

X-ray diffraction with micro- and nanofocused beams is a relatively new technique, bearing lots of potential for improvement. Recent technological advances and new developments have enabled the use of beams confined to sizes of as small as few tens of nanometers, hence very close to the theoretical limits. With these nanofocused beams new valuable information can be derived, for example, from hierarchically structured systems (both biological and synthetic). Due to the high spatial resolution of nanofocused beams the amount of averaging is significantly reduced, otherwise smearing the information content when a beam is used that is larger than the structural length scales. Selected examples from materials and life science are shown earlier, although of course a much wider range of applications is thinkable. One of the biggest challenges in using submicrometer high-flux beams, however, remains the radiation damage that needs to be carefully taken care of especially for biological samples. Here, extensive studies are required to improve the knowledge about the processes leading to radiation damage.

References

1. Bergemann C, Keymeulen H, van der Veen JF (2003) Focusing X-ray beams to nanometer dimensions, *Phys Rev Lett*, **91**, 204801.
2. Fankuchen I, Mark H (1944) X-ray studies of chain polymers, *J Appl Phys*, **15**, 364–370.
3. Als-Nielsen J, McMorrow D (2001) *Elements of Modern X-Ray Physics*, Wiley, New York.
4. Kirkpatrick P, Baez AV (1948) Formation of optical images by X-rays, *J Opt Soc Am*, **38**, 766–773.
5. Snigirev A, Kohn V, Snigireva I, Lengeler B (1996) A compound refractive lens for focusing high-energy X-rays, *Nature*, **384**, 49–51.
6. Kirz J (2001) *X-Ray Data Booklet*, 2nd ed, LNBL, University of Berkeley, Berkeley.
7. Lengeler B, Schroer CG, Kuhlmann M, et al. (2005) Refractive X-ray lenses, *J Phys D: Appl Phys*, **38**, A218–A222.
8. Last A (2014) http://www.X-ray-optics.com, Image reproduced with kind permission of Arndt Last, Institute of Microstructure Technology, Karlsruhe Institute of Technology, 76021 Karlsruhe, Deutschland.
9. Schroer CG, Kuhlmann M, Hunger UT, et al. (2003) Nanofocusing parabolic refractive X-ray lenses, *Appl Phys Lett*, **82**, 1485–1487.
10. Schroer CG, Lengeler B (2005) Focusing hard X-rays to nanometer dimensions by adiabatically focusing lenses, *Phys Rev Lett*, **94**, 054802.
11. Noehammer B, Hoszowska J, Freund AK, David C (2003) Diamond planar refractive lenses for third- and fourth-generation X-ray sources, *J Synchr Rad*, **10**, 168–171.
12. Nazmov V, Reznikova E, Somogyi A, Mohr J, Saile V (2004) Planar sets of cross X-ray refractive lenses from SU-8 polymer, *Proc SPIE*, **5539**, 235–243.
13. Nazmov V, Reznikova E, Last A, et al. (2007) Crossed planar X-ray lenses made from nickel for X-ray micro focusing and imaging applications, *Nucl Instrum Meth Phys Res A*, **582**, 120–122.
14. Chao W, Harteneck BD, Liddle A, Anderson EH, Attwood DT (2005) Soft X-ray microscopy at a spatial resolution better than 15nm, *Nature*, **435**, 1210–1213.

15. Schroer CG (2006) Focusing hard x rays to nanometer dimensions using Fresnel zone plates, *Phys Rev B*, **74**, 033405.
16. Yun W, Lai B, Krasnoperova AA, et al. (1999) Development of zone plates with a blazed profile for hard X-ray applications, *Rev Sci Instrum*, **70**, 3537–3541.
17. Jark W, Matteucci M, Menk RH (2008) On the use of clessidra prism arrays in long-focal-length X-ray focusing, *J Synchrotron Rad*, **15**, 411–413.
18. Kang HC, Yan H, Winarski RP, et al. (2008) Focussing of hard X-rays to 16 nanometers with a multilayer Laue lens, *Appl Phys Lett*, **92**, 221114.
19. Kipp L, Skibowski M, Johnson RL, et al. (2001) Sharper images by focusing soft X-rays with photon sieves, *Nature*, **414**, 184–188.
20. David C, Souvorov A (1999) High-efficiency Bragg-Fresnel lenses with 100 nm outermost zone width, *Rev Sci Instr*, **70**, 4168–4173.
21. Yumoto H, Mimura H, Matsuyama S, et al. (2005) Fabrication of elliptically figured mirror for focusing hard x rays to size less than 50 nm, *Rev Sci Instr*, **76**, 063708.
22. Mimura H, Yumoto H, Matsuyama S, et al. (2007) Efficient focusing of hard x rays to 25 nm by a total reflection mirror, *Appl Phys Lett*, **90**, 051903.
23. Mimura H, Handa S, Kimura T, et al. (2010) Breaking the 10 nm barrier in hard-X-ray focusing, *Nat Phys*, **6**, 122–125.
24. Hignette O, Cloetens P, Rostaing G, Bernard P, Morawe C (2005) Efficient sub 100 nm focusing of hard x rays, *Rev Sci Instr*, **76**, 063709.
25. Morawe C, Osterhoff M (2010) Curved graded multilayers for X-ray nano-focusing optics, *Nucl Instr Meth Phys Res A*, **616**, 98–104.
26. Bilderback DH, Thiel DJ (1995) Microbeam generation with capillary optics, *Rev Sci Instrum*, **66**, 2059–2063.
27. Snigirev A, Bjeoumikhov A, Erko A, et al. (2007) Submicrometer hard X-ray focusing using a single-bounce ellipsoidal capillary combined with a Fresnel zone plate, *J Synchr Rad*, **14**, 227–228.
28. Jarre A, Seeger J, Ollinger C, Fuhse C, David C, Salditt T (2007) X-ray waveguide nanostructure: design, fabrication, and characterization, *J Appl Phys*, **101**, 054306.
29. Neubauer H (2010) private communication, Image reproduced with kind permission of Henrike Neubauer, Georg-August-University Göttingen, Institute for X-ray Physics, 37077 Göttingen, Germany.

30. Cedola A, Lagomarsino S, Di Fonzo S, Jark W, Riekel C, Deschamps P (1998) Submicrometre beams from a hard X-ray waveguide at a third-generation synchrotron radiation source, *J Synchrotron Rad*, **5**, 17–22.
31. Cusack S, Belrhali H, Bram A, Burghammer M, Perrakis A, Riekel C (1998) Small is beautiful: protein micro-crystallography, *Nat Struct Biol*, **5**, 634–637.
32. Riekel C (2000) New avenues in X-ray microbeam experiments, *Rep Progr Phys*, **63**, 233–262.
33. Müller M, Burghammer M, Flot D, et al. (2000) Microcrystallography with an X-ray waveguide, *J Appl Cryst*, **33**, 1231–1240.
34. Krywka C, Döhrmann M, Roth SV, Müller M (2010) The nanofocus endstation of the MINAXS beamline of PETRA III, *AIP Conf Proc*, **1234**, 879–882.
35. Krywka C, Neubauer H, Priebe M, et al. (2012) A two-dimensional waveguide beam for X-ray nanodiffraction, *J Appl Cryst*, **45**, 85–92.
36. Lichtenegger H, Müller M, Paris O, Riekel C, Fratzl P (1999) Imaging of the helical arrangement of cellulose fibrils in wood by synchrotron X-ray microdiffraction, *J Appl Cryst*, **32**, 1127–1133.
37. Massl S, Keckes J, Pippan R (2007) A direct method of determining complex depth profiles of residual stresses in thin films on a nanoscale, *Acta Mater*, **55**, 4835–4844.
38. Poulsen HF (2004) *Three-Dimensional X-Ray Diffraction Microscopy: Mapping Polycrystals and Their Dynamics*, Springer, Berlin, Heidelberg; *Springer Tracts in Modern Physics*, vol. 205.
39. Riekel C, Burghammer M, Müller M (2000) Microbeam small-angle scattering experiments and their combination with microdiffraction, *J Appl Cryst*, **33**, 421–423.
40. Müller M, Burghammer M, Riekel C (2001) Combined scanning microdiffraction and micro small-angle scattering at the microfocus beamline ID13 at the ESRF, *Nucl Instrum Meth A*, **958–961**, 467–468.
41. O'Sullivan AC (1997) Cellulose: the structure slowly unravels, *Cellulose*, **4**, 173–207.
42. Müller M, Czihak M, Vogl G, Fratzl P, Schober H, Riekel C (1998) Direct observation of microfibril arrangement in a single native cellulose fiber by microbeam small-angle X-ray scattering, *Macromolecules*, **31**, 3953–3957.
43. Paris O, Müller M (2003) Scanning X-ray microdiffraction of complex materials: diffraction geometry considerations, *Nucl Instrum Meth B*, **200**, 390–396.

44. Ogurreck, M, Lichtenegger H, Müller M (2010) Analytical description of the scattering of cellulose nanocrystals in tracheid wood cells, *J Appl Cryst*, **43**, 256–263.

45. Müller M, Krasnov I, Ogurreck M, Blankenburg M, Pazera T, Seydel T (2011) Wood and silk: hierarchically structured biomaterials investigated in situ with X-ray and neutron scattering, *Adv Eng Mat*, **13**, 767–772.

46. Kölln K, Grotkopp I, Burghammer M, et al. (2005) Mechanical properties of cellulose fibres and wood – Orientational aspects in situ investigated with synchrotron radiation, *J Synchr Rad*, **12**, 739–744.

47. De la Mora E, Carmichael I, Garman EF (2011) Effective scavenging at cryotemperatures: further increasing the dose tolerance of protein crystals, *J Synchr Rad*, **18**, 346–357.

Chapter 4

Small-Angle X-Ray Scattering

Ulla Vainio
Deutsches Elektronen Synchrotron (DESY), Notkestraße 85, 22607 Hamburg, Germany
Helmholtz-Zentrum Geesthacht, Max-Planck-Str 1, 21502 Geesthacht, Germany
ulla.vainio@hzg.de

Small-angle X-ray scattering (SAXS) is a convenient tool for investigating structures in the nanometer scale in the field of materials science, environmental sciences as well as in biology. Conventional transmission SAXS deals with samples in the "bulk" form, whereas grazing-incidence SAXS (GISAXS) can be used to investigate nanometer scale structures on surfaces. Anomalous SAXS (ASAXS) is a contrast variation method by which one can obtain element specific information about the nanostructures.

The SAXS methods are complementary to electron microscopy and atomic force microscopy techniques. Providing statistically averaged information, SAXS methods are especially useful for obtaining structural information in situ as a function of time, for example, during a chemical process, or as a function of some external parameter such as temperature, pressure, or moisture.

X-Ray Diffraction: Modern Experimental Techniques
Edited by Oliver H. Seeck and Bridget M. Murphy

ISBN 978-981-4303-59-0 (Hardcover), 978-981-4303-60-6 (eBook)
www.panstanford.com

4.1 Introduction

Small-angle X-ray scattering (SAXS) was first observed at end of the 1920s by many researchers fairly simultaneously, but it was the work of A. Guinier in the 1930s which started a rapid development of this new phenomenon into a useful method. During his PhD work which was related to diffraction of metal alloys, Guinier quantified the intense scattering in the centers of his diffraction patterns, at *small angles*, to be related to the nano meter scale structures in the samples [1,2].

In the following decades, plenty of SAXS studies that were made using laboratory X-ray tubes contributed to the development of the theoretical concepts relevant to SAXS. Later, the emergence of synchrotron sources made it possible to study weakly scattering samples and to make time-resolved studies due to the high flux of the synchrotron radiation. The first synchrotron SAXS studies were made already in the 1970s from muscle fibers at the DESY storage ring in Hamburg [3]. By the end of the decade, time-resolved SAXS studies of contracting muscle fibers with 10 ms time resolution were possible [4]. Nowadays, most synchrotrons in the world have dedicated SAXS beamlines.

With synchrotrons becoming readily available, completely new SAXS techniques were developed in the 1980s. H.B. Stuhrmann was the first to extend the concept of anomalous scattering to small angles and used ASAXS to make element-specific measurements of nanometer-scale structures [5]. The ASAXS method would not be feasible without the possibility of tuning the photon energy continuously. At the end of the decade, also the surface-sensitive GISAXS method was invented [6], and soon it became a popular method to complement microscopy studies. While electron microscopy techniques and atomic force microscopy are excellent tools for studying the local structure of surfaces, GISAXS can give statistically averaged information about structures on surfaces without worries about possible artifacts arising from sample preparation.

In this chapter, a basic introduction to SAXS is given. More complete presentations can be found in these three books: *Small Angle Scattering of X-Rays* (1955) by Guinier and Fournet, *Small Angle X-Ray Scattering* (1982) edited by Glatter and Kratky, and *Structure Analysis by Small-Angle X-Ray and Neutron Scattering* (1987) by Feigin and Svergun.

4.2 Experimental Setup

In a typical SAXS experiment, a two-dimensional (2D) X-ray detector is placed a few meters away from the sample to detect scattered X-rays at small angles. Only a small part of the X-rays are scattered by the sample, while most of the incident X-ray beam (primary beam) travels through the sample. Currently available position-sensitive SAXS detectors can get saturated or spoiled if the primary beam hits the detector, therefore a beamstop must be positioned carefully in front of the detector to receive the whole primary beam. Figure 4.1 shows a schematic drawing of a SAXS setup with a wide-angle X-ray scattering (WAXS) option. The basic SAXS setup can be extended and modified to include other methods such as WAXS, fluorescence spectroscopy, infrar ed spectroscopy, and so forth. The sample environment can accommodate setups for various in situ experiments, starting from simple heating and stretching experiments all the way to novel stopped-flow mixing cells or even rheology. In principle, there are no limitations; one only needs to be able to put the X-ray beam through the cell somehow and leave some space for the scattering that comes out.

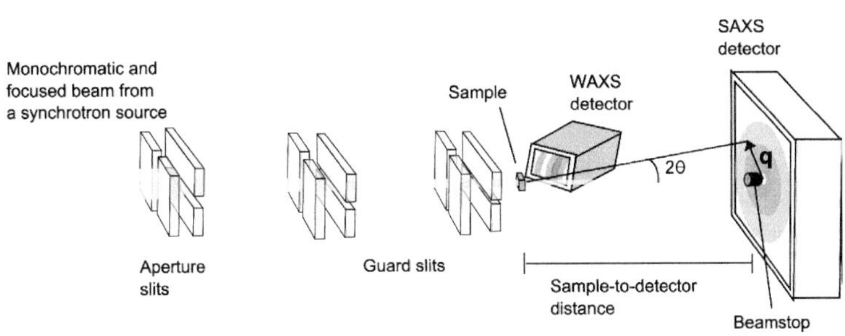

Figure 4.1 A schematic SAXS/WAXS (wide angle X-ray scattering) setup at a synchrotron SAXS beamline. The sample-to-detector distance can usually be varied from less than a meter to several meters. The scattering angle is denoted by 2θ and the scattering vector by $\mathbf{q}$.

The biggest challenge in synchrotron SAXS is to find a way to minimize scattering and artifacts due to the beamline components. Dedicated SAXS beamlines are optimized to provide conditions where this background is as low as possible. Usually, it is also possible to change the sample-to-detector distance at such instruments. The resolution of the SAXS camera is limited by the X-ray beam size at the detector, by pixel size of the detector, by the size of the beamstop, and the sample-to-detector distance.

Correction for background scattering can only be done when the transmission of the sample is known accurately. For this purpose, the beamstop may contain a photodiode by which the primary beam intensity is monitored so that transmission of the sample, that is, the intensity of the direct beam going through the sample compared to intensity without sample, can be measured. Alternatively, a separate photodiode can be moved into the beam for a short period to measure the transmission of the sample before measuring SAXS.

The absorption and scattering of X-rays by air is significant at the X-ray energies at which SAXS is typically measured, therefore usually the beam path before and after the sample is fully in vacuum (or in helium). Even the detector may be placed in vacuum. If the sample needs to be measured in air, the beam path in air is kept as short as possible. Care must be taken that any scattering from the vacuum tube windows before or after the sample does not reach the detector. Therefore, if the detector is placed in air, right after a vacuum window, the beamstop should be placed on the vacuum side to prevent scattering from the vacuum window and the air path that follows. Unfortunately, it is hard to find materials that are nearly transparent to X-rays, nontoxic, and do not scatter but are mechanically strong enough to withstand the atmospheric pressure. Polyimide and beryllium are currently the most frequently used materials for bigger windows.

Every material that comes into contact with the beam is a potential source of background. In addition to scattering from air and window materials, various beamline components, especially the beamstop or the slits, may give a strong inhomogeneous background due to scattering, refraction or fluorescence. Scattering from slits not only spoils parts of the scattering pattern but also limits the smallest achievable scattering angle of the SAXS camera and thus spoils the resolution. Therefore, guard slits are usually used in eliminating slit scattering from the aperture slits. Beamstop material should be selected such that it does not have an absorption edge at the energy range of the beamline, or alternatively the beamstop can be made into a sandwich structure out of different materials. The outer parts of the sandwich structure are selected so that they absorb the fluorescence coming from of the inner parts.

In summary, the background coming from various sources should be minimized before the experiment because correcting the scattering images afterward for various artifacts is usually not straightforward, if at all possible.

As examples of high brilliance SAXS beamlines at third-generation synchrotron sources let us mention just a few that are used for time-resolved studies: ID02 at ESRF in France [7], BioCAT at APS in the US [8], BL45XU-SAXS at Spring-8 in Japan [9], and MiNaXS beamline at PETRA III in Germany that started operation in spring 2011.

4.2.1 Sample Cells and Optimal Sample Thickness

Designing the optimal sample cell for samples that are not thin films but for example powders or liquids is a hard task. For sample cells or cuvettes with windows, thin films of Kapton, Mylar, polystyrene, or mica are often used as window materials [10]. These materials are not perfect because they give scattering at small angles, and Kapton even gives strong diffraction rings. Diamond used for high-pressure sample cells is otherwise good, but gives the so-called Kossel lines in the scattering patterns [11]. In case of liquids, recently developed ultrasonic traps [12] and free liquid jets [13,14] circumvent the problem because in these methods no windows are needed, but unfortunately they work only in special circumstances. Alternatively, the sample can be placed in thin-walled quartz or borosilicate glass capillaries or mark tubes having wall thicknesses of about 10 μm or more. It is then important to make sure that the beam size is much smaller than the size of the capillary in order to avoid total reflection of slit scattering or beam halo from the curved capillary walls. Additionally, the capillary walls can scatter differently depending on the position of the X-ray beam on the capillary and whether the capillary is filled with liquid or not. A flat sample cell is usually a better option than capillaries.

The perfect sample thickness t is such that the sample interacts with the X-rays enough to give a good signal but not so thick that the entire signal is absorbed by the sample. Samples that are too thick may give rise to unwanted multiple scattering events. The optimal thickness can be determined from the linear absorption coefficient μ of the material, which varies with the photon energy. Based on the compromise between scattering volume and absorption, transmission T of $\sim 1/e \approx 0.37$ is desirable. The transmission of the sample is the intensity I of primary beam after going through the sample divided by the intensity I_0 of the beam without sample

$$T = \frac{I}{I_0} = \exp(-\mu t). \qquad (4.1)$$

The values of linear absorption coefficients can be found in various databases, such as the one from NIST [15]. For example, for water at 8 keV, $\mu \approx 10\ \text{cm}^{-1}$ gives an optimal thickness of $1/\mu \approx 1$ mm, while for copper at 16 keV, $\mu \approx 600\ \text{cm}^{-1}$ corresponds to an optimal thickness of 17 μm.

4.2.2 Corrections to Experimental Data

Several corrections need to be carried out before the scattering images (see Fig. 4.2) from a 2D SAXS detector can be analyzed:

1. Subtraction of "dark current" from the image. This is usually a combination of accumulated background radiation (depends on measurement time) and some sort or readout noise of the detector (time independent).
2. Normalization of the 2D image with the detector "flat field" to correct for different sensitivities in different areas of the detector.
3. Division by primary beam intensity measured before the sample to correct for fluctuations in the primary beam intensity.
4. Division by transmission T to correct for absorption of the sample.
5. Subtraction of a background image coming from air, solvent, sample cell, or whatever else is in the beam in addition to the sample particles or material. The background image should be corrected like in steps 1–4.
6. Optional: Normalization to absolute units. This will be explained in the following section.

   In case of isotropic scattering from nonoriented samples, the 2D image can be integrated to get a 1D curve, where intensity is expressed as a function of the length q of scattering vector $\mathbf{q}$. In general, these steps are needed:
7. Determination of beam center in the image.
8. Masking of bad spots, beamstop area, slit scattering, and so forth.

9. Integrating the intensities in each pixel to a certain pixel or q range over angle φ while ignoring the masked areas. The intensities must be normalized by the number of pixels in each integration bin to preserve the relative intensity scale in radial direction.

The center of the primary beam can be determined either through a semitransparent beamstop or attenuator, or by applying an algorithm that finds the center of the scattering rings. Figure 4.2 demonstrates parts of the integration process.

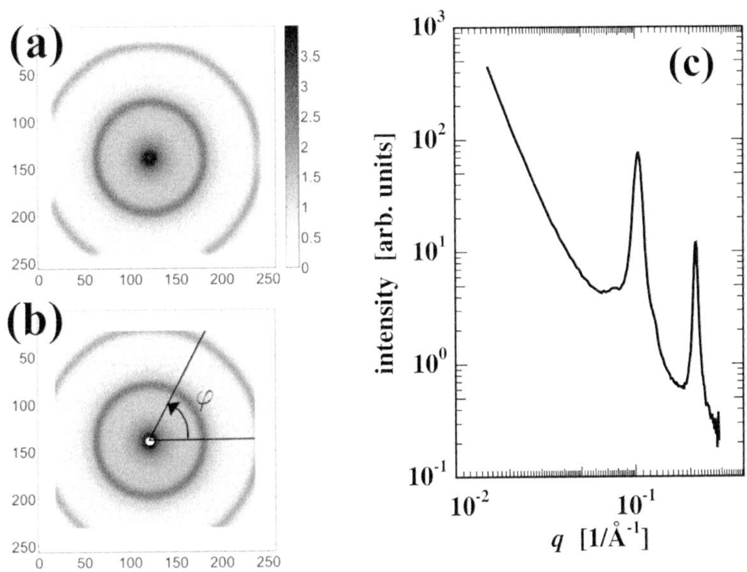

Figure 4.2 (a) A 2D SAXS image of a silver behenate sample on logarithmic intensity scale. The axes show pixel number. Darker areas have more intensity. (b) The same SAXS image after applying a mask for the beamstop area and for the less-sensitive edge areas of the detector. These masked areas are marked here with white color and need to be ignored during integration over the angle φ. (c) The same SAXS data plotted as a function of scattering vector q after integrating φ over $[0, 2\pi]$.

To calibrate the scattering vector length ($q = 4\pi\sin\theta/\lambda$, where θ is half of the scattering angle and λ is the wavelength of the photons, see Section 4.3), one needs to know exactly the distance to the active layer of the detector, the lateral size of the detector pixels and the wavelength (or energy) of X-rays within the relative accuracy that one wishes to achieve for the q scale.

In some types of detectors, for example, gas-filled detectors, the position of the detection layer of the detector is less well defined and also the spatial resolution or "pixel" size may vary from point to point. In any case, it is advisable to calibrate or at least to check the q scale using crystalline materials such as silver behenate (largest period 5.84 nm [16]) or collagen from rat tail tendon (largest period 65 nm). These materials give sharp diffraction peaks at the small angles. Due to stability issues of these materials, so far they have not made it to the list of real *NIST standards*, and indeed, one should note that there might also be differences in the diffraction patterns of materials obtained from different sources.

In synchrotron SAXS studies, smearing of the data is usually not a problem. However, it should be kept in mind that there are always some smearing effects due to the beam size, the limited pixel size of the detector, and the detector response function. The smearing by beam size changes depending on if the focus is at the detector (small smearing) or at the sample (larger smearing).

If smearing is not included in the model or the data is not "de-smeared" before fitting it, this affects, for example, the determination of particle-size distributions, which are then perceived broader than they really are. An analytical treatment of the resolution function is given in Ref. [17].

4.2.3 Absolute Intensity Scale

Intensity is a measure of the scattering power of the sample. Measurements made at different instruments can be easily compared to each other when everybody uses the so-called absolute intensity units. Further, absolute intensities are useful when studying fluctuation in liquids, the specific surface of porous materials or, for example, the molecular weight of proteins.

The scattered intensity is measured in counts, but physically the measured quantity is in fact the differential scattering cross-section $d\sigma/d\Omega$ [cm^2] at the solid angle Ω. Normalizing by the irradiated volume V and multiplying by the number of scattering units N gives the macroscopic scattering cross-section

$$\frac{d\Sigma}{d\Omega}=\frac{N}{V}\frac{d\sigma}{d\Omega}. \tag{4.2}$$

This function has the units[$\mathrm{cm}^2/\mathrm{cm}^3$ = cm^{-1}] and it can be transformed to electron units [e.u./Å^3] by dividing with the classical electron radius squared r_e^2.

There are several ways to normalize the measured intensity to absolute scale. One can either use the information about the beamline [18], or use secondary standards such as water [19,20], polyethylene (especially Lupolen [21]), or glassy carbon [22]. When using polymer standards, one has to be very careful with radiation damage at synchrotron sources.

When using secondary standards, the calibration procedure is fairly simple. After correcting and integrating the scattering from the sample and the secondary standard, the sample intensity $I_1(q)$ is put onto absolute scale by

$$\frac{d\Sigma_1}{d\Omega}(q)=\frac{I_1(q)}{t_1A_1}\cdot\frac{t_sA_s}{I_s(q_0)}\cdot\frac{d\Sigma_s}{d\Omega}(q_0), \tag{4.3}$$

where $I_s(q_0)$ is the measured intensity of the standard at certain $q = q_0$ value, $[d\Sigma_s/d\Omega](q_0)$ is the intensity of the standard in absolute units at the same q value, t_1 and A_1 are the thickness and irradiated area of the sample, and t_s and A_s are the respective parameters for the standard sample. Obviously, if the beam size is the same for both measurements, the area can be omitted.

4.3 Theory

X-ray scattering experiments are usually divided into experiments at small angles (SAXS) and at wide angles (WAXS). The definition of small angle is arbitrary because the scattering angle depends on the energy of the photons, and at synchrotron beamlines, the energy can typically vary. Usually, one can think that SAXS signal is the scattering contribution coming from the sample at $q < 10\ \mathrm{nm}^{-1}$. Throughout this chapter, we will use the definition $q = 4\pi\sin\theta/\lambda$, where θ is half of the scattering angle, for the length of the scattering vector.

All small-angle X-ray scattering techniques (SAXS, GISAXS, ASAXS) rely on common basic assumptions. Small-angle studies are attractive because at small scattering angles nearly all angle-dependent corrections to scattering patterns can be neglected. This is very different from X-ray diffraction (XRD) experiments, where at large angles it is important to take into account the angle dependent corrections for polarization, absorption, geometry, and so on.

It is important to realize that there is no fundamental difference between SAXS and WAXS or XRD. Very often SAXS measurements are made because one needs to find out the periodicity of a certain periodic structure in the nanometer size range. In that case, the diffraction peaks simply happen to be in the "SAXS range". Such is the case for example for lipids that are a vital part of the cell walls of most living beings. In this case, one simply determines the positions of the diffraction peaks to determine the crystalline structure. As in all scattering data, the position of a peak in the scattering image (inverse space) is inversely related to the period d of the structure in real space. Therefore, in SAXS, $d = 2\pi/\hat{q}$, where $\hat{q}$ is the position of the diffraction peak. The larger the structure is in real space, the smaller is the scattering angle to which it scatters.

In addition to diffraction peaks from crystallites, there are a large variety of structures that can contribute to the SAXS signal. The size of these structures is typically in the nanometer range (1–100 nm). To understand the full SAXS signal, one needs to be aware of the different kind of structures that can exist in the nanometer scale in all kinds of materials, ranging from proteins in solution to phase separation in metallic glasses. We start by looking at SAXS signals coming from nanoparticles and crystallites of different shape.

Figure 4.3 demonstrates how differently shaped crystallites give different peak shapes in WAXS. These curves are theoretical and have been calculated from a set of scattering centers using the Debye formula (see Chapter 1, Section 1.2.1), but these structures also have real-life counterparts. For example, clay contains flat sheets and the shape of the diffraction peaks arising from the atomic long-range order within the sheets look like those of the theoretical disk depicted in Fig. 4.3. In fact, the shape of the peak depends, among other things, on the shape and size of the crystals, and remarkably the scattering curves follow in the SAXS region the same power law $I(q) \propto q^{-n}$ that is found in the "tails" of the WAXS peaks. The SAXS signal actually tells us about the shape and size of the crystallite. In principle, the SAXS signal, being centered at $q = 0\ \text{nm}^{-1}$, close to the primary beam, can be thought of as the (000) reflection. Just as one can determine the width of crystallites in certain directions from the widths of the corresponding diffraction peaks by using, for example, the Scherrer formula (see Chapter 1, Section 1.2.2), it is possible in an analogous manner to determine the size and shape of crystallites or nanoparticles from the width and shape of the SAXS signal. In determining the size and shape of nanoparticles and alike, SAXS has the advantage over diffraction studies in that it can also be used to study the shape and size of noncrystalline nanoparticles or macromolecules.

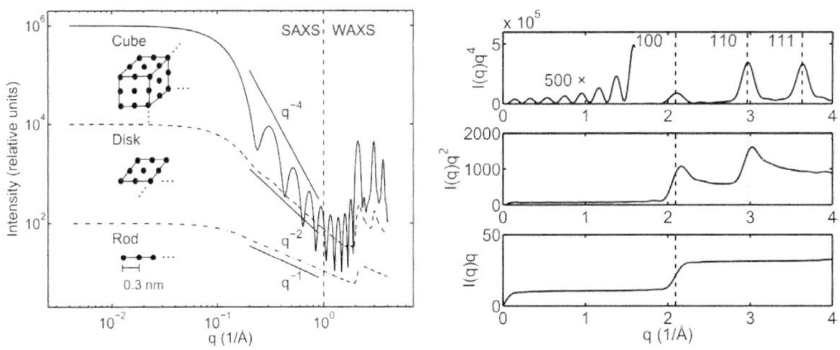

Figure 4.3 Left: Simulated scattering curves of a cube, disk, and a rod, all having edge length of 3 nm either in 3, 2, or 1 dimensions and lattice spacing of 0.3 nm between the scattering centers. Right (*from top to bottom*): intensities on linear scale of the cube, disk, and the rod multiplied by q^n, where n is 4 (Porod plot), 2 (Kratky plot), and 1, respectively. The dotted lines mark the theoretical positions of the reflections, as calculated from $q = 2\pi/d_{hkl}$.

4.3.1 Scattering Length Density

In SAXS, the theoretical models assume the materials to have a homogeneous average electron density in the atomic scale. To calculate the average scattering length density, we have to start by calculating the scattering power of atoms, and the number that describes how strongly an atom scatters X-rays is given by the atomic scattering factor f. It depends on the photon energy and the scattering angle, but luckily at small scattering angles, the angle dependency can be neglected. Thus in SAXS, the atomic scattering factor of an atom is expressed by:

$$f = Z + f'(E) + if''(E), \qquad (4.4)$$

where Z is the number of electrons in the atom, E is the photon energy and $f'(E)$ and $f''(E)$ are the so-called anomalous scattering factors (or dispersion coefficients), which are element specific and change rapidly at the absorption edges of the element. While f' is mostly negative and reaches a minimum at each absorption edge of the element, f'' that is multiplied with the imaginary unit is in fact related to the absorption and rises steeply at each absorption edge.

While the atomic scattering factor is in electron units, we wish to use the scattering amplitude, or scattering length, b [cm], which is compatible with the notation in small-angle neutron scattering and is defined as $b = r_e f$. The classical electron radius is $r_e = 2.818 \cdot 10^{-13}$ cm. The scattering length density of a compound is then calculated by using the following expression:

$$\rho_e = r_e \rho N_A \sum_{n=1}^{N} x_n f_n \bigg/ \sum_{n=1}^{N} x_n M_n \,, \tag{4.5}$$

where x_n is the proportion of element of type n, ρ is the density of the material in g/cm^3, and M_n is the molecular weight of each element in g/mol, and N_A is the Avogadro's constant ($N_A = 6.022 \cdot 10^{23}$ 1/mol). The scattering length density has the units cm^{-2}.

The scattered intensity depends on the difference $\Delta\rho_e$ between two phases rather than the absolute value of the scattering length density. The difference is also named *contrast*. The higher the contrast, the more scattering will be observed. If two structurally different parts in the sample have same scattering length density, these parts cannot be distinguished from each other. When choosing a solvent for macromolecules, one can either increase or decrease the scattering, depending on the density difference between the macromolecules and the solvent.

Because only the scattering length difference matters, the Babinet's principle tells us that certain reciprocity applies here. For a structure composed of phases A and B, identical intensity patterns result when the scattering length densities of phases A and B are interchanged.

4.3.2 Power Laws

In SAXS curves, the intensity $I(q)$ often follows a power law:

$$I(q) \propto q^{-\alpha}, \tag{4.6}$$

where α is the power-law exponent. All powders will give power-law scattering in SAXS. Porous structures with pores of different sizes give power-law scattering and in the case of weakly scattering film samples, such as metal foils, one may even observe power-law scattering coming mainly from the roughness of the sample surface.

Because the power laws often extend all the way to the WAXS region, to large q values, it is important to take into account the background from WAXS. This background is typically of the form $Aq^n + B$, where A, B, and n are positive real numbers [23]. Before fitting the power-law exponent of the SAXS curve, one should therefore make sure that any such background does not exist or has been subtracted from the scattering curve. To be able to fit this WAXS background correctly, the SAXS curves should be measured to large enough q.

Power laws with $\alpha \neq 1, 2$ or 4 can be interpreted as being caused by fractal structures in the samples. With fractals we do not necessarily mean theoretical self-similar fractals, but more like natural "rough" structures. This is in accord with the original idea of Benoît Mandelbrot, the inventor of the fractal concept, who applied the fractal theory for example to clouds and mountains, and other rough natural objects. He was fascinated by the idea that a very complex roughness that extends over many length scales can be described with only one number, the fractal dimension D. Remarkably, this number is directly related to the power-law exponent α.

In SAXS curves, the power laws can be classified into power laws coming from *mass fractals* ($D_m < 3$) and the ones coming from *surface fractals* ($D_m = 3$, $D_s < 3$). The difference between these two fractal types is that in mass fractals, the roughness penetrates throughout the bulk volume (like in a snow flake, or some types of cheese with holes in them), whereas in surface fractals, the roughness is present only at the surface (like in mountain ranges and deep valleys on the surface of a planet, while the planet is not porous inside). In SAXS, the power-law exponent may typically get values in the range $0 < \alpha \leq 4$. Values larger than four are also possible, but they have different origin [24]. The scattering follows the laws:

$$I(q) \propto q^{-Dm} \text{ for mass fractals } (0 < D_m < 3) \tag{4.7}$$

$$I(q) \propto q^{-(2d-Ds)} \text{ for surface fractals } (0 < D_s < 3) \tag{4.8}$$

When dealing with 3D objects, the Euclidian dimension is $d = 3$. The fractal concept has been well presented in the publications by Martin and Hurd [25] and Schmidt [26]. Table 4.1 shows some examples of the fractal dimensions that can be observed in SAXS experiments. Notably, for 3D scattering objects with smooth surfaces one usually observes the "Porod law", $\alpha = 4$. This type of power-law scattering is observed, for example, for spherical nanoparticles or for two-phase systems with smooth surfaces, the structure found in many porous materials.

Table 4.1 Mass fractal dimension D_m, surface fractal dimension D_s, and power-law exponent α for some structural systems in $d = 3$ [26]

Object	D_m	D_s	α	Remarks
Randomly oriented thin rod	1	1	1	
Randomly oriented thin sheet	2	2	2	
Smooth 3D object	3	2	4	Porod law
Mass fractal	$D < 3$	$D < 3$	$0 < \alpha < 3$	
Surface fractal	3	$D < 2$	$3 < \alpha < 4$	

To estimate the length scales for which the structure is fractal, let $q_{\min}$ be the smallest and $q_{\max}$ the largest q value for which the power-law scattering is observed. Now the fractal roughness can be estimated to extend over sizes L, which lie in the interval [27]

$$1/q_{\max} < L < \pi/q_{\min}. \tag{4.9}$$

The power-law scattering is obviously related to roughness and shapes of particles in the sample, but are all structures showing noninteger power-law scattering really fractal? For the fractal interpretation to make sense, the power law should actually extend over a very large q range, preferably over several decades [28]. Power-law scattering can also be caused, for example, by polydispersity of scattering units, a topic that will be covered in Section 4.3.4.4.

4.3.3 Porod Constant

The power-law region at the tail of the SAXS curve is related to the total surface area in the sample. A surface-to-volume ratio of a sample consisting of two phases can be calculated from the SAXS curve if the tail region of the SAXS curve follows the Porod law. In real experiments, the Porod law

$$I(q) = Gq^{-4} + C \tag{4.10}$$

is superimposed on a background (here only constant C, more complicated backgrounds should be subtracted before analysis). Constant G can be obtained by fitting a line to (q^4, Iq^4). G. Porod himself defined his constant in the form

$$K_G = \lim_{q\to\infty} q^4 I(q) \tag{4.11}$$

which at first seems a bit complicated, but one can easily see that constants G and K_G must be the same constant if the Porod law holds. Now, we get the surface-to-volume ratio S/V[m^{-1}] of the sample [29,30]:

$$\frac{S}{V} = \frac{K_G}{2\pi(\Delta\rho_e)^2}\,. \tag{4.12}$$

Here, $\Delta\rho_e$[cm/cm^3] is the difference in scattering length densities of the two phases of the material (see Eq. (4.5)). Note that $I(q)$ here is on the absolute intensity scale, which is normalized by the irradiated volume V of the sample, because we use the definition where $I(q)$ is the macroscopic differential cross section $d\Sigma/d\Omega$, which has units [cm^{-1}].

Further, the specific surface area S_m[m^2/g]—the interface area between two phases normalized by the mass of one of the phases—is calculated

$$S_m = \frac{S}{V\rho}, \tag{4.13}$$

where ρ again is the density of the one of the phases, which in case of a porous two-phase system is in fact the density of the sample. Usually, the value obtained from SAXS for the specific surface area is larger than the one obtained from gas absorption experiments because the gas cannot access closed pores, while SAXS makes no distinction between open and closed pores.

4.3.4 Scattering from Particles

SAXS can be used to determine the shape, size, and polydispersity of nanometer-sized particles. As we will further see, not all of those parameters can be determined at the same time, but usually assumptions must be made either about the shape or about the polydispersity. Unlike microscopy methods, SAXS can give statistically averaged information about all sorts of nanoparticles in all kinds of solvents and conditions, typically in the concentration range from ~1 mg/ml to several tens of mg/ml. The sample volume can be small, the lowest volumes being ~10 μl for stationary sample cells [31]. This makes SAXS an attractive tool for the study of biological macromolecules and their behavior. In this section, we will go through some theoretical concepts that can be applied to the study of both inorganic and organic particles.

Figure 4.4 shows theoretical SAXS curves for two spheres and two spherical core-shell particles. The curves in the figure have been smeared to make them appear more similar to experimental ones by applying a square-shaped filter function. The double logarithmic scale is most commonly used because it emphasizes the details at large q values. The Guinier and Kratky plots are also often used during the analysis of SAXS curves of particles. It is usually a good idea to make qualitative analysis of the curves by plotting them in several different ways before analyzing them quantitatively.

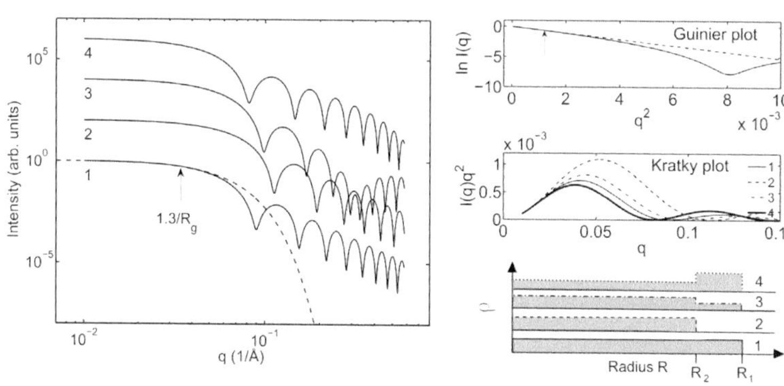

Figure 4.4 Left: Log-log scale representation of theoretical SAXS curves of spheres (solid lines) with radius R_1 = 5 nm (1) and R_2 = 4 nm (2), and core-shell particles (3,4) of same size but having different scattering length densities ρ_e. The curves have been normalized and shifted vertically for clarity. The dashed line shows the fitted Guinier approximation. Guinier plot: SAXS of sphere 1 (solid line) and the Guinier approximation (dashed line) on (q^2, ln(I)) axis. Kratky plot: SAXS from all particles on (q, Iq^2) axis. The vertical arrows show the limit of Guinier approximation for sphere 1. The last plot shows the radial electron density ρ profiles of the four spherically symmetric particles.

Further, in case of macromolecules or nanoparticles in solution, the background coming from the solvent should be subtracted by taking into account the volume fraction of the solvent:

$$I_p(q) = I_{tot}(q) - (1-\Phi)\, I_{sol}(q), \tag{4.14}$$

where Φ is the volume fraction of particles in the solution, $I_p(q)$ is the intensity of the particles, $I_{tot}(q)$ is the total measured intensity of sample solution after correcting for empty cuvette scattering according to steps 1–4 in Section 4.2.2, and $I_{sol}(q)$ is the measured and corrected intensity of the pure solvent. Ideally, one should do this by using absolute intensity units, as demonstrated in Ref. [32].

4.3.4.1 Guinier approximation

The Guinier approximation can be used to derive the radius of gyration of nanoparticles or macromolecules. The approximation is based on the fact that the beginning of every scattering curve at small q has a Gaussian shape (notice the similarity to the shape of diffraction peaks). The Guinier approximation is

$$I(q) = I(0)\cdot\exp\left(-\frac{q^2 R_g^2}{3}\right), \tag{4.15}$$

where R_g is the radius of gyration, which is defined

$$R_g^2 = \frac{\int_V \rho_e(\mathbf{r})|\mathbf{r}|^2\, dV}{\int_V \rho_e(\mathbf{r})\, dV}, \tag{4.16}$$

where $\rho_e(\mathbf{r})$ is the scattering length density at distance $(\mathbf{r})$ from the center of the particle. A concise derivation of the Guinier approximation can be found in Ref. [33]. Note that A. Guinier uses definition $s = q/(2\pi)$. The Guinier approximation (Eq. (4.15)) holds for particles of all shapes. Figure 4.4 demonstrates the validity of this approximation for spheres. As a rule of thumb, the approximation holds when $q < 1.3/R_g$ [34].

The radius of gyration can easily be found from experimental data by plotting the data in coordinates$(q^2, \ln(I))$ and fitting a straight line to the points. With particles of known shape, the radius of gyration can be used for calculating the size of the particles. For example for spheres with radius R, radius of gyration is $R_g = (3/5)^{1/2}R$.

4.3.4.2 Form factor

As in Young's double-slit experiment with visible light, there are two things that contribute to the scattering pattern in small-angle scattering. One contribution comes from the particle or pore size and shape that results in the kind of scattering as depicted in Fig. 4.4. This contribution is called the *form factor* $P(q)$, and it is analogous to Young's experiment with a single slit where the width of the scattering pattern is inversely related to the slit width in real space. The other contribution comes from the arrangement of the scattering units (particles or pores) and is called the *structure factor* $S(q)$. A special case of the structure factor with intense maxima happens when scattering centers are arranged in a lattice, having long-range order.

The Fourier transform of the scattering length density difference between the particle and its surroundings gives an amplitude

$$F(\mathbf{q}) = \int \Delta\rho_e(\mathbf{r})\exp(-\mathbf{q}\cdot\mathbf{r})dV \,, \tag{4.17}$$

which is squared to give the form factor

$$P(\boldsymbol{q}) = |F(\boldsymbol{q})|^2. \tag{4.18}$$

As the simplest case, let us consider a homogeneous sphere with radius R and scattering length density $\rho_e = \rho_1$ surrounded by a medium for which $\rho_e = \rho_0$. As ρ_e is centrosymmetric, it can be expressed as a function of the distance r from the center of the sphere

$$\rho_e(r) = \rho_1,\ \mathrm{r} \le R$$

$$\rho_e(r) = \rho_0,\ \mathrm{r} > R \tag{4.19}$$

Remembering that the dot product can be written as

$$\mathbf{q}\cdot\mathbf{r} = qr\cos\theta \tag{4.20}$$

and that the volume element in spherical coordinates is

$$dV = d\varphi \cdot \sin\theta d\theta \cdot r^2 dr, \tag{4.21}$$

we can integrate Eq. (4.17) to give the form factor amplitude of a sphere of volume V:

$$F(q) = V(\rho_1 - \rho_0)\cdot 3\frac{\sin(qR) - qR\cos(qR)}{(qR)^3}. \tag{4.22}$$

For spherical monodisperse particles, the form factor and structure factor are decoupled so that the total intensity is a multiplication of the two factors:

$$\frac{d\sigma}{d\Omega}(q) = P(q)S(q). \tag{4.23}$$

In dilute systems, we may approximate $S(q) = 1$. When long-range order gives sharp diffraction maxima in $S(q)$, it can sometimes happen that the sharp minima characteristic to the form factors of spherical particles (see Fig. 4.4) suppress some of the diffraction maxima of the structure factor.

For particles with small anisotropy (deviation from spherical symmetry) or polydispersity, Eq. (4.23) can be rewritten to a different form using the so-called *decoupling approximation* [35]. A complete review of form factors for different kind of particle shapes is given by Pedersen [36]. The fitting procedure of SAXS curves is explained in the review in great detail. Note that the definition of form factor amplitude is different from the one presented here.

4.3.4.3 Structure factor

When particle concentration in solution becomes high enough, the so-called *concentration effect* is observed in the scattering curves. In low concentrations of particles, the structure factor $S(q)$ is close to one because the particles are free to move everywhere in the solution. However, once the concentration of particles starts to increase, they are more and more likely to spend time close to each other. This gives rise to a maximum in the structure factor and at smaller q values $S(q)$ starts to get lower. To see the concentration effect, one can measure solutions of differing particle concentrations and normalize the intensities by the volume fraction of particles $I(q)/\Phi$ or by their concentration $I(q)/c$. Then the intensities for different concentrations of particles should be on the same scale. A small flattening of the intensity curve should be observable at low q when the concentration increases, but the high-q part of the curve should stay nearly the same. By extrapolating as a function of concentration for each q value separately, one can obtain the shape of the "ideal" SAXS curve in zero particle concentration and thus eliminate the concentration effect.

In case of charged nanoparticles or macromolecules, the structure factor is always present and it has a sharp maximum. This so-called *polyelectrolyte effect* harms the determination of particle shape because one cannot simply get rid of it by extrapolating to zero concentration. Instead, one can try to reduce the charges of the molecules, for example, by adding salt to the solution.

Sometimes it is interesting to study the interactions between particles. If particle's shape stays the same in all concentrations, one can extract the structure factor from the SAXS curve by dividing the intensity by the form factor obtained from the SAXS curves for low particle concentrations. In principle, this works only for monodisperse spherical particles, but in case of highly charged particles, the particle shape becomes less important and the method can be applied to such systems.

Using liquid state theory, it is possible to derive analytical structure factors for certain special types of systems from the interaction potential between the particles. Notably, for noninteracting monodisperse spherical colloids a simple analytical solution can be found for the structure factor. The Percus–Yevick approximation applied to the direct correlation function in Ornstein–Zernike integral equation gives this solution, which is known as the hard-sphere structure factor [36,37]. This system is a special case of the mean spherical approximation (MSA).

For charged colloidal systems the rescaled MSA (RMSA) gives a semianalytical solution, which is also known as the Hayter–Penfold structure factor [38]. The original papers ignore the so-called penetrating background of counter-ions, but it can be taken into consideration by introducing corrections for effective charge and effective screening as explained by R.J. Hunter [37].

In addition to structure factors related to repulsive potentials, there are also analytical structure factors derived for attractive potentials. For finding analytical structure factors for other systems, such as sticky hard spheres or polydisperse systems, the review by Pedersen [36] is a good starting point, whereas the book from Hunter [37] explains the theory behind these calculations. For example, the study by Zhang et al. [39] shows how a collection of different structure factors could be applied to map the phase diagram of bovine serum albumin (BSA) protein in aqueous solution.

When the long-range order of scattering units is sufficient the structure factor starts to resemble that of diffraction patterns familiar from XRD. The typical structural arrangements and theories related to them can be found for example in Förster et al. [40].

4.3.4.4 Polydispersity

So far, we have ignored the possible polydispersity of particles. Biological macromolecules of same type all have the same size and shape if we ignore aggregation effects, but synthesized nanoparticles are hardly ever fully monodisperse. Therefore, sometimes one actually wants to find out the particle-size distribution from the SAXS data. For noninteracting particles ($S(q) = 1$, dilute solution), the SAXS intensity is simply an integral over the form factors of the particles of different sizes:

$$\frac{d\sigma}{d\Omega}(q) = \int_0^\infty N(R)P(q,R)dR. \tag{4.24}$$

Here the number size distribution $N(R)$ is normalized so that $\int N(R)dR = 1$. In some cases, it is also interesting to look at the volume size distribution $V(R)$, which is calculated from $N(R)$ by multiplying it with the particle volume. It has a maximum at larger particle sizes than $N(R)$. Because the intensity scales with volume squared, the presence of a few larger particles makes it more difficult to observe smaller particles with SAXS. The volume size distribution presents better which particle sizes give strongest contribution in the SAXS curve.

There are several analytical particle-size distributions $N(R)$ that are based on the experimental data obtained for different types of nanoparticles. The shape of the size distribution depends on the growth process of the nanoparticles. The most commonly used distributions are log-normal [41] or Gaussian.

If the distribution is too complex to be described by a single analytical distribution function or a combination of them, one can fit a distribution without functional form using, for example, Monte Carlo [42] or maximum entropy [43] approaches.

When moving from dilute polydisperse systems to more concentrated ones, particles of all sizes interact with particles of all other sizes, and there is no single-structure factor that can describe this interaction. One way to circumvent heavy calculations is to apply the *local monodisperse approximation*, [44] where we simply assume that the particles only interact with other particles of equal size. Mathematically, this is expressed with the simple equation:

$$\frac{d\sigma}{d\Omega}(q) = \int_0^\infty N(R)P(R,q)S(R,q)dR \cdot \qquad (4.25)$$

4.3.4.5 Distance distribution function

In biological SAXS studies, one of the major goals is to determine the shape of macromolecules or macromolecule complexes. In case of such monodisperse systems, there is one simple tool in SAXS analysis that gives a good starting point for such studies. The distance distribution function $p(r)$ of a particle is a probability function that gives the weighted probability of finding distances r inside the particle. It is characteristic of the particle shape (see Fig. 4.5) and can be easily derived from the scattered intensity.

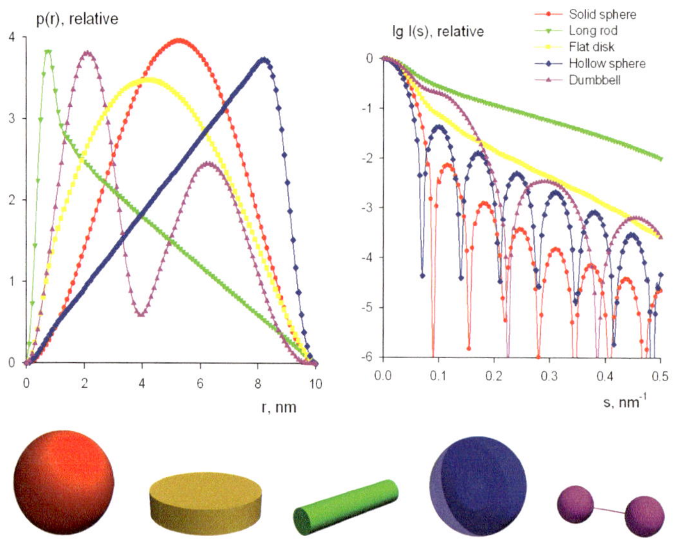

Figure 4.5 Scattering intensities and distance distributions of geometrical bodies of homogeneous electron density. The $p(r)$ function goes to zero at the maximum dimension of the particle (figure from Ref. [34]).

Rearranging the integrals of the scattered intensity we expressed by using Eqs. (4.17) and (4.18), we get the following:

$$\frac{d\sigma}{d\Omega}(q) = \iint \Delta\rho_e(\mathbf{r}_1)\Delta\rho_e(\mathbf{r}_2)\exp[-i\mathbf{q}\cdot(\mathbf{r}_1 - \mathbf{r}_2)]dV_1 dV_2 \,. \qquad (4.26)$$

When defining $\mathbf{r} = \mathbf{r}_1 - \mathbf{r}_2$, we have an autocorrelation of the scattering length density difference related to the particle

$$\gamma(\mathbf{r}) = \int \Delta\rho_e(\mathbf{r}_1)\Delta\rho_e(\mathbf{r}_1 - \mathbf{r})dV_1 \,. \tag{4.27}$$

Now we apply the assumptions that (1) the particles are oriented isotropically (i.e., there is no preferred direction) and (2) there is no long-range order; and take an orientation average of the Fourier term:

$$< \exp(-i\mathbf{q}\cdot\mathbf{r}) >= \frac{\sin(qr)}{qr} \,. \tag{4.28}$$

The intensity is now written conveniently as a function of the autocorrelation function:

$$\frac{d\sigma}{d\Omega}(q) = \int 4\pi r^2 \gamma(r) \frac{\sin(qr)}{qr} dr \,. \tag{4.29}$$

The inverse transform of which gives us the $p(r)$ function:

$$p(r) = r^2 \gamma(r) = \int q^2 \frac{d\sigma}{d\Omega}(q) \frac{\sin(qr)}{qr} dq \,. \tag{4.30}$$

The $p(r)$ function can be directly derived from the experimental intensity by using Eq. (4.30). In practice, one must first extrapolate the intensity at low q to $q = 0$, for example, by the Guinier law or other simple approximation and the high q to a very large q value where the intensity is nearly zero using, for example, the Porod law. To determine the shape of the $p(r)$ function, absolute intensity units are not necessary; here, we have used the differential scattering cross section $d\sigma/d\Omega$ instead of usual $I(q)$ just for consistency.

Obviously, direct applying of Eq. (4.30) is problematic because the intensity has to be extrapolated. Therefore, indirect methods to determine the $p(r)$ function have also been developed [45,46].

If the particles are very polydisperse, the $p(r)$ function analysis does not work and the shape of the particles will look more elongated than they really are.

4.3.4.6 Kratky plot and Porod invariant

The Kratky plot, already introduced briefly in Fig. 4.3, is a special way of presenting SAXS intensity. This simple plot can be a powerful tool in visualization of SAXS data while determining the folding state or conformation of proteins as has been shown, for example, by Shi et al. [47]. Figure 4.6 demonstrates how the Kratky plot nicely shows the state of the protein (unfolded or globular).

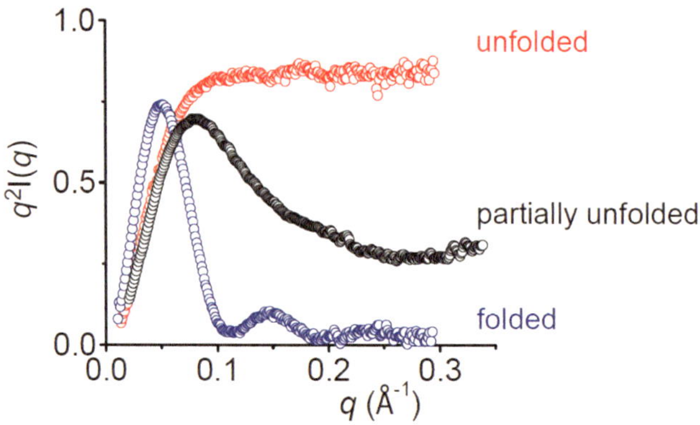

Figure 4.6 Kratky plots of unfolded, partially unfolded and folded proteins (figure from Ref. [48]). Note that unfolded proteins that are basically free Gaussian random chains follow the power law $I(q) \propto q^{-2}$.

From Kratky plot one can also calculate the so-called *Porod invariant*, which is basically an integral over the Kratky plot:

$$Q = \int_0^\infty q^2 I(q) dq \,. \tag{4.31}$$

In order to do this in practice, one must first extrapolate the data to zero at low q and extend the large q values by another extrapolation, like when applying Eq. (4.30).

The Porod invariant depends on the mean square fluctuation of electron density $< \eta^2 >$ in the sample but not on the specific structure causing it; thus, the name "invariant" (do not mix with Porod *constant*). For example, same globular macromolecule in different conformations should give same Q. It has also other interesting properties and some parameters can be derived from SAXS curves without absolute intensity units by introducing the invariant into the equation. One important example concerns determination of the particle volume, which can be derived from SAXS data by using the Porod invariant and extrapolated intensity at $q = 0$. The meaning and use of the invariant is explained, for example, by G. Porod in the book of Glatter and Kratky.

4.3.5 Scattering from Fluctuations

Fluctuations in atomic composition in bulk samples are difficult to measure with microscopy methods, therefore this is an area where SAXS can really show its power. It gives statistically averaged data without much sample preparation. For example for metallic glasses, the combined SAXS/WAXS technique has made possible in situ measurements of spinodal decomposition and phase separation [49]. In case of metallic glasses, the fluctuations present themselves in the SAXS curve in the form of a clear maximum because there exists a period in the element density fluctuations. SAXS can be used for studying the density fluctuations in supercritical liquids by applying the Ornstein–Zernike theory [50]. Further, fluctuations from random assemblies that can be described with a characteristic correlation length give rise to a hump in the SAXS curve that can be fitted using the Debye–Bueche random phase model [51].

4.3.6 Generalized Scattering Functions

Sometimes the studied system is not clearly described with simple form factors and structure factors or correlation lengths, but fortunately there are other ways to fit scattering curves of more complicated systems. Here, we mention just a few of them.

In the 1990s, G. Beaucage created an approach to fit more or less featureless scattering curves by introducing the so-called *unified* or *global scattering functions* that combine the Guinier approximation and power-law scattering by relying on a set of approximations [52]. These functions can be used to fit the whole scattering curve without having to make assumptions about the actual structure of the sample. Recently, some corrections to this approach have been proposed by B. Hammouda [53].

An interesting development has been to use *clipped Gaussian random field (GRF)* method to model ordered inhomogeneous nanostructure morphologies [54]. By choosing the right clipping function, one can make a model and fit nearly any structure one can imagine. The difficulty with this method lies in finding that correct clipping function.

Because of their properties in Fourier transforms, Gaussians are also used to describe the electron density profile of complicated but spherically symmetric systems. This approach has been applied, for example, to slightly polydisperse microemulsions [55] and synaptic vesicles [56].

4.4 Radiation Damage

Since the emergence of third-generation synchrotron sources, SAXS is no longer a nondestructive method. Radiation damage is a serious issue, but most of the radiation damage studies have so far been mainly concerned with protein crystallography samples. In the intense photon beam of undulator beamlines at third-generation synchrotron sources, local temperature increase of several degrees Kelvin can occur in the sample within seconds when the incident photon flux at the sample is around 10^{13} photons/s/mm^2. Even in case of cryocooled samples, one can expect temperature increase of several degrees [57].

In SAXS studies, radiation damage is also created due to ionization and free radicals that are formed during irradiation. It is well known in the biological SAXS community that strong synchrotron beams cause unwanted aggregation of biological macromolecules in solution [58]. Therefore, in the bioSAXS experiments, radical scavengers such as glycerol, ethylene glycol, sucrose or dithiothreitol (DTT) are typically added to the solvent in small amounts to prevent aggregation of the studied macromolecules.

One way to overcome radiation damage in solution samples is to use a *flow-through cell* or a *free-liquid jet,* where fresh liquid continuously flows through the synchrotron beam. These methods require a lot of sample and are mainly useful when studying molecules or nanoparticles that can be produced in large quantities.

In the case of solid samples, especially polymers, radiation damage usually leads to a decrease in the scattering signal in a few seconds. Therefore, one should always check for radiation damage before making the "real" experiments by exposing the sample in different positions for different periods of time. The suitable exposure time is then a compromise between good scattering statistics and the least amount of radiation damage effects in the scattering pattern.

4.5 BioSAXS

BioSAXS, or X-ray solution scattering as it is sometimes called, is normal transmission SAXS from biological samples, especially proteins, DNA, RNA, or their complexes in solution. What makes it special compared with other SAXS techniques is that special computational methods have been developed to combine information from protein crystallography and NMR with that coming from SAXS so that conformation changes or shapes of complex structures of macromolecules or macromolecule assemblies in solution can be studied. In addition, SAXS can be used as a screening technique to check the quality of biological samples. Two excellent reviews of biological SAXS experiments are those of Putnam et al. [48] and Lipfert and Doniach [59].

The current popularity of bioSAXS can mostly be attributed to the tremendous efforts put into the development of highly automated beamlines dedicated for SAXS measurements of biological samples and further to the development of freely available and highly dedicated software packages for analysis of biological SAXS data.

4.6 GISAXS

The SAXS technique for studying structures on surfaces and thin films is usually called grazing-incidence or glancing-incidence small-angle X-ray scattering (GISAXS). Even though SAXS and GISAXS have a lot in common, there are differences that should not be overlooked. In transmission SAXS, we rely on the kinematical scattering theory where multiple scattering events are ignored and Born approximation is used, while in GISAXS multiple scattering events can become significant. Therefore, scattering in GISAXS geometry is usually simulated using the distorted wave Born approximation (DWBA) [60]. For comprehensive reviews of GISAXS, the reader is advised to see the reviews by Renaud et al. [61], P. Müller-Buschbaum [62], and R. Lazzari [63].

Figure 4.7a shows a schematic GISAXS setup. In order to understand the structural features in GISAXS patterns, let us go through the terminology that is commonly used. The most prominent feature in the GISAXS pattern is the specular peak if no beamstop is used to mask it. This peak is observed when the exit angle α_f equals the incident angle α_i. It is the reflection of the primary beam from the sample surface. The smoother the sample surface, the stronger the specular reflection. The critical angle α_c for total external reflection of the studied material also affects the strength of this reflection. The critical angle depends on the density of the material and on the X-ray wavelength. At a photon energy of 9 keV, α_c is of the order 0.15° for polymers [62].

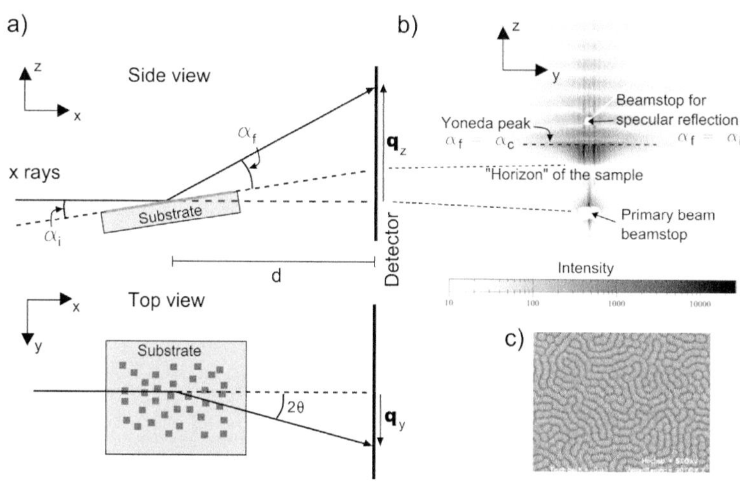

Figure 4.7 (a) A schematic presentation of a GISAXS setup. All the angles have been exaggerated to make it clearer. The sample-to-detector distance *d* is typically much longer than the substrate. (b) A GISAXS pattern of a PS-*b*-PMMA diblock-co-polymer-template sample on which 6 nm of Fe has been sputtered. The measurement was made at beamline BW4 at DORIS III storage ring. (c) A FESEM image of the sample surface. The width of the worm-like structure is approximately 100 nm. Measured data shown (b and c) are through courtesy of K. Schlage (DESY).

At incident angles $\alpha_i < \alpha_c$, X-rays can hardly penetrate to the material and thus only the top layers at the surface of the material are probed, whereas the specularly reflected beam is very strong. When $\alpha_i > \approx \alpha_c$ more information is obtained from buried layers of the sample close to the surface and the sample scatters strongly. At incident angles $\alpha_i > \alpha_c$, the intensity in the scattering pattern due to the surface component decreases while some features of the scattering pattern are maybe more easily resolved.

Figure 4.7b shows a GISAXS pattern of the structure in Fig. 4.7c. The worm-like lamellar long-range order in the lateral direction parallel to the surface of the sample gives the diffraction peaks in *y* direction. In this direction, GISAXS is like transmission SAXS and the same theories apply to it. In *z* direction, where the layered substrate–polymer–iron structure in sample surface normal gives the modulation, the GISAXS signal is more similar to reflectivity measurements.

An indication of good quality GISAXS samples is a maximum of the off-specular diffuse scattering observed close to the critical angle $\alpha_f = \alpha_c$. This maximum that is observed for flat and well-defined sample layers is often called the Yoneda peak (or sometimes Vineyard peak, named after G.H. Vineyard who was the first one to apply the DWBA to surface diffraction of X-rays [64]). This enhancement in intensity is caused by increased total internal reflections inside the sample close to the critical angle.

Compared to transmission SAXS, the definition of the scattering vector $\mathbf{q} = (qx, qy, qz)$ is a little more complicated in GISAXS geometry [63]:

$$\begin{aligned} q_x &= \frac{2\pi}{\lambda}(\cos\alpha_f \cos 2\theta - \cos\alpha_i) \\ q_y &= \frac{2\pi}{\lambda}(\cos\alpha_f \sin 2\theta) \\ q_z &= \frac{2\pi}{\lambda}(\sin\alpha_f - \sin\alpha_i). \end{aligned} \tag{4.32}$$

When the angles are small, the curvature of the Ewald sphere and the length of the scattering vector in forward direction q_x can actually be neglected.

In GISAXS, the precise alignment of the sample with respect to the incoming beam is crucial. The correct height where half of the beam is blocked by the sample at incident angle $\alpha_i = 0$ can be found by iteratively scanning the sample height and rocking the sample (scanning the incident angle). The X-rays are monitored either with a point detector on a goniometer arm or with a detector that is aligned with the primary beam behind the sample, typically at the primary beamstop. The rocking is necessary to find the sample surface parallel to the beam. The accurate value of angle α_i can be calculated from the position of the specular reflection on the 2D detector.

GISAXS experiments are often combined with other techniques, especially X-ray reflectivity, but the big advantage of GISAXS as a static method, where sample does not need to be moved, is that it is well suited for in situ studies, such as studying growth processes of thin metal films on different surfaces during vacuum deposition [65]. In addition to studying nanoparticles and other structures on solid substrates, one can even apply GISAXS to liquid surfaces with a special arrangement where the beam is tilted slightly downward, for example, using a mirror.

4.7 ASAXS

Anomalous SAXS (or resonant SAXS) is a contrast variation method, the theory of which is similar to that used for contrast variation in small-angle neutron scattering (SANS), and it bears some resemblance to the multiwavelength anomalous dispersion (MAD) experiments in macromolecular crystallography. In SANS, contrast variation is typically achieved by measuring several structurally identical samples that have different deuterium content. In principle, the same is possible in SAXS by varying the electron density of parts of the sample for example in solution scattering [48]. However, several difficulties are involved in such studies because the conformation of macromolecules might be different in different solvents and chemical bonding might change when the elements are exchanged.

In ASAXS, only one sample is needed. The chemical composition of the sample remains the same, but by recording SAXS images at several photon energies, below an absorption edge of an element the scattering power of the element in the sample is different in each image and this gives the contrast variation effect. The atomic scattering factor of an element changes strongly close to the edge, as shown in Fig. 4.8a.

At L3 edges, the contrast variation effect is stronger than at K edges, and therefore to achieve maximum contrast variation, it is good to choose an L3 edge whenever possible. However, even then ASAXS is by no means a simple method. One major obstacle is the energy-dependent background. The closer to the edge one measures

SAXS, the more background one gets from fluorescence and resonant Raman scattering. Since energy-sensitive 2D detectors with very fine energy resolution are not yet available, measurements above the absorption edge are usually not done because of the strong fluorescence background which is not easily separated from the actual signal.

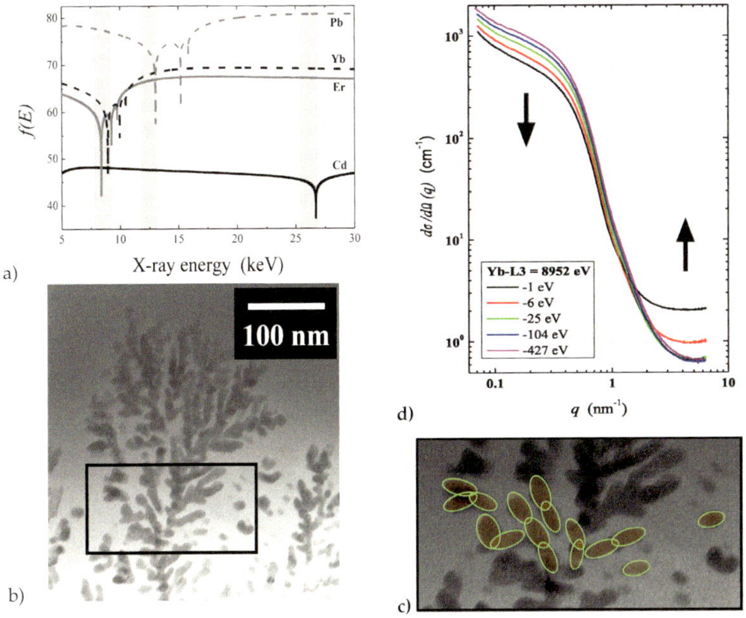

Figure 4.8 (a): Anomalous scattering factor as a function of energy for elements Pb, Yb, Er, and Cd. (b) and (c): TEM images of glasses similar to the one that was studied using ASAXS. (d): Measured scattering curves at five different energies close to the Yb L3 edge. At small q values, the arrow indicates the decrease of the scattering contrast when the energy increases close to the particular absorption edge. At large q values, the arrow pointing upward shows the dramatic increase in energy-dependent background when approaching the absorption edge. All figures are from Haas et al. [72].

In ASAXS measurements, knowing the accurate position of the absorption edge is important because the magnitude of the atomic scattering factor of the element needs to be known accurately if quantitative analysis is to be done. The edge position can be measured by making an X-ray absorption near-edge structure (XANES) scan of the sample. After taking a derivative of the absorption curve or linear absorption coefficient, which can be derived from it, the edge position is usually identified as the first maximum of the curve as explained in [66]. From the absorption scan, the anomalous scattering coefficients f' and f'' (Eq. (4.2)) can be determined using the Kramers–Kronig relation [67].

The absolute intensity of the SAXS patterns is of importance in ASAXS measurements. A secondary absolute intensity calibration sample is usually measured at each photon energy where SAXS from the sample is measured. This ensures that energy dependency of monitors or detectors are eliminated and the intensity variation is purely due to contrast variation in the sample.

In the simplest case, where only one element has its absorption edge at the studied energy range, the SAXS intensity can be written as the weighted sum of three *partial structure factors* (PSFs):

$$I(q, E) = |f_a(E)|^2 S_{aa}(q) + (f_a^*(E)f_b + f_a(E)f_b^*)S_{ab}(q) + |f_b|^2 S_{bb}(q), \quad (4.33)$$

where a refers to the anomalous component and b to the rest of the sample. This formula can be derived, for example, from the Debye formula by simply regrouping the terms belonging to different atoms. It should be noted that here the term "structure factor" is not the same as the one used for scattering from particles. The partial structure factors contain both the normal form factor and the structure factor. If the cross-term $S_{ab}(q)$ is zero, as may sometimes be the case, the element specific scattering can simply be obtained by subtracting from each other two measurements made at different energies. Further, there are many quantitative ways to analyze ASAXS data [67–69]. For a complete review, see the one from D. Tachev [70].

The ASAXS method is a powerful tool when there is doubt if some feature in the SAXS pattern comes from a certain structure or not. One of the early applications was for ion conducting polymers where the application of ASAXS to sulfonated polystyrene ionomer showed that the upturn in SAXS patterns of these materials is related to the agglomerates of sulfonate groups rather than other structural features [71]. Recently, Haas et al. [72] have shown how ASAXS can be used as a quantitative tool to study the chemical composition of nanostructures. Haas et al. studied nanoparticles inside a promising material for optoelectronic devices. The oxyfluoride glass ceramic contained a multitude of compounds: SiO_2 (21.1%), B_2O_3 (6.5%), Al_2O_3 (7.0%), PbF_2 (21.0%), CdF_2 (14.3%), YbF_3 (11.0%), ErF_3 (0.5%), PbO (11.0%), and CdO (7.6%). All the heavy elements of this material have absorption edges in the energy range 8–27 keV, which is achievable on several ASAXS beamlines.

It was known from TEM studies that the glass consists of regions of different density (Fig. 4.8b,c), and even though the chemical composition of the whole material was well known, it was not clear which elements are present in the nanoparticles and which are in the surrounding glass matrix. ASAXS measurements were made close to four absorption edges: Er L3 (8358 eV), Yb L3 (8944 eV, Fig. 4.8d), Pb L3 (13 035 eV), and Cd K (26 711 eV). The nanoparticles were estimated as polydisperse spheroids and a nonlinear regression of all scattering curves obtained at all X-ray energies was done simultaneously. From the fit, the chemical composition of the nanoparticles inside the glass matrix was determined to be $Pb_{17}Yb_{17}Er_2F_{64}$, and surprisingly cadmium was found only in the glass matrix and not in the nanoparticles.

Contrast variation is possible also in GISAXS geometry. This technique is so new that thus far no consensus has been reached if it should be named anomalous grazing-incidence SAXS (AGISAXS) or grazing-incidence anomalous SAXS (GIASAXS). The method is promising, especially for low atomic number elements that cannot be studied easily in transmission mode due to their absorption edges being located at soft X-ray region that requires very thin samples. Further, there is reason to expect more contrast enhancement in the GISAXS geometry compared to transmission SAXS. The first quantitative GIASAXS study with soft X-rays, by Okuda et al. [73], has shown both theoretically and experimentally that contrast enhancement at Si K edge at 1.83 keV is significant for capped germanium nanodots on silicon substrate.

Acknowledgment

The author wishes to thank Mika Torkkeli, Jan Perlich, Kai Schlage, Oliver Seeck, and Teemu Ikonen for their helpful comments that have improved this book.

References

1. Guinier A (1939) La Diffraction des Rayons X aux Très Faibles Angles: Applications à l'Etude des Phénomènes Ultra-microscopiques, *Ann Phys (Paris)*, **12**, 161–236.
2. Guinier A (1969) 30 years of small-angle X-ray scattering, *Phys Today*, **22**, 25–30.
3. Holmes KC (2010) 50 years of fiber diffraction, *J Struct Biol*, **170**, 184–191.
4. Huxley HE, Faruqi AR, Bordas J, Koch MHJ, Milch JR (1980) The use of synchrotron radiation in time-resolved X-ray diffraction studies of myosin layer-line reflections during muscle contraction, *Nature*, **284**, 140–143.
5. Stuhrmann HB (1981) Anomalous small-angle scattering, *Q Rev Biophys*, **14**, 433–462.
6. Levine JR, Cohen JB, Chung YW, Georgopoulos P (1989) Grazing-incidence small-angle X-ray scattering: new tool for studying thin film growth, *J Appl Cryst*, **22**, 528–532.
7. Narayanan T, Diat O, Bösecke P (2001) SAXS and USAXS on the high brilliance beamline at the ESRF, *Nucl Instrum Meth A*, **467–468**, 1005–1009.
8. Fischetti R, Stepanov S, Rosenbaum G, et al. (2004) The BioCAT undulator beamline 18ID: a facility for biological non-crystalline diffraction and X-ray absorption spectroscopy at the advanced photon source, *J Appl Cryst*, **11**, 399–405.
9. Fujisawa T, Katsuaki I, Oka T, et al. (2000) Small-angle X-ray scattering station at the SPring-8 RIKEN beamline, *J Appl Cryst*, **33**, 797–800.
10. Lurio L, Mulders N, Paetkau M, Jemian PR, Narayanan S, Sandy A (2007) Windows for small-angle X-ray scattering cryostat, *J Synchrotron Rad*, **14**, 527–531.
11. Nishikawa Y, Fujisawa T, Inoko Y, Moritoki M (2001) Improvement of a high pressure cell with diamond windows for solution X-ray scattering of proteins, *Nucl Instrum Meth A*, **467–468,** 1384–1387.
12. Delißen F, Leiterer J, Bienert R, Emmerling F, Thünemann AF (2008) Agglomeration of proteins in acoustically levitated droplets, *Anal Bioanal Chem*, **392**, 161–165.
13. Haberkorn H, Franke D, Frechen Th, Goesele W, Rieger J (2003) Early stages of particle formation in precipitation reactions—quinacridone and boehmite as generic samples, *J Colloid Interface Sci*, **259**, 112–126.

14. Polte J, Erler R, Thünemann AF, Emmerling F, Kraehnert R (2010) SAXS in combination with a free liquid jet for improved time-resolved in situ studies of nucleation and growth of nanoparticles, *Chem Commun,* **46**, 9209–9211.

15. Chantler CT (1995) Theoretical form factor, attenuation and scattering tabulation for Z = 1–92 from E = 1–10 eV to E = 0.4–1.0 MeV, *J Phys Chem Ref Data,* **24**, 71.

16. Huang T, Toraya H, Blanton TN, Wu Y (1993) X-ray powder diffraction analysis of silver behenate, a possible low-angle diffraction standard, *J Appl Cryst,* **26**, 180–184.

17. Pedersen JS, Posselt D, Mortensen K (1990) Analytical treatment of the resolution function for small-angle scattering, *J Appl Cryst,* **23**, 321–333.

18. Russell TP, Lin JS, Spooner S, Wignall GD (1988) Intercalibration of small-angle X-ray and neutron scattering data, *J Appl Cryst,* **21**, 629–638.

19. Orthaber D, Bergmann A, Glatter O (2000) SAXS experiments on absolute scale with Kratky systems using water as a secondary standard, *J Appl Cryst,* **33**, 218–225.

20. Cunsolo A, Formisano F, Ferrero C, Bencivenga F, Finet S (2009) Pressure dependence of the large-scale structure of water, *J Chem Phys,* **131**, 194502.

21. Kratky O, Pilz I, Schmitz PJ (1966) Absolute intensity measurement of small angle X-ray scattering by means of a standard sample, *J Colloid Interface Sci,* **21**, 24–34.

22. Zhang F, Ilavsky J, Long GG, Quintana JPG, Allen AJ, Jemian PR (2010) Glassy carbon as an absolute intensity calibration standard for small-angle scattering, *Metall Mater Trans A,* **41A**, 1151–1158.

23. Balta-Calleja FJ, Vonk CG (1989) *X-Ray Scattering of Synthetic Polymers,* Elsevier, Amsterdam, pp. 241–247; *Polymer Science Library* 8.

24. Schmidt PW, Avnir D, Levy D, Höhr A, Steiner M, Röll A (1991) Small-angle X-ray scattering from surfaces of reversed-phase silicas: power-law scattering exponents of magnitudes greater than four, *J Phys Chem,* **94**, 1474–1479.

25. Martin JE, Hurd AJ (1987) Scattering from fractals, *J Appl Cryst,* **20**, 61–78.

26. Schmidt PW (1991) Small-angle scattering studies of disordered, porous and fractal systems, *J Appl Cryst,* **24**, 414–435.

55. Gradzielski M, Langevin D, Magid L, Strey R (1995) Small-angle neutron scattering from diffuse interfaces. 2. Polydisperse shells in water–*n*-alkane–$C_{10}E_4$ microemulsions, *J Phys Chem*, **99**, 13232–13238.

56. Castorph S, Riedel D, Arleth L, et al. (2010) Structure parameters of synaptic vesicles quantified by small-angle X-ray scattering, *Biophys J*, **98**, 1200–1208.

57. Kuzay TM, Kazmierczak M, Hsieh BJ (2001) X-ray beam/biomaterial thermal interactions in third-generation synchrotron sources, *Acta Cryst D*, **57**, 69–81.

58. Kuwamoto S, Akiyama S, Fujisawa T (2004) Radiation damage to a protein solution, detected by synchrotron X-ray small-angle scattering: dose-related considerations and suppression by cryoprotectants, *J Synchrotron Rad*, **11**, 462–468.

59. Lipfert J, Doniach S (2007) Small-angle X-ray scattering from RNA, proteins, and protein complexes, *Annu Rev Biophys Biomol Struct*, **36**, 307–327.

60. Lazzari R (2002) IsGISAXS: a program for grazing-incidence small-angle X-ray scattering analysis of supported islands, *J Appl Cryst*, **35**, 406–421.

61. Renaud G, Lazzari R, Leroy F (2009) Probing surface and interface morphology with grazing incidence small angle X-ray scattering, *Surf Sci Rep*, **64**, 255–380.

62. Müller-Buschbaum P (2009) A basic introduction to grazing incidence small-angle X-ray scattering, in *Applications of Synchrotron Light to Scattering and Diffraction in Materials and Life Sciences* (Ezquerra TA, Garcia-Gutierrez MC, Nogales A, Gomez M, ed), Springer, Berlin Heidelberg, pp. 61–89.

63. Lazzari R (2009) Grazing incidence small-angle X-ray scattering from nanostructures, in *X-Ray and Neutron Reflectivity: Principles and Applications* (Daillant J, Gibaud A, ed), Springer, Berlin Heidelberg, pp. 283–342.

64. Robinson IK, Tweet DJ (1992) Surface X-ray diffraction, *Rep Prog Phys*, **55**, 599–651.

65. Kaune G, Ruderer MA, Metwalli E, et al. (2009) In Situ GISAXS study of gold film growth on conducting polymer films, *ACS Appl Mater Interfaces*, **1**, 353–360.

66. Kraft S, Stümpel J, Becker P, Kuertgens U (1996) High resolution X-ray absorption spectroscopy with absolute energy calibration for the

determination of absorption edge energies, *Rev Sci Instrum*, **67**, 681–687.

67. Stuhrmann HB, Goerigk G, Munk B (1991) Anomalous X-ray scattering, in *Handbook on Synchrotron Radiation, Vol. 4* (Ebashi S, Koch M, Rubenstein E, ed), North-Holland, Amsterdam, pp. 555–580.
68. Haubold H-G, Gebhardt R, Buth G, Goerigk G (1994) Structural characterization of compositional and density inhomogeneities by ASAXS, in *Resonant Anomalous X-Ray Scattering* (Materlik G, Sparks CJ, Fischer K, ed), Elsevier Science, Amsterdam, pp. 295–304.
69. Simon JP, Lyon O (1994) Anomalous small angle X-ray scattering in materials science, in *Resonant Anomalous X-Ray Scattering* (Materlik G, Sparks CJ, Fischer K, ed), Elsevier Science, Amsterdam, pp. 305–322.
70. Tatchev D (2008) Structure analysis of multiphase systems by anomalous small-angle X-ray scattering, *Philos Mag*, **88**, 1751–1772.
71. Ding YS, Hubbard SR, Hodgson KO, Register RA, Cooper SL (1988) Anomalous small-angle X-ray scattering from sulfonated polystyrene ionomer, *Macromolecules*, **21**, 1698–1703.
72. Haas S, Hoell A, Wurth R, Rüssel C, Boesecke P, Vainio U (2010) Analysis of nanostructure and nanochemistry by ASAXS: accessing phase composition of oxyfluoride glass ceramics doped with Er^{3+}/Yb^{3+}, *Phys Rev B*, **81**, 184207.
73. Okuda H, Kato M, Ochiai S, Kitajima Y (2009) Anomalous grazing incidence small-angle scattering of capped Ge nanodots at the Si K absorption edge, *Appl Phys Express*, **2**, 126501.

Chapter 5

The X-Ray Standing Wave Technique: Fourier Analysis with Chemical Sensitivity

Jörg Zegenhagen
European Synchrotron Radiation Facility, 6 rue Jules Horowitz, F-38000 Grenoble, France
Present address: Diamond Light Source Ltd, Harwell Science and Innovation Campus, Didcot, Oxfordshire, OX11 0DE, UK
jorg.zegenhagen@diamond.ac.uk

5.1 Introduction

The X-ray standing wave (XSW) technique [1,2] adds spatial resolution to many traditional X-ray spectroscopy techniques. Instead of a transient X-ray wave, it utilizes an X-ray interference field (XIF), that is, a stationary X-ray wavefield or standing wave, as a source of excitation. The foundations for the method were laid by Boris W. Batterman [3] over 40 years ago, and it has meanwhile found widespread use in condensed matter research. Commonly, the XSW is formed by the superposition of two coherent plane waves. Thus, a plane X-ray wave, that is, a sufficiently collimated and monochromatic X-ray beam, is needed. This has become much easier with the help of synchrotron radiation [4] in particular with the availability of modern third-generation sources. The formation of a

X-Ray Diffraction: Modern Experimental Techniques
Edited by Oliver H. Seeck and Bridget M. Murphy

ISBN 978-981-4303-59-0 (Hardcover), 978-981-4303-60-6 (eBook)
www.panstanford.com

second, coherent wave is (until now) based on the elastic scattering of X-rays. Most frequently employed is Bragg reflection from a single crystal, which needs to be treated by the dynamical theory of X-ray diffraction (DTXD) [5]. Also employed is total reflection from mirror surfaces [6] or Bragg reflection from multilayer [7]. The signal of interest is the inelastic scattering excited by the stationary X-ray wave, that is, by the spatially modulated intensity.

When an X-ray photon is absorbed by an atom, an electron is excited and may be emitted as a "photoelectron" [8]. The probability for the photo excitation process is in the dipole approximation[1] directly proportional to the strength of the X-ray wavefield at the center of the atom. (No angular momentum is transferred to the electron by the photon.) Monitoring the intensity of the photoelectron signal as a function of the movement of the XSW provides information of the structural arrangement of the emitting atoms. After the initial excitation, the excited atom relaxes rapidly, leading to a cascade of subsequent emission processes. In case of a photo-excited core hole, the hole is filled very rapidly (of the order of fs) by an electron from an upper shell. The de-excitation energy can be carried away by a photon $\hbar\omega_F$ or by an electron kicked out from another shell with lower binding energy. This so-called Auger electron is emitted (to a good approximation) with the kinetic energy $E_{\text{kin}} = E_{B1} - E_{B2} - E_{B3}$, where E_{B1} is the binding energy of the photo-emitted electron; E_{B2} is the binding energy of the electron, which fills this hole; and E_{B3} is the binding energy of the electron that is then emitted. The atom is subsequently left with two (core) holes. The process is called a B1B2B3 Auger process (e.g., KLL, KLM, and LMM) according to the involved electronic shells. These two de-excitation processes, that is, the emission of an Auger electron or a fluorescence photon are competing. For light elements, the Auger relaxation process is dominant and for heavy atoms the decay by fluorescence is more probable. With further de-excitation of the atom, other relaxation channels open up (vibration, rotation, etc.).

Since the strength of the X-ray photo absorption is proportional to the X-ray intensity, the intensity of the photoelectron emission and subsequent decay channels depends on the position of the atoms with

[1]In the dipole approximation, in the transition matrix element of the photo absorption process, the operator of the photon plane wave is expanded in a Taylor series $\exp(-i2\pi\mathbf{k}\cdot\mathbf{r}) = 1 - 2\pi i\mathbf{k}\cdot\mathbf{r} + \ldots$, and only the first term $\exp(-i2\pi\mathbf{k}\cdot\mathbf{r}) = 1$ is retained.

respect to the XIF. The emission is maximum or minimum depending on whether the atoms are located at the maximum (antinodes) or the minimum (nodes) of the X-ray standing wave. Thus, by shifting the wavefield in space, accurate structural information is obtained when detecting signals resulting directly from the photoelectric absorption (primarily photo electrons, but also Auger electrons, or X-ray fluorescence). The intensity of the photoelectron emission and all relaxation channels depends on the X-ray intensity, and therefore reacts to the movement of the XSW. However, it is important to realize that generally only the photoelectrons carry with certainty the information about the initial absorption process.[2] For all other signals, it should be carefully checked whether and to what extent they could have been the result of excitation by secondary radiation. X-ray fluorescence, inelastic scattered photons, or photo- and Auger electrons excited by the incident photon will again give rise to emission signals such as Auger electrons and X-ray fluorescence. The lower the needed excitation energy in comparison with the initial energy of the X-ray photon, the higher the likelihood that such a signal may not be the result of the primary photon excitation.

Just like X-ray diffraction (XRD), the XSW technique is in fact a Fourier technique [10], but there are some significant differences. The XSW technique probes the correlation of the atoms with the XIF, whereas X-ray diffraction techniques probe the autocorrelation of the electron density. Moreover, diffraction techniques suffer from the so-called phase problem since only the magnitude of the (complex) structure factor can be recorded and the phase information is lost.[3] With an XSW measurement, one can determine the amplitude and phase of an atomic distribution function. Thus, in certain cases, the result of just one XSW measurement, just one phase value can provide real space information. Further, element specific information is difficult to obtain by diffraction, while the XSW method allows analyzing the structure of materials element specific. When using high-resolution spectroscopic techniques, XSW analysis is even sensitive to the chemical environment and electronic properties. On

[2]Note the limitations of the dipole approximation. Higher order multiple terms may need to be taken into account in the XSW analysis when detecting differential photoelectron signals [9].

[3]For completeness we should note that there are some ways to regain this information [11].

the contrary, the number of Fourier components that are accessible to the XSW technique is intrinsically restricted. In case of producing the XIF by Bragg diffraction, the possible Fourier components are determined by the substrate-allowed reflections, that is, the diffraction vectors $\mathbf{H} = \mathbf{Q}/2\pi$ giving rise to Bragg reflections. This imposes strong limitations on the maximum length scale which can be probed by the XSW technique. It is restricted by the largest lattice plane spacing $d_H = H^{-1}$ with $H = |\mathbf{H}|$, which can excite a Bragg reflection. In other words, the XSW technique can only provide structural information on the scale of the unit cell of the crystal, which is producing the Bragg reflection. Longer length scales are accessible when employing multilayer or mirrors as plane wave generators, but Fourier analysis can then only be carried out in one dimension.

In the following, I will first concisely discuss the formation of the XIF, the formalism of the XSW analysis, and the relationship of the XSW technique with XRD. Then, I will review briefly as a particularly illustrative example the XSW analysis of 4 percent Mn in ultrathin films of GaMnAs on GaAs(001). Recording ~10–20 XSW "structure factors" of the Mn, images of Mn in the GaAs unit cell can be reconstructed by direct Fourier inversion. They immediately reveal three different Mn lattice locations. Based on the identified positions, refinement of the site occupation of the Mn leads to a highly accurate determination of the fraction of Mn located at three distinct positions in the GaAs unit cell for differently treated samples.

5.2 Formation of an XSW

A planar X-ray wavefield is created by the superposition of two plane X-ray waves:

$$\mathbf{E}_0 = \mathbf{e}_0 E_0 \exp[2\pi i(\nu_0 t - \mathbf{K}_0 \cdot \mathbf{r})] \tag{5.1}$$

and

$$\mathbf{E}_\mathbf{H} = \mathbf{e}_\mathbf{H} E_\mathbf{H} \exp[2\pi i(\nu_0 t - \mathbf{K}_\mathbf{H} \cdot \mathbf{r})] \tag{5.2}$$

Here, $\mathbf{e}_0$ and $\mathbf{e}_\mathbf{H}$ are the polarization vectors and $\mathbf{K}_0$ and $\mathbf{K}_\mathbf{H}$ are the X-ray propagation vectors with

$$|\mathbf{K}_0| = |\mathbf{K}_\mathbf{H}| = |\mathbf{K}| = K = \lambda^{-1}, \tag{5.3}$$

where λ is the X-ray wavelength. Since $E_\gamma = h\nu_0$ and $c = \lambda\nu_0$, the X-ray energy E_γ and λ are related via $\lambda E_\gamma = hc = 1.23984$ keV·nm. The two waves are coherent with identical frequencies.

Thus, one can relate the two E-field vectors by amplitude and phase factors via

$$E_{\mathbf{H}} = \sqrt{R}E_0 \exp(i\varphi) \tag{5.4}$$

with

$$R = I_{\mathbf{H}}/I_0 = |E_{\mathbf{H}}|^2/|E_0|^2. \tag{5.5}$$

Further, the propagation direction of the two plane waves is related by

$$\mathbf{K}_{\mathbf{H}} = \mathbf{K}_0 + \mathbf{H}. \tag{5.6}$$

Thus, it is straightforward to show that for the total wavefield intensity (normalized to the incident intensity I_0) one obtains the following expression:

$$I_{\mathbf{H}} = 1 + R + 2\sqrt{R}\cos(\nu - 2\pi\mathbf{H}\cdot\mathbf{r}) = 1 + R + 2\sqrt{R}\cos(\nu - 2\pi z_{\mathbf{H}}/d_{\mathbf{H}}). \tag{5.7}$$

Here, $\mathbf{r}$ is the distance with respect to a chosen reference point and thus $z_{\mathbf{H}}$ is the projection of this distance along $\mathbf{H}$, that is, normal to the wavefield planes. The phase ν (not to be confused with the frequency ν_0) between the two plane waves depends of course on the origin chosen for $\mathbf{r}$. When using a standing wave created by single-crystal Bragg diffraction, the origin of $\mathbf{r}$ is the same as the origin chosen for the structure factor of the crystal which produces the XSW. Further, the wavefield spacing $d_{\mathbf{H}}$ corresponds, in this case, exactly to the chosen diffraction plane spacing and $\mathbf{H} = [hkl]$ denotes the chosen diffraction vector with $|\mathbf{H}| = H = 1/d_{\mathbf{H}}$, where (hkl) are the Miller indices of the diffraction planes.

Equation (5.7) describes a planar wavefield, which is modulated in the direction of $\mathbf{H}$ with the spacing $d_{\mathbf{H}}$. The position of the maxima and minima in the direction of $\mathbf{H}$ is determined by the value of ν.

5.3 XSW Analysis

If the dipole approximation[4] for the photoabsorption process can be applied, the intensity I of the X-ray excited photoelectrons, Auger

---

[4]See footnote 1 on page 1.

electrons, or X-ray fluorescence from a single atom A reflects the X-ray intensity at its position **r**. Thus, the corresponding normalized yield is given by the following equation:[5]

$$Y_{\mathbf{H}} = I / I_0 = 1 + R + 2\sqrt{R}\cos(v - 2\pi\mathbf{H}\cdot\mathbf{r}). \quad (5.8)$$

The normalization constant I_0, called "off-Bragg intensity", is determined by incident X-ray flux, absorption cross section, and so forth. When the standing wave is created by Bragg reflection characterized by a reciprocal lattice vector **H**, R denotes the reflectivity. The phase v between the incident and the reflected X-ray wave changes by π when passing the Bragg reflection. Because of the dot product $\mathbf{H}\cdot\mathbf{r}$, the magnitude of the yield excited by the standing wave depends only on the distance $z_{\mathbf{H}}$ of the atom, which is the component of **r** along the direction of **H**. The origin of **r** is most conveniently chosen the same as the one chosen for the structure factor. The magnitude of the yield is independent of the two coordinates normal to **H.** The value of v can be calculated at each point on the reflectivity curve with the help of the DTXD. Thus, the exact location of the standing wave planes with respect to the unit cell of the crystal is known precisely for each value of v. It depends on the value of the complex structure factor for the chosen reflection [12].

Obviously, since $|\mathbf{H}| = H = 1/d_{\mathbf{H}}$, the yield from an atom at position $z_{\mathbf{H}}$ and $z_{\mathbf{H}} + nd_{\mathbf{H}}$ is identical, since $\cos(v - 2\pi z_{\mathbf{H}}/d_{\mathbf{H}}) = \cos(v - 2\pi z_{\mathbf{H}}/d_{\mathbf{H}} - 2\pi n)$. This feature of the XSW analysis is usually referred to as the modulo-d ambiguity. The quantity $|\mathbf{H}| = H = 1/d_{\mathbf{H}}$ has the dimension of an inverse length, and thus the dot product $\mathbf{H}\cdot\mathbf{r}$ is a dimensionless number, that is, can be written as $\mathbf{H}\cdot\mathbf{r} = P^{\mathbf{H}}$. The dimensionless parameter $P^{\mathbf{H}}$ is historically called the coherent position, which represents a position normal to the wavefield planes, normalized by the plane distance $d_{\mathbf{H}}$ and, because of the above said, $0 \le P^{\mathbf{H}} < 1$.

In reality, a large number N_A of specific atoms A will be excited by the XSW. If all identical atoms are exposed to the same X-ray intensity, the total yield is simply the summation of the yield of all individual atoms, that is, can be written after normalization as follows:

$$\mathrm{Y}_{\mathbf{H}} = N_A^{-1}\sum_{j}^{N_A}\left\{1 + R + 2\sqrt{R}\cdot\cos(v - 2\pi P_{Aj})\right\}. \quad (5.9)$$

[5]See footnote 2 on page 2.

The result of the summation over cosine function is a cosine function. However, if the phases P_{Aj} are different, the result will exhibit an amplitude $f_A^{\mathbf{H}} < 1$ and a "mean" phase $P_A^{\mathbf{H}}$, that is

$$Y_{\mathbf{H}} = 1 + R + 2\sqrt{R} \cdot f_A^{\mathbf{H}} \cos(v - 2\pi P_A^{\mathbf{H}}) \tag{5.10}$$

with

$$P_A^{\mathbf{H}} = Hz_{A\mathbf{H}} \tag{5.11}$$

where $z_{A\mathbf{H}}$ represents the **H**-projected "average" position, of the entity of atoms, which, however, does not correspond to the arithmetic average of the z_{Aj}. Because the number N_A of atoms is very large, Eq. (5.9) can be rewritten by defining a distribution function $G(\mathbf{r})$ using the properties of the delta function via

$$G(\mathbf{r}) = N_A^{-1} \sum_j^{N_A} \delta(\mathbf{r} - \mathbf{r}_j) \tag{5.12}$$

and exchanging the summation with an integral, we can write

$$Y_{\mathbf{H}} = \int_{\mathbf{r}} \mathrm{d}\mathbf{r} G(\mathbf{r}) \left\{ 1 + R + 2\sqrt{R} \cdot \cos(v - 2\pi \mathbf{H} \cdot \mathbf{r}_{Aj}) \right\}. \tag{5.13}$$

The function $G(\mathbf{r})$ can be interpreted as a quasicontinuous, normalized particle density function of the particular atomic species under consideration. The two parameters $f_A^{\mathbf{H}}$ and $P_A^{\mathbf{H}}$ in Eq. (5.10), called coherent fraction and coherent position, respectively are in fact the Fourier transform of $G(\mathbf{r})$, that is,

$$G(\mathbf{H}) = G^{\mathbf{H}} = f^{\mathbf{H}} \mathrm{e}^{-\mathrm{i}2\pi \mathbf{H} \mathrm{r}} = f^{\mathbf{H}} \mathrm{e}^{-\mathrm{i}2\pi P^{\mathbf{H}}}. \tag{5.14}$$

This can be easily shown by converting the cosine function into exponentials:

$$\cos(v - 2\pi \mathbf{H} \cdot \mathbf{r}) = \cos(v - 2\pi P^{\mathbf{H}}) = [\mathrm{e}^{-i(v - 2\pi P_{\mathbf{H}})} + \mathrm{e}^{i(v - 2\pi P_{\mathbf{H}})}]/2. \tag{5.15}$$

Using $\int_{\mathbf{r}} \mathrm{d}\mathbf{r} G(\mathbf{r}) = \mathbf{1}$ we convert Eq. (5.13) into

$$Y_{\mathbf{H}} = \int_{\mathbf{r}} d\mathbf{r} G(\mathbf{r}) \{1 + R + \sqrt{R} \cdot \left(\exp[i(v - 2\pi \mathbf{H} \cdot \mathbf{r}_{Aj})] + \exp[-i(v - 2\pi \mathbf{H} \cdot \mathbf{r}_{Aj})] \right) \}$$

$$= 1 + R + \sqrt{R} \cdot \left\{ e^{iv} \int d\mathbf{r} G(\mathbf{r}) \exp[i(-2\pi \mathbf{H} \cdot \mathbf{r}_{Aj})] + e^{-iv} \int d\mathbf{r} G(\mathbf{r}) \exp[i(2\pi \mathbf{H} \cdot \mathbf{r}_{Aj})] \right\}$$

$$= 1 + R + 2\sqrt{R} \cdot \mathrm{Re}[e^{iv} G(\mathbf{H})] \tag{5.16}$$

and consequently we obtain Eq. (5.10):

$$Y_{\mathbf{H}} = 1 + R + 2\sqrt{R} \cdot f_A^{\mathbf{H}} \cos(v - 2\pi P_A^{\mathbf{H}}).$$

5.4 XSW Structure Factor versus XRD Structure Factor

The two parameters $f^{\mathbf{H}}$ and $P^{\mathbf{H}}$ (coherent fraction and coherent position, respectively) are related to the XRD structure factor $F_{\mathbf{H}}$, which represents (neglecting dispersion corrections) the Fourier transform of the total electron density function $\rho_e(\mathbf{r})$. By contrast, $f_A^{\mathbf{H}}$ and $P_A^{\mathbf{H}}$ are amplitude and phase, respectively of the (complex) Fourier coefficient of an particle density function specific for an element A, which is selected by spectroscopy. In fact, even a particular chemical species of an element can be selected by spectroscopy.

The analogy and differences of XSW and XRD results had already been pointed out [2]. The two parameter coherent fraction $f^{\mathbf{H}}$ and coherent position $P^{\mathbf{H}}$ can be directly compared with the more familiar X-ray structure factor, defined by the following equation:

$$F_{\mathbf{Q}} = \sum_{j=1}^{n} c_j F_{j\mathbf{Q}} \tag{5.17}$$

with $\sum_{j=1}^{n} c_j = 1$ and

$$F_{j\mathbf{Q}} = N_j^{-1} \sum_{n=1}^{N_j} (f_{j0} + f_j' + i f_j'') \exp(-i\mathbf{Q} \cdot \mathbf{r}_n). \tag{5.18}$$

Here, $\mathbf{Q}$ is a reciprocal lattice vector, the index j counts the different elements and f_{j0}, f_j' and f_j'' are the atomic form factor, dispersion correction, and absorption coefficient of a particular element labeled by j. With $2\pi\mathbf{H} = \mathbf{Q}$, the XSW "structure factor" $G^{\mathbf{H}}$ can be identified as follows:

$$G^{\mathbf{H}} = F_{j\mathbf{Q}} / (f_{j0} + f_j' + f_j'') = F^{\mathbf{H}} \exp(2\pi i P^{\mathbf{H}}), \tag{5.19}$$

where the element j is selected by spectroscopy. Thus, an XSW measurement directly determines an atomic distribution whereas an X-ray diffraction experiment measures the electron density distribution.

Frequently, $f^{\mathbf{H}}$ and $P^{\mathbf{H}}$ are considered as representative of the order and mean position, respectively of the contemplated species. However, some caution is here advised. Just like a particular X-ray structure factor, a particular XSW Fourier component can be extinct and when a particular coherent fraction $f^{\mathbf{H}}$ (i.e., a particular Fourier component) of the particle distribution is zero, this does not prove that there is no order. A simple example would be that two specific sites $\mathbf{r}_{A1}$ and $\mathbf{r}_{A2}$ are occupied such that for the chosen diffraction vector $\mathbf{H}\cdot\mathbf{r}_{A2} = P^{\mathbf{H}}_{A2} = \mathbf{H}\cdot\mathbf{r}_{A1} + 0.5 = P^{\mathbf{H}}_{A1} + 0.5$. The according XSW yield is then given by $Y_{\mathbf{H}} = 1 + R + 2\sqrt{R}f^{\mathbf{H}}\cos(v - 2\pi P^{\mathbf{H}}) = 1/2\Sigma_{k=1}^{2}\{1 + R + 2\sqrt{R}\cdot\cos(v - 2\pi P^{\mathbf{H}}_{\mathrm{Ak}})\}$. Adding two cosine functions that are out of phase by π leads consequently to $f^{\mathbf{H}} = 0$.

5.5 XSW Fourier Analysis: Imaging of Mn In GaAs

In many applications, small quantities of impurities or dopant atoms determine the properties of materials, in particular in the case of semiconductors. Recent examples are dilute magnetic semiconductors [13], that is, semiconductors doped with small quantities of magnetic impurities such as GaAs doped with Mn. Their potential use for spintronic devices relies on the ability to precisely control the site distribution the impurity. We recently used the XSW for this purpose. A three-dimensional Fourier analysis of ~4 percent Mn in (Ga, Mn)As thin films allowed to directly visualize the location of the Mn atoms in the unit cell of GaAs [14] (see Fig. 5.1).

The XSW measurements were conducted at beamline ID32 [15] at the European Synchrotron Radiation Facility. Figure 5.2a schematically shows the experimental setup and the principle of the measurement. The 10 keV X-radiation from the undulator was selected by a pair of Si111 crystals. A Si555 channel cut crystal acts as a high-resolution monochromator and reduces the energy spread of the beam to about 15 meV. Together with the small angular spread of the undulator beam, this setup allowed to properly resolve the reflectivity curves of GaAs for all used reflections. The principle of the technique is illustrated in Fig. 5.2b.

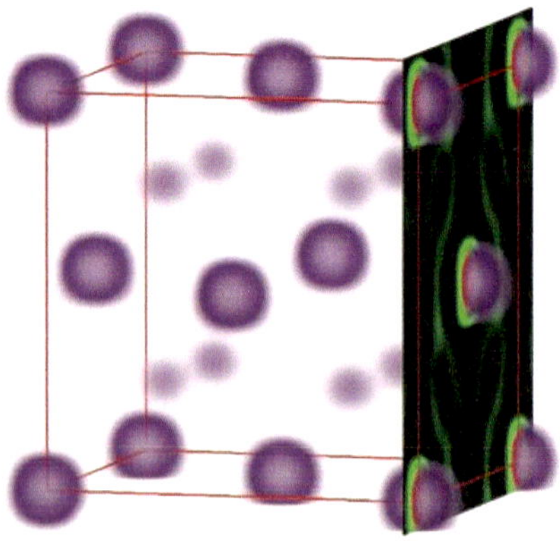

Figure 5.1 The image of Mn in the unit cell of GaAs reconstructed from 18 Fourier components for an as-grown 3-nm $Ga_{0.96}Mn_{0.04}As$ film, where 82 percent of the Mn is found to substitute the Ga. The Mn distribution along a {100} plane is also shown.

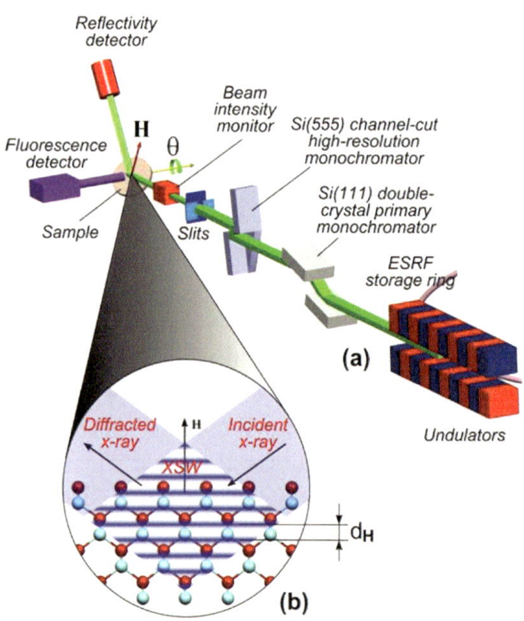

Figure 5.2 Principle and result of an XSW imaging measurement and the experimental setup at a third generation Schematic representation of the XSW with period d_H induced by the GaAs(004) reflection.

As the incident angle (θ) traverses the range of the Bragg reflection, the XSW, which has the periodicity of the Bragg plane spacing $d_{\mathbf{H}}$, moves by half of $d_{\mathbf{H}}$ normal to the Bragg planes. This movement modulates the intensity of the fluorescence X-rays emitted from Mn in a way that is characteristic of its atomic position within the GaAs unit cell.

We analyzed three 23 nm $Ga_{0.952}Mn_{0.048}AS$ films, two of which had received a certain postgrowth treatment, that is, were annealed in inert or hydrogen atmosphere. Figure 5.3 shows as examples the reflectivity curves and the Mn K_{α} fluorescence yields modulated by the GaAs(111), (1–13), (004), and (422) XSWs for the 23 nm $Ga_{0.952}Mn_{0.048}AS$ film. The distinctive modulations result from the movement of the XSWs, which probe the Mn distribution along the different crystallographic directions. For each substrate reflection, specified by a reciprocal lattice vector **H,** the XSW measurement determines two quantities $f^{\mathbf{H}}$ and $P^{\mathbf{H}}$ of the Mn distribution.

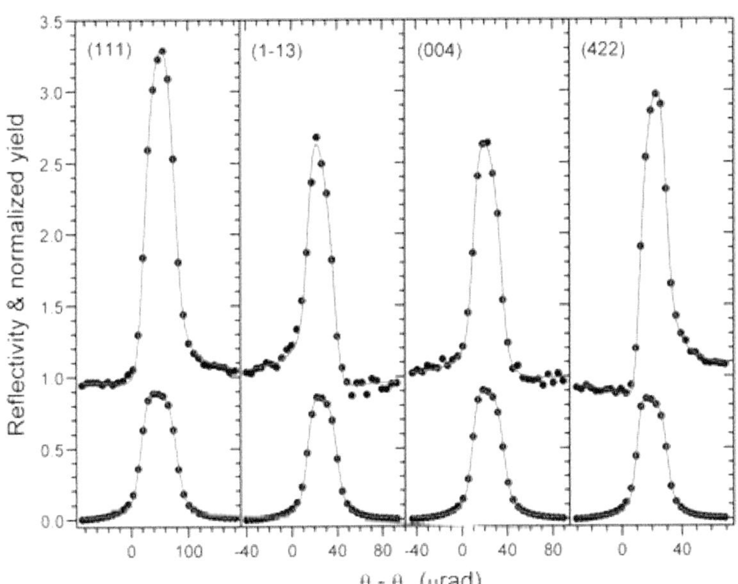

Figure 5.3 Data (dots) and best fits (lines) of the reflectivity (bottom) and XSW-induced Mn K_{α} fluorescence yields (top) for the GaAs(111), (1–13), (004) and (422) reflections for a post-annealed 23 nm $Ga_{0.952}Mn_{0.048}AS$ film.

For the Mn in a (Ga,Mn)As film its distribution, $\rho_a(\mathbf{r})$, projected into the substrate GaAs unit cell, is described by a Fourier series of $2\pi\mathbf{H}$ as $\rho_a(\mathbf{r}) = (1/V_{uc})\Sigma_{\mathbf{H}}G^{\mathbf{H}}\exp(-2\pi i\mathbf{H}\cdot\mathbf{r})$, where V_{uc} is the volume of the unit cell. With $\rho_a(\mathbf{r})$ being normalized such that $\int_{V_{uc}} \rho_a(\mathbf{r})d\mathbf{r} = 1$, it can be shown that the complex coefficients $G^{\mathbf{H}}$ are related to $f^{\mathbf{H}}$ and $P^{\mathbf{H}}$ as $G^{\mathbf{H}} = f^{\mathbf{H}}\exp(2\pi iP^{\mathbf{H}})$. Each XSW measurement thus determines, in contrast to diffraction, not only the amplitude but also the phase of the **H**-th Fourier coefficient of $\rho_a(\mathbf{r})$, which can then be readily reconstructed [16], with a sufficient number of measured $f^{\mathbf{H}}$ and $P^{\mathbf{H}}$, by the expansion

$$\rho_a(\mathbf{r}) = (1/V_{uc})\left\{1 + 2\sum_{\mathbf{H}} f^{\mathbf{H}}\cos\left(2\pi\left(P^{\mathbf{H}} - \mathbf{H}\cdot\mathbf{r}\right)\right)\right\}, \tag{5.20}$$

considering that $f^{\bar{\mathbf{H}}} = f^{\mathbf{H}}$ and $P^{\bar{\mathbf{H}}} = -P^{\mathbf{H}}$.

For the reconstructions, we used the $f^{\mathbf{H}}$ and $P^{\mathbf{H}}$ values that were determined from the results of the XSW measurements of 18 nonequivalent reflections. Utilizing the crystal symmetry of GaAs, this amounted to 50 Fourier components (50 $f^{\mathbf{H}}$ and $P^{\mathbf{H}}$ values). The resulting images (see Fig. 5.1) showed with striking clarity that the majority of the Mn substitutes for the Ga (Mn_{Ga} site). The maxima of the Mn distributions all occurred slightly above the bulk Ga site due to the ~ 0.4 percent larger *c* lattice constants of the (Ga, Mn)As layers. However, the images also showed that Mn occupies other positions than the Mn_{Ga} site. Small fractions of the Mn could be identified on two different interstitial positions. These are the open spaces in the center of a Ga or As tetrahedrons, that is, the Ga and As interstitial sites, respectively.

The study showed that the exact fraction of the Mn at the different sites strongly depends on the postgrowth treatment. To quantify the Mn site distribution for each sample precisely, a structural model based on the reconstructed image was introduced and refined to best reproduce the measured Fourier coefficients. The result [14] revealed characteristic differences for the three differently treated samples. The fraction of Mn substituting for Ga was measured to be 75 percent for an as-grown sample. However, the fraction increased after annealing to 89 percent at the expense of interstitial Mn. This was surprising, since it had been commonly assumed that low temperature (LT) annealing does not affect the Mn_{Ga} population [17,18] because of the rather high activation energy needed for interstitial Mn to move to the Ga site. However, the observed increase of Mn_{Ga} upon annealing could be due to the filling of Ga vacancies, created during either the growth or the annealing process. The refinement further showed that the as-grown sample contained a small fraction of disordered Mn, which was largely annihilated by the post growth annealing and by hydrogenation.

5.6 Summary

In summary, amplitude and phase of the Fourier transform of atomic distributions are determined by the XSW technique with chemical sensitivity and high accuracy for very dilute systems. In contrast to diffraction experiments, using only some ten Fourier components, with the help of the phases high-resolution images can be reconstructed. For the presented case of a dilute magnetic semiconductor, the complete Mn site distributions in 23 nm thick $Ga_{0.952}Mn_{0.048}AS$ films with concentrations as low as 0.3 percent at the corresponding lattice sites could be revealed. Structural refinements based on the images yielded the site occupation of the Mn with an error of less than ±0.1 percent [14]. This precise structure analysis by the XSW imaging technique provided invaluable information related to the growth parameters and to the structure and electronic/magnetic properties of this dilute magnetic semiconductor material.

References

1. Batterman BW (1969) Detection of foreign atom sites by their X-ray fluorescence scattering, *Phys Rev Lett*, **22**, 703–705.
2. Zegenhagen J (1993) Surface structure determination with X-ray standing waves, *Surf Sci Rep*, **18**, 199–271.
3. Batterman BW (1964) Effect of dynamical diffraction in X-ray fluorescence scattering, *Phys Rev*, **133**, A759.
4. Materlik G, Zegenhagen J (1984) X-ray standing wave analysis with synchrotron radiation applied for surface and bulk systems, *Phys Lett*, **104** A, 47–50.

5. Laue M (1941) *Röntgenstrahl-Interferenzen*, Akademische Verlagsgesellschaft, Beccker & ErlerKolm.-Ges., Leipzig; (1960) *Röntgenstrahl-Interferenzen*, Akademische Verlagsgesellschaft, Frankfurt.
6. Bedzyk MJ, Bommarito GM, Schildkraut JS (1989) X-ray standing waves at a reflecting mirror surface, *Phys Rev Lett*, **62**, 1376–1379.
7. Barbee TW, Warburton WK (1984) Evanescent and standing wave. Fluorescence from a multilayer structure, *Mater Lett*, **3**, 17–25; Bedzyk MJ, Bilderback DH, Bommarito GM, Caffrey M, Schildkraut JS (1988) X-ray standing waves: a molecular yardstick for biological membranes, *Science*, **241**, 1788–1791.
8. Einstein A (1905) Über einen die Erzeugung und Verwandlung des Lichtes betreffenden heuristischen Gesichtspunkt, *Ann Phys*, **322**, 132–148.
9. Vartanyants I, Zegenhagen J (1999) Photoelectric scattering from an X-ray interference field, *Solid State Commun*, **113**, 299–320.
10. Hertel N, Materlik G, Zegenhagen J (1985) X-ray standing wave analysis of bismuth implanted in Si(110), *Z Phys B*, **58**, 199–204.
11. Hauptman HA, Karle J (1953) *Solution to the Phase Problem I: The Centrosymmetric Crystal* (ACA Monograph No. 3 Polycrystal Book Service).
12. Bedzyk MJ, Materlik G (1985) Two-beam dynamical diffraction solution of the phase problem: a determination with X-ray standing wave fields, *Phys Rev B*, **32**, 6456–6463.
13. Jungwirth T, Sinova J, Mašek J, Kučera J, MacDonald AH (2006) Theory of ferromagnetic (III,Mn)V semiconductors, *Rev Mod Phys*, **78**, 809–864.
14. Lee T-L, Bihler C, Schoch W, et al. (2010) Fourier transform imaging of impurities in the unit cells of crystals: Mn in GaAs, *Phys Rev B*, **81**, 235207–235216.
15. Zegenhagen J, Detlefs B, Lee T-L, et al. (2010) X-ray standing waves and hard X-ray photoelectron spectroscopy at the insertion device beamline ID32, *J Electron Spectrosc Relat Phenom*, **178–179**, 258–267.
16. Cheng L, Fenter P, Bedzyk MJ, Sturchio NC (2003) Fourier-expansion solution of atom distributions in a crystal using X-ray standing waves, *Phys Rev Lett*, **90**, 255503–255506.

17. Zhao LX, Staddon CR, Wang KY, et al. (2005) Intrinsic and extrinsic contributions to the lattice parameter of GaMnAs, *Appl Phys Lett*, **86**, 071902–071904.

18. Jungwirth T, Wang KY, Mašek J, et al. (2005) Prospects for high temperature ferromagnetism in (Ga,Mn)As semiconductors, *Phys Rev B*, **72**, 165204–165216.

for the diffuse scattering of neutrons, where energy and momentum transfers are coupled, the above treatment is not valid.

Equation (6.9) defines the so-called S-matrix introduced by M. Born in 1942 [26] and already contained in I. Waller's dissertation, as M. Born mentions. The S-matrix decomposition has been applied before for the analysis of Debye–Waller factors [27–29].

For the special case of monoatomic crystals (with one atom per unit cell), we get the following:

$$Z(\mathbf{Q}) = f(\mathbf{Q})\exp(-W(\mathbf{Q}))\cdot\hat{\mathbf{I}} \tag{6.10}$$

$$I(\mathbf{Q}) \propto f^2(\mathbf{Q})\exp(-2W(\mathbf{Q}))\cdot\mathbf{Q}^T\cdot S(\mathbf{q})\cdot\mathbf{Q}. \tag{6.11}$$

As $S(\mathbf{q})$ and $\mathbf{D}(\mathbf{q})$ for a monoatomic crystal are real and symmetric, the minimum $\mathbf{Q}$-set for a complete $\mathbf{D}(\mathbf{q})$ reconstruction at arbitrary $\mathbf{q}$ assumes six observations at most. Reciprocal space mapping with a 2D detector allows overdetermination for the system of Eq. (6.9) for any $\mathbf{q}$ thus providing the phonon dispersion, density of states, and all the derived values. The polarization correction has to be applied prior to the treatment.

For nonmonoatomic crystal $S(\mathbf{q})$ at an arbitrary $\mathbf{q}$ remains just Hermitean even if the dynamical matrix takes a special shape (i.e., if every atom sits at a center of inversion) and part of the information is irreversibly lost in a kinematic scattering experiment; information on the missing elements of the S-matrix still can be recovered from coherent X-ray scattering.

Some use of the above formalism can also be made in the long-wave limit, where the dominating scattering comes from acoustic phonons with the linear dispersion $\omega \propto qV$ ($q = |\mathbf{q}|$, V is the sound velocity). Collecting data in all directions for multiple Γ-points (Bragg nodes) allows the determination of the full set of elastic moduli from a diffraction experiment.

6.3 Experimental Technique

All IXS instruments are based on the triple-axis spectrometer as developed by Brockhouse for INS. The three axes comprise the very high energy resolution monochromator (first axis), the sample goniometry (second axis), and the crystal analyzer (third axis). A detailed account of the instrumental developments and the two different types of very high energy resolution monochromators is given in Ref. [17]. The X-ray beam from the undulator source is premonochromatized by a silicon or a diamond (1 1 1) double-crystal monochromator to a relative bandwidth of $\Delta E/E \approx 2 \cdot 10^{-4}$. Due to the high heat load produced by the intense undulator beam, the silicon crystals have to be cooled down to cryogenic temperatures of about 125K, where silicon displays a maximum in the thermal conductivity and a minimum in the linear thermal expansion coefficient. This allows the thermal deformation of the crystal to be kept below the limits where photon-flux losses occur. In the case of diamond, conventional water cooling is sufficient thanks to its thermal properties and its weak photoelectric absorption. The necessary very high energy resolution is then obtained either by a monochromator, operating at a Bragg angle very close to 90° (ESRF and SPring-8), or by a multiple crystal in-line monochromator (APS). The highly monochromatized X-rays are then focused onto the sample position by focusing optics. The scattered photons are energy analyzed by a spherical perfect silicon crystal analyzer, operated in Rowland geometry, and at a Bragg angle of 89.98°. In the case of the most widely spread spectrometers, the momentum vector **Q** is selected by rotating the analyzer arm in the horizontal plane. Most of the IXS spectrometers are equipped with more than one analyzer so that spectra at several momentum transfers can be recorded simultaneously. ID28 at the ESRF is currently equipped with nine crystal analyzers, arranged with a spacing of 0.75° in the horizontal scattering plane. The X-rays diffracted from the analyzer crystal are recorded by solid-state detectors, which provide a very low background.

Table 6.1 Characteristics of the ID28 spectrometer in the main operation modes

Reflection (Si)	(8 8 8)	(9 9 9)	(11 11 11)	(12 12 12)
Wavelength [Å]	0.7839	0.6968	0.5701	0.5226
Energy [eV]	15816	17793	21747	23725
Q_{max} [nm^{-1}]	67.7	76.2	93.2	100
Resolution (meV)	6.0	3.0	1.5–1.7	1.3–1.4

In the case of backscattering monochromators, energy scans are conveniently done by changing the lattice constant *via* changing the temperature T: $\Delta E/E = \Delta d/d = \alpha(T)\,\Delta T$, where $\alpha(T) = \alpha_0(T - 25.5°\text{C}) + \beta\Delta T[\text{K}]$ is the thermal expansion coefficient of silicon, equal to $2.58{\cdot}10^{-6}\ \text{K}^{-1}$ at room temperature [30]. In order to obtain an energy step of about one-tenth of the energy resolution, that is, $\Delta E/E \sim 10^{-6}$, it is necessary to control the monochromator crystal temperature with a precision of about 0.5 mK [31]. Key characteristics of the ID28 ESRF instrument are represented in Table 6.1; the scheme of its layout is given in Fig. 6.1. A more detailed account of the experimental setup is given in [18].

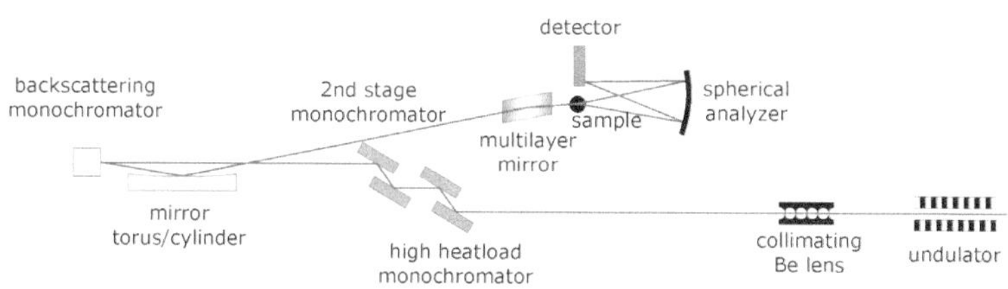

Figure 6.1 Schematic optical layout of the inelastic scattering beamline II (ID28) at the European Synchrotron Radiation Facility. Figure taken from Ref. [18].

6.4 Mapping of Phonon Dispersion Surfaces

The use of a multianalyzer system for single-crystal work allows the efficient mapping of the bidimensional dispersion surfaces, provided that the spectrometer settings are selected to cover the region of interest with sufficient density. The data throughput can be significantly improved, and further allows the unambiguous assignment of phonon branches/surfaces around low symmetry points. One of the first examples of such studies concerned the 2D phonon dispersion relations around the K point of graphite single crystals [19]. As the dispersion perpendicular to the graphene layers in graphite is practically zero for the phonon modes around K, graphite can be utilized as an approximant of graphene.

The experiments were performed with a total energy resolution of ~3.3 meV, and nine IXS spectra were recorded simultaneously, thus allowing an efficient coverage of the relevant portion of the phonon–dispersion branches. A set of ~100 data points was obtained in the $q_x - q_y$ plane in a region of ±0.6 Å^{-1} around the K point, further extended by the application of symmetry operations (Fig. 6.2).

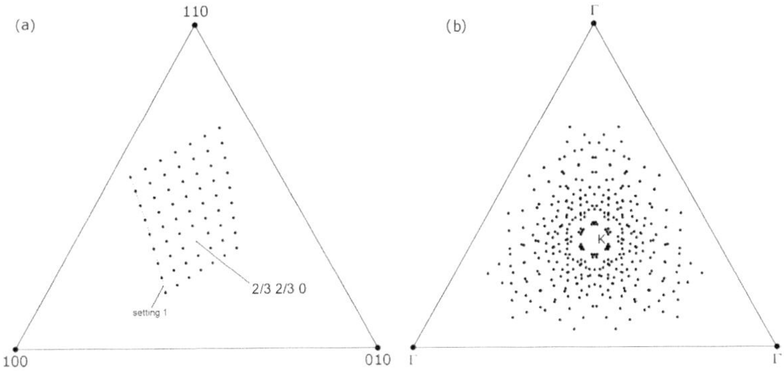

Figure 6.2 (a) Mesh in the reciprocal space of graphite, corresponding to a subset of seven experimental settings, taken with nine analyzers simultaneously; the dotted line represents the section of the Ewald sphere. (b) Expanded mesh, obtained by applying the symmetry operations for the K point.

The phonon mode assignment was carried out by explicit comparison of the measured and calculated IXS intensities in the 2D map and used to disentangle overlapping phonon modes. From the experimental data points 3D surfaces of the phonon–dispersion relations are interpolated using spline functions (Fig. 6.3a,b). The interpolated phonon maps were used for further data analysis.

The phonon trigonal warping effect for the LA, LO, and TO modes around the K point is illustrated from the phonon equienergy contours shown in Fig. 6.3c along with the GW calculations (Green's function G of the screened Coulomb interaction W). Notably, the experiments and calculations are in excellent agreement to each other which further supports the performed mode assignment.

This work provided not only a correct assignment of the phonon branches and hence the phonon trigonal warping, but it has also provided an all-experimental determination of the electron–phonon coupling constant of 166 eV/Å^2, in excellent agreement to GW calculations. These results are fundamental for understanding angle-resolved photoemission, double-resonance Raman and transport measurements of graphene-based systems. In a more general context, it illustrates that 2D phonon mapping can be the key for understanding phonon properties in a plethora of other materials: recent applications comprise studies of chromium [32], quartz [33] and ferroelectric perovskites [34].

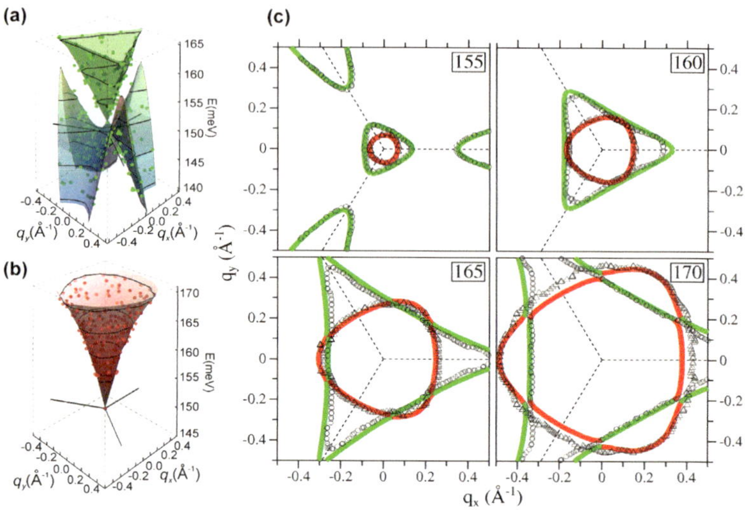

Figure 6.3 The interpolated experimental 2D phonon dispersion relation for (a) the longitudinal optic and acoustic phonon branches and (b) the transverse optical phonon branch that displays a Kohn anomaly. Dots denote the experimental data (c) Equienergy contours from the interpolated phonon surfaces for the three highest phonon branches around K for energies 155–170 meV. Circles and triangles denote an interpolation to the raw experimental data in (a) and (b). Lines are GW calculations. The present color code follows the diabatic notation: branches with the same color have similar phonon patterns. Figure taken from Ref. [19].

6.5 Combining IXS and TDS

As has been already mentioned in the Introduction, IXS experiments are time-consuming and even exploiting the efficient mapping of phonon sheets as described above, typical experiment times are in the order of days. TDS offers in principle a very attractive alternative as scattering patterns can be recorded on the timescale of seconds, using large-area detectors, and full 3D-reconstructions are a matter of a few hours at the most. On the contrary, it is evident from the inspection of Eq. (6.6) that information of the phonon energy and the assignment of the phonon branch is "encoded" in the observed **Q**-dependent scattering intensity. Though the full inversion of TDS cases is only possible for monoatomic crystals (one atom per unit cell), and the quantitative treatment of data is much less straightforward than for INS and IXS, TDS constitutes a powerful complement to IXS studies. It allows, on the one hand, the localization of interesting portions of reciprocal space and the rapid study of changes in the lattice dynamics as a function of pressure, temperature and other parameters in the case of TDS. On the other hand, the separation of static and dynamics contributions to the diffuse scattering can be investigated by IXS. In the following, some applications shall be discussed in detail.

6.5.1 Visualization of the Fermi Surface of Zinc

The Fermi surface—the surface in electron momentum space of a metal that separates occupied and empty electronic states—is one of the most important theoretical concepts in solid-state physics. Indeed, the shape of the Fermi surface controls the electric, magnetic, and thermal properties of materials. Interestingly, one of the earliest techniques proposed to study the Fermi surface was to search for abrupt variations in the dispersion of phonons [35]. These so-called Kohn anomalies originate from the singularity at $\mathbf{q} = 2\mathbf{k}_F$ in the static dielectric function of the conduction electrons, and therefore the Fourier transform of the pair-interaction potential, reflecting changes in the charge screening by the conduction electrons when the phonon wave vector $\mathbf{q}$ exceeds the spanning across the Fermi surface $\mathbf{q} = 2\mathbf{k}_F$.

While numerous inelastic neutron scattering (INS), and later inelastic X-ray scattering (IXS) studies, documented these Kohn anomalies, only a few studies were devoted to actually mapping these Kohn anomalies in 2D or 3D [36–39]. As the phonon frequencies are typically determined point by point and branch by branch along the high-symmetry directions, the reconstruction of the full Fermi surface in 3D would be extremely time-consuming with this approach. However, the dispersion anomalies can be identified in a model-independent fashion via characteristic intensity variations of the TDS. Still, there has been little work in this area [40–42] most of which is limited to 1D-anomalies, and to a few high-symmetry points or high-symmetry planes.

In a recent work of metallic zinc synchrotron, X-ray diffraction was utilized in conjunction with area detectors to visualize 3D lens-shaped diffuse objects [20]. By complementing these measurements with an inelastic X-ray scattering study and ab initio calculations, it could be demonstrated that the observed diffuse scattering is of phonon-related nature and that the lens-shaped objects mimic the shape of the Fermi surface via electron–phonon interactions.

The TDS studies were performed in two runs, the first on the Swiss–Norwegian beam lines (SNBL) at the ESRF, and the second on beamline X06a at the SLS. The use of an energy discriminating pixel detector (PILATUS 6M [43]) was found indispensable to suppress the strong zinc fluorescence contribution. Figure 6.4a shows the reconstructed experimentally derived diffuse scattering intensity in the HHL and H0L planes.

Besides the TDS from long-wavelength acoustic phonons, dominating in the close vicinity of the Bragg reflections, additional well-defined features are clearly visible. Six-fold (*HK*2 layer) or (approximately) three-fold (*HK*3 layer) intensity variations (Fig. 6.4b,c) around Bragg spots reflect both the crystallographic structure and the anisotropic dispersion of acoustic phonons. The expected monotonously varying TDS intensity is additionally modulated, clearly showing the lens-like objects with an axial symmetry centered on intense Bragg spots. Their boundaries are characterized by sharp intensity variation and are easy to locate. Furthermore hollow lens-like objects of smaller size (i.e., centered around (1/2 1/2 3) are observed. As no structural disorder is expected for a chemically pure zinc single crystal, the observed lens-like diffuse objects are related to portions of the Fermi surface. These are visualized via nonmonotonic intensity variations in the TDS, and correspond to the occurrence of a Kohn anomaly in the phonon dispersion.

The contrast is particularly favorable when the phonon wave vector translates, or nests, parallel pieces of the Fermi surface. A value of $q \approx 3.14$ Å^{-1} for the radius of curvature of the large lens-like objects (i.e., centered around (0 0 2*n*) spots) is derived, resulting in a value of $q/2 \approx k_F \approx 1.57$Å^{-1}, in close agreement with the corresponding value for a free-electron model (1.573 Å^{-1} if two conduction electrons are attributed to each zinc atom). A 3D representation of constant-intensity surfaces is shown in Fig. 6.5, where lens-like objects with six-fold modulation can be easily identified along the c^* axis.

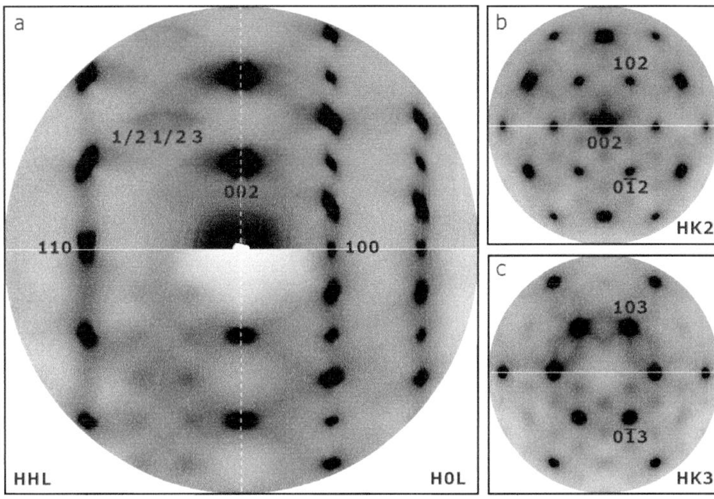

Figure 6.4 Reciprocal space maps for zinc. Sections along the c^* axis: (a) *HHL* and *H0L*, and perpendicular to it: (b) *HK*2 and (c) *HK*3. The upper part of panels show the experimental pattern, and the lower part the calculated pattern. Adapted from Ref. [20].

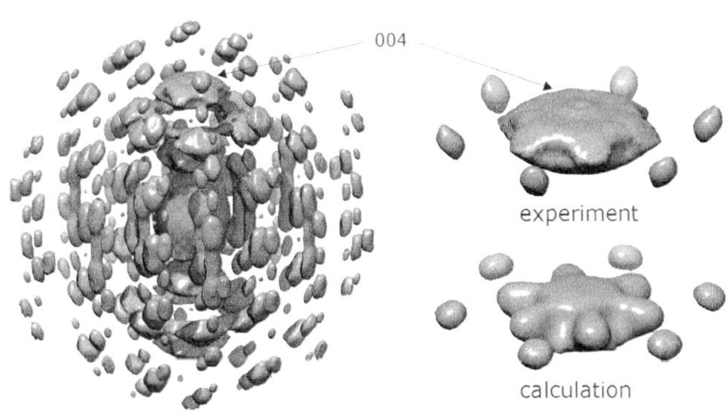

Figure 6.5 Constant-intensity surfaces for diffuse scattering of zinc: general view of the experimental dataset and comparison of experiment and theory for the region around the (004) Bragg spot; for visualization purposes, a $\cos^2(2\theta)$ factor is introduced Adapted from Ref. [20].

The experimental observations are further supported by ab initio calculations of the Fermi surface (Fig. 6.6). The large lens-like objects correspond to the inner part of the Fermi surface around the Γ point. The shape of this part of the Fermi surface mimics the corresponding surface of dispersion anomaly locations (Kohn surface [38]—in the case of zinc it has a well-defined geometrical meaning) through its homothetic transformation with the scaling factor 2 ($q = 2k_F$). This offers the opportunity to visualize the shape of the Fermi surface in a straightforward way. The hollow lens-like objects of smaller size, observed in *HHL* sections (i.e., centered around (1/2 1/2 3)) are off-center sections of the large "lenses".

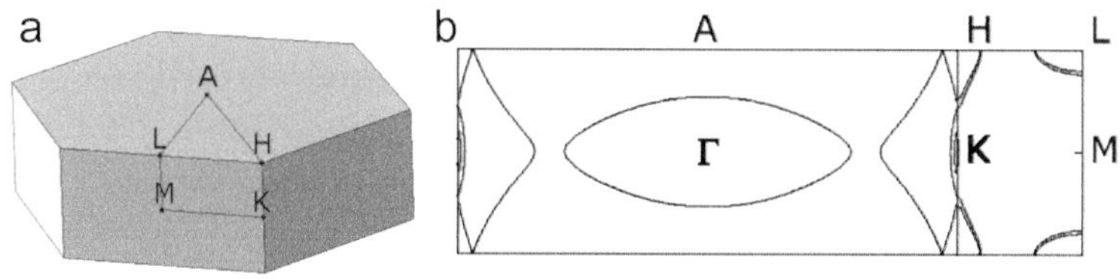

Figure 6.6 (a) Brillouin zone of zinc with denoted special points. (b) High symmetry section of the Fermi surface obtained by ab initio calculations. Figure adapted from Ref. [20].

6.5.2 Giant Kohn Anomaly in $ZrTe_3$

Dynamical modulations of the spin or charge density are a ubiquitous phenomenon in crystalline metals and form an important ingredient of solid state theory [44]. In many materials, they remain dynamical at all temperatures and contribute to the rich phenomenology of electronic, magnetic, and thermal transport and ordering. In general, the modulation wave vector **q** can be incommensurate with the underlying lattice and is derived from a nesting geometry of the Fermi surface that provides strong screening at $\mathbf{q} = 2\mathbf{k}_F$. Such a nested Fermi surface is found, in particular, in low-dimensional crystal structures with chainlike atomic arrangements.

While the charge density wave (CDW) ground state and its many fascinating properties are well described by theory, the phase transition is accessible only by experiment. The combined diffuse scattering (DS)/IXS experimental study of the fluctuations and the soft-mode phonon in $ZrTe_3$ elucidated the properties of the charge density wave in the important regime close to and above the phase transition, where the density waves are of dynamical nature [21]. The Peierls modulation $\mathbf{q}_p = (0.07,0,0.333)^T$ [45] is transverse to the prismatic chains existing in this compound with a small a^* component and a tripling of the unit cell along the layering c^* direction. Resistivity measurements show an anomaly due to the transition at $T_P = 63$ K, and the system remains metallic below T_P [46]. Electronic structure calculations [47,48] and a detailed photoemission study [49] explain this behavior by nesting in a small electron pocket of the highly directional Te $5p_x$ band running along these Te–Te chains, while other sheets of the Fermi surface remain unaffected by the transition.

Phonon dispersions were measured with an energy resolution $\Delta E = 1.7$ meV, and the momentum resolution was $\Delta Q = 0.05$ Å^{-1} in the horizontal $a^* - b^*$ plane and 0.15 Å^{-1} in the vertical c^* direction. The temperature dependence of the Kohn anomaly measured next to (4 0 −1) spot along the modulation vector is traced in Fig. 6.7. In the figure, the a^* component of $\mathbf{Q}$ is indicated in relative lattice units. It shows a dip at $\mathbf{q}_P$ deviating from a sinusoidal model dispersion. The displacement pattern of the soft mode is of mostly transverse character with respect to the Te–Te chains whose Te $5p_x$ orbitals form the quasi-1D Fermi surface.

Already at room temperature ($4.6 \cdot T_P$), a slight dip is visible, barely outside the experimental error bars. As the temperature is lowered, the dip gets more pronounced, until the phonon frequency drops rapidly to zero at T_P. Mean field theory predicts this freezing of the phonon as a logarithmic divergence that would correspond to a square-root power law close to the phase transition [44]. Inspection of the data on a double-logarithmic scale did not reveal well-defined power-law behavior. Over a broad range in temperature, the closest correspondence to a power law was approximately $(T - T_P)^{1/8}$ (see inset of Fig. 6.7). Thus, it can be concluded that the soft mode freezes in the fashion of a giant Kohn anomaly, and suggests that the phonon frequency drops to zero at the phase transition.

The deviation from mean field behavior stimulated the investigation of the thermodynamic signature of the phase transition by the study of the diffuse scattering ($T > T_P$) and superstructure intensity ($T < T_P$) at $\mathbf{q}_P$. The combined diffuse and inelastic X-ray scattering study allowed disentangling the static and dynamical contributions to the diffuse intensity. On inspection, the IXS spectra show that the dominant contribution is the central peak (zero energy loss). Fig. 6.8a compares the intensity of the central peak and the phonon (symbols) with the energy integrated signal (line) at $T = 73$ K > T_P. The central peak is not only dominant, but its momentum dependence corresponds to that of the diffuse scattering signal. Thus the diffuse scattering is a quasistatic diffraction feature which is representative of the (fluctuating) order of the CDW. Scans of the integrated scattering intensity along $\mathbf{q}_P$ are displayed in Fig. 6.8b and show the emergence of a diffuse scattering feature. At high temperature, a broad peak is partially due to TDS, but the emergence of a sharp feature with decreasing width at lower T is due to the longer correlation length of the CDW order. Below T_P, this peak does not sharpen any further, although it remains slightly broadened with respect to the experimental resolution. Its intensity grows rapidly on lowering the temperature below T_P.

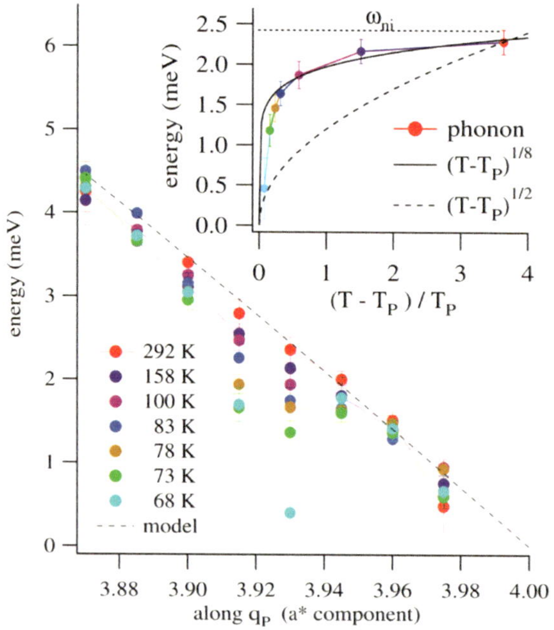

Figure 6.7 $ZrTe_3$ phonon dispersion recorded from $\tau = (40-1)$ in the direction of $\mathbf{q}_P$ at various temperatures $T > T_P$ above the transition. The inset shows the phonon energy $\hbar\omega$ at $\mathbf{q}_P$ as a function of reduced temperature $(T - T_P) / T_P$. The solid and dashed lines represent a 1/8 and a 1/2 power law, respectively. Figure taken from Ref. [21].

The temperature evolution of the intensity and the width along b^* were also studied (Fig. 6.8c). The intensity and width show power-law scaling with reduced temperature. The critical exponents derived from these dependencies are $2\beta = 0.26 \pm 0.06$ for the intensity, that is taken here as a measure of the order parameter squared, and $\nu = 0.85 \pm 0.2$ for the width, that is proportional to the inverse correlation length. The critical exponent $\beta = 0.13 \pm 0.03$ is even smaller than the ones observed in other CDW materials (0.32 ± 0.05 for blue bronze [50] and 0.23 ± 0.02 for $(TaSe_4)_2I$ [51]). Thus, a mean field description of the phase transition is not appropriate and fluctuation corrections are needed.

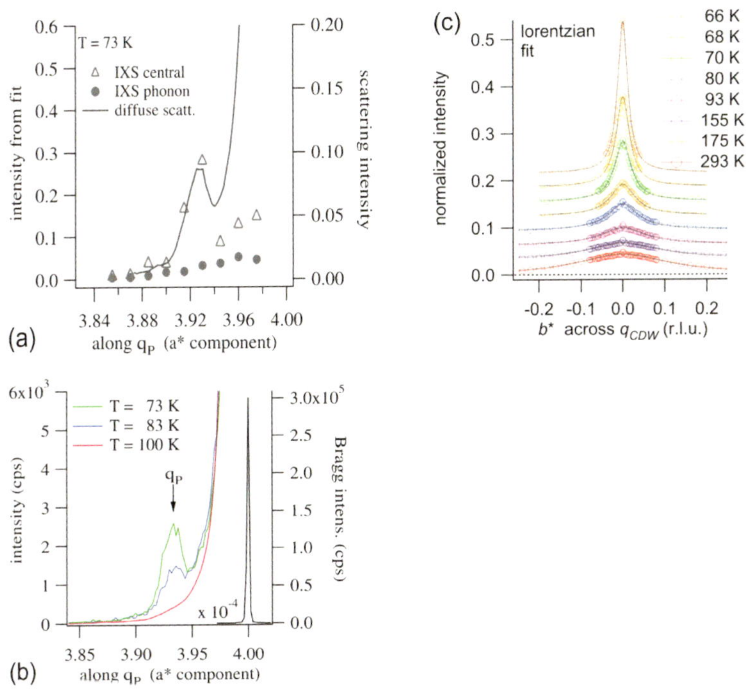

Figure 6.8 Diffuse scattering in $ZrTe_3$ along $\mathbf{q}_P$. (a) Symbols represent the intensities extracted from data fitting of the IXS spectra: central peak (triangles), phonon contribution (filled circles). The solid line is the total diffuse scattering intensity as measured with an energy-integrating detector. (b) Diffuse scattering intensities at three selected temperatures as indicated. (c) Scans of diffuse scattering across $\mathbf{q}_P$ along b^* at various temperatures above T_P. Figure taken from Ref. [21].

6.5.3 Diffuse Scattering and Correlated Disorder in Manganese Analogue of Prussian Blue

Prussian Blue (PB), probably the first coordination compound ever synthesized by mankind [52], together with its metal substituted analogues attract nowadays permanently growing interest due to the unique combinations of physicochemical and structural properties such as high-temperature magnetism, magnetic pole inversion, photochromism, light-induced phase transformations, and other switching phenomena. These properties arise from superexchange interactions, charge transfer, and spin-state instabilities, all of them obviously dependent on the chemical composition [53,54]. One of the most attractive and technologically promising features provided by some of PB analogues is a possibility to manipulate optical and magnetic properties by external stimuli—such as temperature, pressure, light, magnetic, and electric fields; this option persists down to a nanoscale.

The average spatial arrangement of interacting ions determining the physical properties is seemingly simple for PB-like compounds. The structure of a generic PB analogue $A\{[B(CN)_6]_x[(H_2O)_6]_{1-x}\}\cdot yH_2O$ (with A and B being metal cations) consists of two interpenetrating face-centered nearly cubic sublattices. One sublattice is built from A cations, whereas the other sublattice is a mixture of $[B(CN)_6]^{\delta-}$ anions and $(H_2O)_6$ clusters [55]—see Fig. 6.9a. The real structure of most of the PB-like compounds is therefore inherently disordered and complex: "unit cells" have different chemical composition, and the distribution of these unit cells is correlated and forms short-range order—the long-range ordering is rejected by the absence of long-range order.

Building up on previous X-ray and neutron diffuse scattering studies [55–58] synchrotron single crystal diffraction data were collected at the BM1A station of the SNBL at the ESRF with a MAR345 image plate detector. Follow-up experiments were performed at X06SA beamline at SLS with a PILATUS 6M detector. The IXS data, recorded with an energy resolution of 3.0 meV, indicate that in the proximity of strong diffuse features the elastic contribution is largely dominating and the inelastic part can be safely neglected.

Visual inspection of diffuse scattering maps (see Fig. 6.9b) suggests that the basic pattern may be parameterized by three families of intersecting diffuse rods. Such a suggestion is supported by the fact that experimental intensity distribution at the intersection of rods is mostly additively. Following the approach of Krivoglaz [59], the diffuse intensity could be approximated in terms of Fourier components of concentration fluctuations, $c_{\mathbf{q}}$:

$$I_{DIF} = N\left\langle \left|c_{\mathbf{q}}\right|^2 \right\rangle \left(f_{\mathbf{q}}^A - f_{\mathbf{q}}^B\right)^2 \tag{6.12}$$

Here, N plays the role of a scale factor and the molecular form-factors f_q^A and f_q^B stand for the structure factors of the A and B structural units. f_q^A and f_q^B can be calculated from the atomic coordinates reported for the average structure [55]; here and below the momentum transfer $\mathbf{q}$ is taken in r.l.u. Noteworthy, at variance with binary alloys, both substituted object, $[Mn(CN)_6]^{3-}$ and substituent $(H_2O)_6$ are not spherically symmetric. The system of intersecting rods is parameterized by the following equation:

$$\left\langle \left|c_{\mathbf{q}}\right|^2 \right\rangle \approx p(\mathbf{q}) \otimes \sum_{\alpha,\beta,\gamma} \left[\sum_{n_{\alpha,\beta,\gamma}=-\infty}^{+\infty} \delta(q_\alpha + 2n_\alpha) \cdot [\delta(q_\beta + 2n_\beta + 1) + \delta(q_\gamma + 2n_\gamma + 1)] \right] \tag{6.13}$$

Here, $\alpha \neq \beta \neq \gamma$ enumerate directions in reciprocal space, n_α, n_β, n_γ are integer and $p(\mathbf{q})$ is a function accounting for a nonzero width of diffuse rods that can attain ~1/6–1/8 r.l.u. (FWHM)—a simple Lorentzian has been used in the modeling.

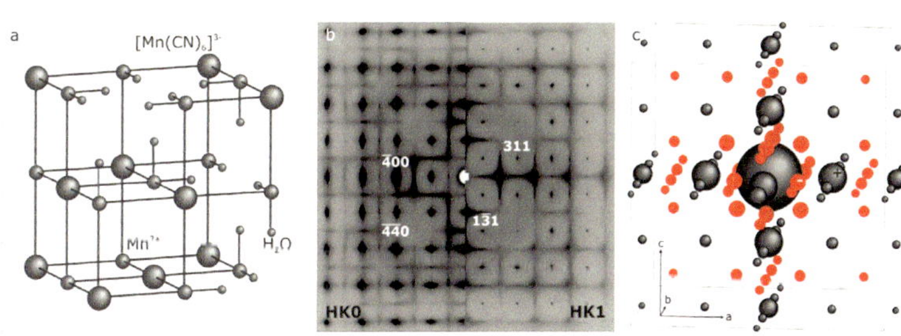

Figure 6.9 (a) Schematically represented fragment of Mn–PB structure; (b) experimental reciprocal space cuts for Mn–PB in HK0 and HK1 planes; (c) fragment of the 3D self-correlation function of Mn–PB; the sphere radius is proportional to the function value in the corresponding node, the sign is associated with the color pattern as denoted in figure.

The real space pattern corresponding to Eq. (6.12) is an autocorrelation function of the deviation from the average structure. A similar function has been recently termed 3D difference pair-distribution function or PDF, 3D Δ-PDF [60]. This quantity is readily available from 3D diffuse data and can be expressed via correlation parameters introduced by Cowley and Warren [61,62] and also be used to evaluate some energetic characteristics of disordered alloys [63]. Despite the fact that this simple description does not take into account many details of the diffuse pattern, it has the advantage of an analytical form whose Fourier transformation can be easily found. The final Δ-PDF function unifying three systems of diffuse rods is written as follows:

$$G(\mathbf{r}) = A' \cdot P(\mathbf{r}) \cdot \sum_{\alpha,\beta,\gamma} \left[\sum_{n_{\alpha,\beta,\gamma}=-\infty}^{+\infty} \delta(r_\alpha) \cdot \left(\delta(r_\beta + n_\beta)\delta(r_\gamma + n_\gamma) - \delta(r_\beta + n_\beta + 1/2)\delta(r_\gamma + n_\gamma + 1/2) \right) \right], \tag{6.14}$$

where A′ is the scaling factor and $P(\mathbf{r})$ is a decay function, which is a Fourier transform of $P(\mathbf{q})$. Figure 6.9c shows the autocorrelation pattern, it is checkerboard-like in [XY0]/[0YZ]/[X0Z] layers and vanishes in octants between them. Such a distribution manifests a tendency toward alternation of $[Mn^{3+}(CN)_6]^{3-}$ ions and $(H_2O)^6$ clusters along the [110] cubic directions and clustering of similar components along the [100] direction, while any correlation is lost along the out-of-plane directions.

It is quite probable that the set of correlations is general for Prussian Blue family, therefore it will directly provide real space description of crystals, which in turn will elucidate the variety of its physicochemical properties in more details – compared with the average structure description.

6.5.4 Powder Wide-Angle IXS

An often encountered major obstacle to obtain the desired complete set of phonon dispersion curves is that conventional approaches require single crystals in order to be able to scan along specific directions in reciprocal space. On the contrary, many materials can only by synthesized as polycrystalline samples, and thus all the directional information is lost. While the formalism described in Section 6.2 is still valid, for the isotropic aggregate the experimental parameter is now $|\mathbf{Q}|$ and not vector $\mathbf{Q}$:

$$S(Q, E, T) = g(Q, E')F(E, E', T). \tag{6.15}$$

where

$$g(Q,E') = \left\langle \left| \sum_d f_d(\mathbf{Q}) e^{-W_d(\mathbf{Q}) + i\mathbf{Q}\cdot\mathbf{r}_d} \left[\mathbf{Q}\cdot\boldsymbol{\sigma}_d^j(\mathbf{q})\right] M_d^{-1/2} \right|^2 \delta(E' - E_{\mathbf{Q},j}) \right\rangle \tag{6.16}$$

$$F(E,E',T) = \frac{1}{E'} \cdot \left[(<n(E,T)>+1)\delta(E-E') + <n(E,T)>\delta(E+E')\right] \tag{6.17}$$

and $\langle\ldots\rangle$ denotes the sampling over both the sphere of radius Q and the phonon modes j.

In the long wavelength limit, the displacements of different atom species become equal in amplitude, collinear, and parallel to the eigenvectors of the corresponding elastic waves, and the scattering is completely defined by the macroscopic elasticity of the crystal. The large Q-region allows the reconstruction of the (generalized) vibrational density-of-states (VDOS) from the inelastic scattering spectra under conditions of (i) correct directional averaging in polycrystalline samples and (ii) an appropriate reciprocal space sampling [64–67].

In order to increase the information content of polycrystalline IXS spectra, the intermediate Q-range must be explored, where the spectral shape varies with momentum transfer due to—though relaxed—selection rules. Therefore, additional constraints (beyond the VDOS and averaged longitudinal sound velocity) are provided for the lattice dynamics. Such an approach was successfully applied to beryllium and textured graphite, and it was shown that a combination of force field model calculations and IXS can provide full lattice dynamics, including phonon dispersion curves, from powder measurements [23]. However, the use of force fields is rather problematic, as these have a very limited transferability. This limitation can be overcome by the combination of parameter-free quantum mechanical calculations and IXS as this approach is applicable to all materials. A benchmark study was devoted to the lattice dynamics of stishovite, a high-pressure polymorph of SiO_2 [24].

The IXS data on the polycrystalline sample were recorded with 3.0 meV energy resolution spanning a momentum transfer range from 3.9 to 55 nm^{-1}. In this context, the simultaneous collection of nine IXS spectra at different momentum transfers is of particular advantage as only the magnitude of the momentum transfer is of relevance. Lattice dynamics calculations were performed using the CASTEP code [68] employing density functional perturbation theory for phonons. Directional averaging was performed over a 20 × 10 polar mesh (1/16 of sphere) for momentum transfers inside the Brillouin zone centered around (0 0 0); the mesh size was progressively increased with Q.

Representative experimental data are shown in Fig. 6.10, where they are compared to the predicted values obtained from the DFT calculations. An outstanding agreement between the two data sets is achieved if a scaling factor 1.05 is applied to the calculated phonon energies. The necessity of applying a scaling factor was expected, as DFT-GGA calculations generally give too low phonon frequencies. The remarkable agreement over a wide range of momentum transfer indicates the validity of the model for *any* crystallographic direction, as orientational averaging naturally probes the low-symmetry directions as well. Relative intensities of phonons with the same momentum transfer Q are naturally embedded in the experimental spectrum thus providing the instrument to check the consistency of the phonon eigenvectors within the model. The phonon dispersion for high symmetry directions thus appears as a by-product.

As a decisive test of the proposed approach the polycrystalline results were confronted with data obtained from a single crystal in the "conventional fashion." Figure 6.11 displays the obtained phonon dispersions along the three main high-symmetry directions. Excellent agreement between the experimental and calculated dispersion is restored after the theoretical phonon energies by the same factor of 1.05, which was established from the experiment on the polycrystal.

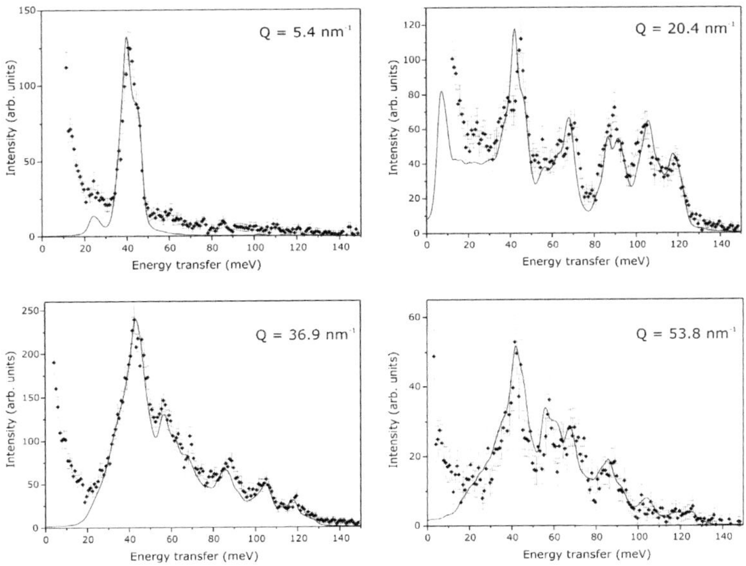

Figure 6.10 Representative polycrystalline IXS spectra of stishovite recorded at: $Q = 5.4\ \text{nm}^{-1}$, $20.4\ \text{nm}^{-1}$, $36.9\ \text{nm}^{-1}$, and $53.8\ \text{nm}^{-1}$ (circles with error bars) compared to ab initio calculations (solid lines). A scale factor of 1.05 has been applied to the energies obtained by DFT–GGA calculations. Figure taken from Ref [24].

Once the (rescaled) lattice dynamics model is validated, it provides immediately the energy distribution function $g(E)$, or VDOS, which in turn gives important insight into the physical properties of materials, since it allows the derivation of numerous thermodynamic properties. Further, the complete elasticity tensor can be derived either from the initial slope of the (rescaled) acoustic phonon dispersions along the principal crystallographic directions or from the DFT-GGA calculation by the stress-strain method [69]. In the present case, an improved agreement with BLS [70–72] results was obtained after rescaling the sound velocity, density, and cell shape by factors derived from comparison of calculations to experimental data. In particular, the integral properties such as the bulk modulus and Debye velocity are in very good agreement, thus confirming the predictive power of DFT-GGA stress-strain calculations. The derived anisotropy of longitudinal waves in stishovite is equal to 1.33(1) reproducing BLS results within 3 percent.

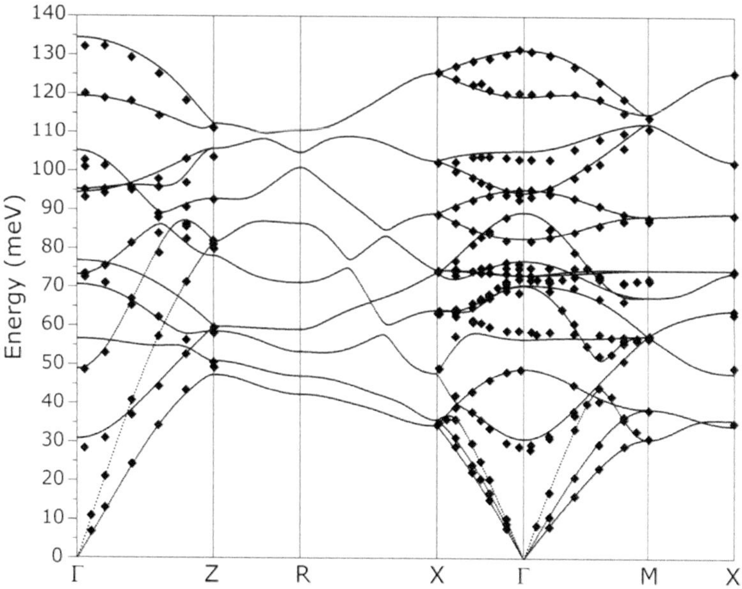

Figure 6.11 Dispersion relations for stishovite along high symmetry directions. Single crystal IXS data (points) are comparedwith ab initio calculations (lines). A scaling factor of 1.05 is applied to the calculated energy scale. Figure taken from Ref. [24].

The very fact of getting the elastic anisotropy from a polycrystalline data makes the employed approach attractive, for example, in Earth science. When the anisotropy of individual phases is known, the macroscopic anisotropy observed in many parts of the Earth can be related to the preferable orientation pattern.

6.5.5 Conclusions and Outlook

The selected examples above have illustrated recent trends in the study of lattice dynamics by X-rays, and have more specifically shown how data collection times might be reduced and which caveats are related to this. In fact, a full determination of the vibrational properties would necessitate the extraction of both phonon eigenfrequencies and eigenvectors for arbitrary points by appropriate sampling of the 4D ($\mathbf{Q}$,E) space. The dimensionality of the problem can be reduced in different ways: (i) integration over Q—reduction to 1D energy distribution; (ii) integration (weighted) over solid angle keeping |**Q**| variable—reduction to 2D momentum-energy map; and (iii) integration over energy transfer—reduction to 3D momentum map.

In the first case, the vibrational density of states—true or generalized—can be retrieved from powder IXS data taken at large momentum transfers within the incoherent approximation. In order to increase the information content of polycrystalline IXS spectra, the intermediate Q-range needs to be explored, thus going to the second case. Here, the spectral shape varies with Q and strong constraints are provided for lattice dynamics. The benchmark case of stishovite showed that a combination of parameter-free quantum mechanical calculations and IXS can be used to recover the full description of lattice dynamics from polycrystalline data. The third case corresponds to 3D mapping of TDS. Typical data collection times in this case can range from a few minutes to hours (instead of several days or weeks in the case of inelastic X-ray or neutron scattering); moreover, using a much simpler experimental setup. While the direct extraction of lattice dynamics date is possible only in very specific cases, it provides important discriminating feedback to quantum mechanical calculations—particularly efficient if coupled to poly- and single-crystal IXS.

It can be expected that TDS studies, combined with IXS studies and/or theoretical modeling, will offer new unique research opportunities in fields ranging from Earth and Planetary Science over solid-state physics and chemistry to nanotechnological applications. To cite only a few major expected breakthroughs:

- Determination of elastic properties of elements and minerals in conditions of high pressure and high temperature using diamond anvil cell and laser heating techniques.
- Study of the coupling of electronic and/or magnetic fluctuations with phonons in strongly correlated electron systems.
- Vibrational properties of nanomodulated and low-dimensional systems.
- Time-resolved (ms $\rightarrow$ s) TDS experiments in the study of phase transformations, that is, the microscopic decomposition kinetics in ionic solids.
- Determination of the shape of the Fermi surface in metallic systems by a directional tracing of Kohn anomalies.

References

1. Lonsdale K (1948) *Crystals and X-Rays*, G. Bell, London.
2. Wooster WA (1962) *Diffuse X-Ray Reflections from Crystals*, Clarendon, Oxford.
3. Xu R, Chiang C (2005) Determination of phonon dispersion relations by X-ray thermal diffuse scattering, *Z Kristallogr*, **220**, 1009–1016.
4. Holt M, Sutton M, Zschack P, Hong H, Chiang T-C (2007) Dynamic fluctuations and static speckle in critical X-ray scattering from $SrTiO_3$, *Phys Rev Lett*, **98**, 065501.
5. Holt M, Zschack P, Hong H, Chou MY, Chiang T-C (2001) X-ray studies of phonon softening in $TiSe_2$, *Phys Rev Lett*, **86**, 3799–3802.
6. Wong J, Wall M, Schwartz AJ, et al. (2004) Imaging phonons in a fcc Pu-Ga alloy by thermal diffuse X-ray scattering, *Appl Phys Lett*, **84**, 3747–3749.
7. Moss SC (1969) Imaging the Fermi surface through diffuse scattering from concentrated disordered alloys, *Phys Rev Lett*, **22**, 1108–1111.
8. Reichert H, Klein O, Bugaev V, et al. (2003) Absence of $2k_F$ splitting in the diffuse scattering from Cu_3Au at the (001) surface, *Phys Rev Lett*, **90**, 185504.
9. Ruocco G, Sette F (1999) The high-frequency dynamics of liquid water, *J Phys: Condens Matter*, **11**, R259–R293.
10. Burkel E (2000) Phonon spectroscopy by inelastic X-ray scattering, *Rep Prog Phys*, **63**, 171–232.
11. Ruocco G, Sette F (2001) High-frequency vibrational dynamics in glasses, *J Phys: Condens Matter*, **13**, 9141–9164.
12. Scopigno T, Ruocco G, Sette F (2005) Microscopic dynamics in liquid metals: the experimental point of view, *Rev Mod Phys*, **77**, 881–933.
13. Fiquet G, Badro J, Guyot F, et al. (2004) Application of inelastic X-ray scattering to measurements of acoustic wave velocities in geophysical materials at very high pressure, *Phys Earth Planet In*, **143–144**, 5–18.
14. Giordano VM, Krisch M, Monaco G (2009) Phonon spectroscopy at high pressure by inelastic X-ray scattering, *J Synchr Radiation*, **16**, 707–713.
15. Murphy BM, Müller M, Stettner J, et al. (2008) Investigating surface dynamics with inelastic X-ray scattering, *J Phys: Cond Matt*, **20**, 224001.

16. Wehinger B, Krisch M, Reichert H (2011) High-frequency dynamics in the near-surface region studied by inelastic X-ray scattering: the case of indium, *New J Phys*, **13**, 023021.
17. Krisch M, Sette F (2007) Inelastic X-ray scattering from phonons, in *Light Scattering in Solids IX, Novel Materials and Techniques* (Cardona M, Merlin R, ed), Springer Verlag Berlin Heidelberg, pp. 313–370.
18. Bosak A, Fischer I, Krisch M (2010) Phonon spectroscopy of polycrystalline materials using inelastic X-ray scattering, in *Thermodynamics Properties of Solids. Experiment and Theory* (Chaplot M, Choudhoury N, ed), Wiley, Weinheim, pp. 123–158.
19. Grüneis A, Serrano J, Bosak A, et al. (2009) Phonon surface mapping of graphite: disentangling quasi-degenerate phonon dispersions, *Phys Rev B*, **80**, 085423.
20. Bosak A, Hoesch M, Krisch M, et al. (2009) 3D imaging of the Fermi surface by thermal diffuse scattering, *Phys Rev Lett*, **103**, 076403.
21. Hoesch M, Bosak A, Chernyshov D, Berger H, Krisch M (2009) Giant Kohn anomaly and the phase transition in charge density wave $ZrTe_3$, *Phys Rev Lett*, **102**, 086402.
22. Chernyshov D, Bosak A (2010) Diffuse scattering and correlated disorder in manganese analogue of Prussian Blue, *Phase Transitions*, **83**, 115–122.
23. Fischer I, Bosak A, Krisch M (2009) Single crystal lattice dynamics derived from polycrystalline inelastic X-ray scattering spectra, *Phys Rev B*, **79**, 134302.
24. Bosak A, Fischer I, Krisch M, et al. (2009) Lattice dynamics of stishovite from powder inelastic X-ray scattering, *Geophys Res Lett*, **36**, L19309.
25. Bosak A, Chernyshov D (2008) On model-free reconstruction of lattice dynamics from thermal diffuse scattering, *Acta Cryst A*, **64**, 598–600.
26. Born M (1942) Theoretical investigations on the relation between crystal dynamics and X-ray scattering, *Rep Prog Phys*, **9**, 294–333.
27. Groenewegen PPM, Huiszoon C (1972) Debye-Waller B values for some NaCl-type structures and interionic interaction, *Acta Cryst A*, **28**, 166–169.
28. Scheringer C (1972) On the interpretation of anisotropic temperature factors, *Acta Cryst A*, **28**, 512–515.
29. Willis BTM, Pryor AW (1975) *Thermal Vibrations in Crystallography*, Cambridge University Press, Cambridge.

30. Verbeni R, D'Astuto M, Krisch M, et al. (2008) Energy calibration of a high-resolution inelastic X-ray scattering spectrometer, *Rev Scient Instr*, **79**, 083902.

31. Verbeni R, Sette F, Krisch M, et al. (1996) X-ray monochromator with 2 × 10^{-8} energy resolution, *J Synchr Radiat*, **3**, 62–64.

32. Lamago D, Hoesch M, Krisch M, et al. (2010) Measurement of strong phonon softening in Cr with and without Fermi-surface nesting by inelastic X-ray scattering, *Phys Rev B*, **82**, 195121.

33. Bosak A, Krisch M, Chernyshov D, et al. (2012) New insights into the lattice dynamics of α-quartz, *Z Kristallogr*, **227**, 84-91

34. Vakhrushev S, Bosak A, Chernyshov D, Korablev-Dyson M, Krisch M, unpublished.

35. Kohn W (1959) Image of the Fermi surface in the vibration spectrum of a metal, *Phys Rev Lett*, **2**, 393–394.

36. Sharp RI (1969) The lattice dynamics of niobium II. Kohn anomalies in niobium, *J Phys C*, **2**, 432–443.

37. Stedman R, Nilsson G (1965) Observations on the Fermi surface of aluminum by neutron spectrometry, *Phys Rev Lett*, **15**, 634–637.

38. Weymouth JW, Stedman R (1970) Fermi surface of aluminum from Kohn anomalies, *Phys Rev B*, **2**, 4743–4751.

39. Iyengar PK, Venkataraman G, Gameel YH, Rao KR (1968) *Proceedings Symposium Neutron Inelastic Scattering*, vol. I, Copenhagen, Vienna IAEA, pp. 195–201.

40. Paskin A, Weiss RJ (1962) Image of the Fermi surface in lead, *Phys Rev Lett*, **9**, 199–200.

41. Costello JM, Weymouth JW (1969) Thermal diffuse X-ray scattering measurements of the Fermi surface of chromium, *Phys Rev*, **184**, 694–701.

42. Yamada Y, Tsang JC, Subba-Rao GV (1975) Imaging of the Fermi surface of 1T-$Ta_{0.9}Hf_{0.1}S_2$ by X-ray diffuse scattering, *Phys Rev Lett*, **34**, 1389–1392.

43. Brönnimann C, Eikenberry EF, Henrich BB, et al. (2006) The PILATUS 1M detector, *J Synchrotron Radiat*, **13**, 120–130.

44. Grüner G (1994) *Density Waves in Solids*, Perseus Publishing, Cambridge, MA.

45. Eaglesham DJ, Steeds JW, Wilson JA (1984) Electron microscope study of superlattices in $ZrTe_3$, *J Phys C*, **17**, L697–L698.

46. Takahashi S, Sambongi T, Okada S (1983) Conduction properties of $ZrTe_3$, *J Phys (Paris), Colloq*, **44**, C3-1733–C3-1736.

47. Felser C, Finckh E, Kleinke H, Rocker F, Tremel W (1998) Electronic properties of $ZrTe_3$, *J Mater Chem*, **8**, 1787–1798.

48. Stöwe K, Wagner F (1998) Crystal structure and calculated electronic band structure of $ZrTe_3$, *J Solid State Chem*, **138**, 160–168.

49. Yokoya T, Kiss T, Chainani A, Shin S, Yamaya K (2005) Role of charge-density-wave fluctuations on the spectral function in a metallic charge-density-wave system, *Phys Rev B*, **71**, 140504.

50. Girault S, Moudden AH, Pouget JP (1989) Critical X-ray scattering at the Peierls transition of the blue bronze, *Phys Rev B*, **39**, 4430–4434.

51. Requardt H, Kalning M, Burandt B, Press W, Currat R (1996) Critical X-ray scattering at the Peierls transition in the quasi-one-dimensional system $(TaSe_4)_2I$, *J Phys Condens Matter*, **8**, 2327–2336.

52. Anonymous (1710) Notitia Coerulei Berolinensis nuper inventi, *Miscellanea Berolinensia ad incrementum scientiarum*, **1**, 377–378.

53. Verdaguer M, Galvez N, Garde R, Desplanches C (2002) Electrons at work in Prussian blue analogues, *Electrochem Soc Interface*, **11**, 28–32.

54. Ohkoshi S-I, Hashimoto K (2002) New magnetic functionalities presented by Prussian Blue analogues, *Electrochem Soc Interface*, **11**, 34–38.

55. Buser HJ, Schwarzenbach D, Petter W, Ludi A (1977) The crystal structure of Prussian Blue: $Fe_4[Fe(CN)_6]_3 \cdot xH_2O$, *Inorg Chem*, **16**, 2704–2710.

56. Franz P, Ambrus C, Hauser A, et al. (2004) Crystalline, mixed-valence manganese analogue of prussian blue: magnetic, spectroscopic, X-ray and neutron diffraction studies, *J Am Chem Soc*, **126**, 16472–16477.

57. Kumar A, Yusuf SM, Keller L (2006) Neutron diffraction study of molecular magnetic compound $Ni_{1.125}Co_{0.375}[Fe(CN)_6] \cdot 6.4H_2O$, *Physica B: Condens Matter*, **358–386**, 444–446.

58. Kumar A, Yusuf SM, Keller L, Yakhmi JV, Srivastava JK, Paulose PL (2007) Variation of structural and magnetic properties with composition in the $(Co_xNi_{1-x})_{1.5}$ $[Fe(CN)_6] \cdot zH_2O$ series, *Phys Rev B*, **75**, 224419.

59. Krivoglaz MA (1992) *Diffraction of X-Rays and Thermal Neutrons in Imperfect Crystals*, Springer, Berlin.

60. Kobas M, Weber T, Steurer W (2005) Structural disorder in the decagonal Al-Co-Ni, I. Patterson analysis of diffuse X-ray scattering data, *Phys Rev B*, **71**, 224205–224220.

61. Cowley JM (1950) X-ray measurement of order in single crystals of Cu_3Au, *J Appl Phys*, **21**, 24–30.
62. Walker CB, Chipman DR (1971) Theory of long-range order in β-CuZn, *Phys Rev B*, **4**, 3104–3106.
63. Mezger M, Reichert H, Ramsteiner IB, et al. (2006) Temperature and concentration dependence of the effective pair interaction parameters in Ni-Pd from high-energy X-ray diffuse scattering, *Phys Rev B*, **73**, 184206–184220.
64. Needham LM, Cutroni M, Dianoux AJ, Rosenberg HM (1993) A study of the vibrational spectrum of amorphous and crystalline SeTe samples by inelastic neutron scattering, *J Phys: Condens Matter*, **5**, 637–646.
65. De Wette FW, Rahman A (1968) Inelastic scattering of neutrons by polycrystals, *Phys Rev*, **176**, 784–790.
66. Squires GL (1978) *Introduction to the Theory of Thermal Neutron Scattering*, Cambridge University Press, Cambridge.
67. Bosak A, Krisch M (2005) Phonon density of states probed by inelastic X-ray scattering, *Phys Rev B*, **72**, 224305.
68. Clark SJ, Segall MD, Pickard CJ, et al. (2005) First principles methods using CASTEP, *Z Kristallogr*, **220**, 567–570.
69. Winkler B, Hytha M, Warren MC, Milman V, Gale JD, Schreuer J (2001) Calculation of the elastic constants of the Al_2SiO_5 polymorphs andalusite, sillimanite and kyanite, *Z Kristallogr*, **216**, 67–70.
70. Weidner DJ, Bass JD, Ringwood AE, Sinclair W (1982) The single-crystal elastic moduli of stishovite, *J Geophys Res*, **87**, 4740–4746.
71. Brazhkin V, McNeil L, Grimsditch M, Bendeliani N, Dyuzheva T, Lityagina L (2005) Elastic constants of stishovite up to its amorphization temperature, *J Phys: Condens Matter*, **17**, 1869–1875.
72. Jiang F, Gwanmesia GD, Dyuzheva TI, Duffy TS (2009) Elasticity of stishovite and acoustic mode softening under high pressure by Brillouin scattering, *Phys Earth Planet In*, **172**, 235–240.

Chapter 7

Magnetic X-Ray Scattering

S. P. Collins
Diamond Light Source Ltd, Diamond House, Harwell Science & Innovation Campus, Didcot, Oxfordshire, OX11 ODE, UK
steve.collins@diamond.ac.uk

In this chapter, we outline key aspects of magnetic X-ray diffraction. Common uses and limitations of the technique are described. As an example, we consider nonresonant magnetic scattering from a relatively simple antiferromagnetic structure, and the techniques for calculating and measuring its angle dependence. Finally, we discuss some studies where resonant magnetic X-ray scattering plays a unique role, and consider future trends and developments.

7.1 Introduction

X-rays have, over the past century, played a dominant role in studies pertaining to the positions of atoms and the distribution of electronic charge in crystalline materials. Studies of magnetic properties with the same atomic resolution have fallen predominantly within the realm of neutron scattering. This is for good reason: neutrons carry magnetic moments that interact with the local atomic magnetic fields

X-Ray Diffraction: Modern Experimental Techniques
Edited by Oliver H. Seeck and Bridget M. Murphy

ISBN 978-981-4303-59-0 (Hardcover), 978-981-4303-60-6 (eBook)
www.panstanford.com

Magnetic scattering, even with little or no resonance enhancement, is straightforward if the sample is in other ways well optimized. The fact that larger magnetic moments help considerably is of no surprise: doubling the moment leads to a quadrupling of the diffraction intensity. However, the sample quality and sharpness of the peaks is even more important, due to the very high resolution of most synchrotron beamlines. In fact, there are very few examples where the peaks are instrumentally narrow, and therefore the peak intensity scales with the inverse of the volume of reciprocal space occupied by the magnetic reflection.

Halving the width in all directions in reciprocal space will give an intensity increase of a factor of eight—a greater benefit than doubling the moment. The importance of crystal quality in the few tens of microns where the beam will penetrate cannot be overemphasized. Treating the sample by mechanical polishing and/or chemical etching can make a huge difference, although not always in the direction expected!

Putting numbers to these properties is very difficult as the smallest measurable moment will depend, for example, on the strength of the resonance and sample quality, etc. Light materials with a high fraction of magnetic ions, high crystal symmetry and a small unit cell are good. The presence of ions with absorption edges just below the photon beam energy is bad. In the end, one just has to try an experiment.

In the next few sections, we consider an example measurement, on $FeBO_3$. We outline some of the relevant information already available on the material, describe the experimental equipment and measurements, and compare the measured signals with calculations based on the nonresonant magnetic scattering cross-section and documented magnetic structure.

7.5 $FeBO_3$: Introduction

This material is one of a class that is of significant current interest. It is a rhombohedral system with iron ions polarized perpendicular to the *c*-axis, almost oppositely from one layer to the next. The moments, however, are slightly canted, giving rise to a weak ferromagnetic moment in the a–b plane. The direction of the ferromagnetic moment is thought to be determined by the where T_N is the ordering temperature, β is the critical exponent for magnetization with zero external field and we assume that the intensity scales with the square of the magnetic moment. A least squares fit of Eq. (7.17) to the experimental data gives β = 0.305(2)—slightly different from the value of 0.345 for the 3d X–Y model, which one might expect to describe the current system as the moments are free to rotate in the a - b plane. The slight discrepancy might be explained by the fact that an applied filed is thought [34] to strengthen the antiferromagnetic coupling close to the phase transition.

One might ask, what has been learnt about $FeBO_3$ by carrying out these measurements? It is a useful illustration of many important principles, and it has given strong support to the magnetic structure determined from neutron and Mossbauer techniques. More importantly, it forms the basis for future work on a highly topical subject: the Dzyaloshinsky–Moriya exchange interaction that is responsible for weak ferromagnetism in $FeBO_3$. If the sign of the Dzyaloshinsky–Moriya interaction is reversed, then the patterns of magnetic moments on the odd and even layers (Fig. 7.2) are reversed. This would change the sign of the magnetic scattering amplitude but not its intensity, hence the present measurements are not sensitive to such a change. However, interference with another scattering process might reveal the amplitude of the magnetic scattering and the elusive sign of the Dzyaloshinsky–Moriya interaction [41]. This is a subject of on-going research.

7.9 Magnetic Scattering and Polarization

The polarization of the X-ray beam, that is, the direction of the electric vector of the electromagnetic wave, is of central importance to magnetic scattering. This is because magnetic scattering, and most forms of resonant scattering, have a complex polarization dependence that couples to sample properties, unlike the simple polarization dependence for nonresonant charge scattering (Fig. 7.5) which is independent of the sample. Magnetic scattering facilities typically exploit both linear and circular polarization states.

Most hard X-ray beamlines produce beams that are very highly linearly polarized in the horizontal direction. If the scattering

The presence of various atomic multipoles in a material, and their energies, depend on the symmetry of the atomic sites and local potential [48]. A more interesting situation arises when multipoles order spontaneously to reduce the symmetry of the system. A phase transition is driven by interactions between multipoles of a particular order (rank), which is then said to be the "order parameter" for the phase transition. Phase transitions of this kind are important as they can change the electronic properties of the system [49]. One of the most powerful techniques for observing them is resonant X-ray scattering.

Some aspects of multipole order are strikingly clear even from raw intensity data, such as the dependence on polarization and azimuthal rotation ψ (about the scattering vector). The intensity variation with angle, in the absence of any absorption corrections, etc., can be written as a finite series of sine and cosine curves:

$$I = \sum_{n=0}^{2K} a_n \sin(n\psi) + b_n \cos(n\psi) \tag{7.18}$$

and so, for example, we see that the six-fold pattern in Fig. 7.8 (top) requires $K \geq 3$, which corresponds to $2^3 = 8$ (octapole) order. Moreover, a $K = 3$ resonant scattering tensor is not accessible via pure electric dipole transitions ($K_{\max} = 2$), and therefore it is clear even from the raw data that higher-order multipole transitions are involved.

While the identification of the rank of atomic multipoles is often unambiguous due to the angle dependence, the same cannot be said of the determination of the symmetry of the tensor with respect to spatial and time reversal. These require a much more detailed analysis of the symmetries and the nature of the ordering phase transitions. In particular, the assignment of resonant scattering tensors to magnetoelectric multipoles necessitates an unusually stringent burden of proof [51].

7.11 Future Directions

The past two decades have seen an explosion in the use of synchrotrons for studies of magnetic materials, brought about largely by the advent of "third-generation" light sources. Combined with advances in optics, the quality of X-ray beams at modern synchrotron beamlines is now so good that the main limitation (by far) is the technology for X-ray detectors, where magnetic scattering requires high dynamic range, spatial, and energy resolution.

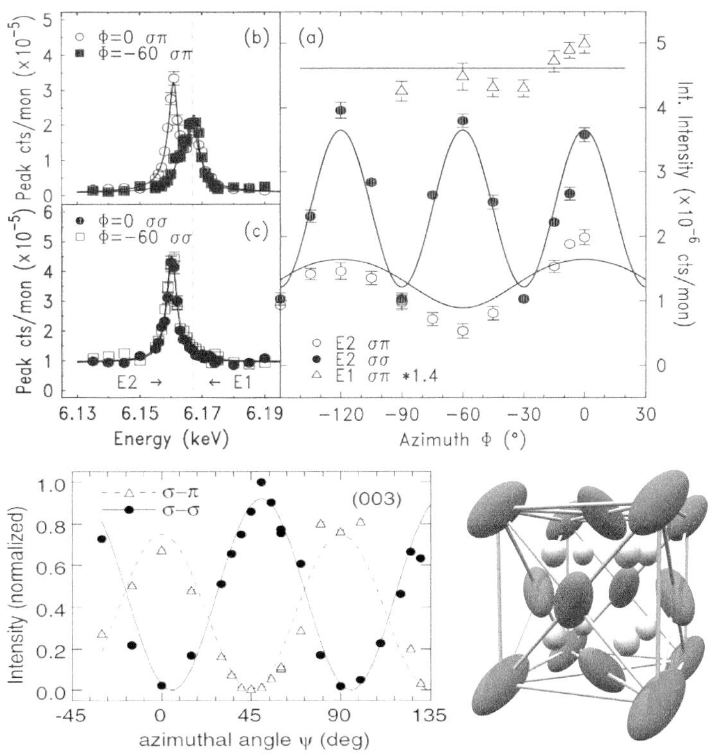

Figure 7.8 Ordering of atomic multipoles. Bottom right: The pattern of electric quadrupoles in NpO_2 revealed by Np M-edge dipole resonant scattering [16], producing a four-fold azimuthal intensity curve (Bottom left). It is believed that the quadrupole order is actually driven by the interaction between magnetic triakontadipoles (K = 5) [50], although these are not evident from the resonant scattering data. In other systems, higher order atomic multipoles can be observed directly via electric quadrupole transitions, leading to characteristic high-frequency angle dependences. Top: the six-fold intensity pattern in $Ce_{0.7}La_{0.3}B_6$ providing direct evidence for octapole ordering [17].

The use of soft X-rays has emerged as a standard tool for studies of magnetism, and features prominently in many of the most active areas of growth. Measurements of magnetic excitations and their dispersion, from minute sample volumes, can be achieved through the use of resonant inelastic X-ray scattering (RIXS), where each new instrument sets new standards for energy resolution. Lensless imaging of magnetic structures by coherent scattering offers the prospect of observing the spatial patterns of natural domains [52] and artificial structures and their evolution with time (see also Chapters 12 and 13). Here, the low photon energies provide not only enhanced magnetic sensitivity but also very long coherence lengths.

Possibly the most exciting prospect for the coming years is to study magnetic properties at ultrashort time scales using new techniques such as pump-probe methods. First results have already been published. A pump-probe experiment has been performed to determine the dynamics of demagnetization of a CoPd film following rapid demagnetization by a laser "pump" pulse. The evolution of Co spin and orbital moments following the pulse is monitored by XMCD using a 100 fs soft X-ray "probe" pulse. The results indicate that the orbital moment responds faster to the demagnetization than the spin moment [53].

Even better time resolution can be achieved by exploiting the next ("Fourth") generation of accelerator-based X-ray sources—free-electron laser (Fig. 7.9). With a highly intense and coherent soft X-ray pulse of 30fs duration the spatial magnetic structure of a sample can detected via a characteristic speckle pattern [54]. The possibility of studying the dynamics of magnetic structures on femtosecond timescales is a very exciting one, with important implications both for fundamental materials physics and future device application.

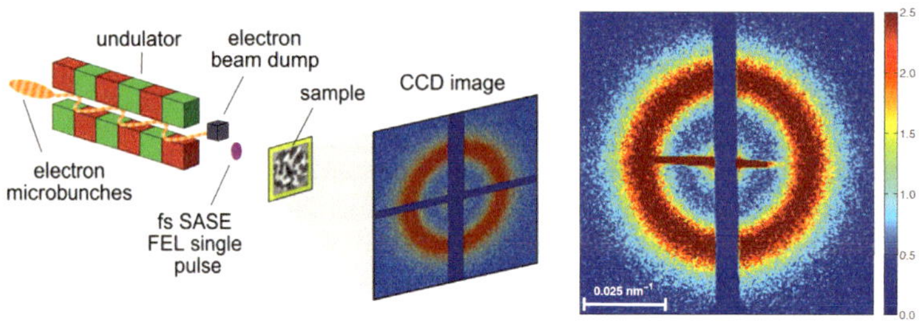

Figure 7.9 The future: Ultrafast soft X-ray diffraction from a Co/Pt multilayer. The diffraction pattern (right) was obtained from a single 30 fs pulse at the FLASH free-electron laser source [54].

Acknowledgment

The author gratefully acknowledges assistance from, and valuable discussions with, G Nisbet, C Vecchini, A Bombardi, and V E Dmitrienko.

References

1. Squires GL (1978) *Introduction to the Theory of Thermal Neutron Scattering*, Cambridge University Press, Cambridge.
2. De Bergevin F, Brunel M (1972) Observation of magnetic superlattice peaks by X-ray diffraction on an antiferromagnetic NiO crystal, *Phys Lett*, **39**A, 141–142.
3. De Bergevin F, Brunel M (1981) Diffraction of X-rays by magnetic materials. I. General formulae and measurements on ferro- and ferrimagnetic compounds, *Acta Cryst*, A**37**, 314–324.
4. Lovesey SW, Collins SP (1996) *X-Ray Scattering and Absorption by Magnetic Materials*, Oxford, Clarendon.
5. Gibbs D, Moncton DE, D'Amico KL, Bohr J, Grier BH (1985) Magnetic X-ray scattering studies of holmium using synchrotron radiation, *Phys Rev Lett*, **55**, 234–237.
6. Inami T, Honma T, Tsutsui S, et al. (2007) Resonant X-ray magnetic diffraction experiments on SmB_2C_2, *J Magn Magn Mater*, **310**, 748–750.
7. Mannix D, Stunault A, Bernhoeft N, et al. (2001) Resonant enhancements at nonmagnetic ions: new possibilities for magnetic X-ray scattering, *Phys Rev Lett*, **86**, 4128–4131.
8. Gibbs D, Harshman DR, Isaacs ED, McWhan DB, Mills D, Vettier C (1988). Polarization and resonance properties of magnetic X-ray scattering in holmium, *Phys Rev Lett*, **10**, 1241–1244.
9. Langridge S, Lander GH, Bernhoeft N, et al. (1997) Separation of the spin and orbital moments in antiferromagnetic UAs, *Phys Rev B*, **55**, 6392–6398.
10. Collins SP, Laundy D, Guo GY (1993) Spin and orbital magnetic X-ray diffraction in $HoFe_2$, *J Phys: Condens Mat*, **5**, L637–L642.
11. Chen CT, Idzerda YU, Lin H-J, et al. (1995) Experimental confirmation of the X-ray magnetic circular dichroism sum rules for iron and cobalt, *Phys Rev Lett*, **75**, 152–155.

12. Langridge S, Stirling G, Lander GH, et al. (1994) X-ray study of the critical magnetic scattering in NpAs, *Europhys Lett*, **25**, 137–142.

13. Bohr J, Gibbs D, Axe JD, et al. (1989) Diffraction studies of rare earth metals and superlattices, *Physica B*, **159**, 93–105.

14. Pascut GL, Coldea R, Radaelli P, et al. (2011) Direct observation of charge order in triangular metallic $AgNiO_2$ by single-crystal resonant X-ray scattering, *Phys Rev Lett*, **106**, 157206.

15. Nakamura K, Arima T, Nakazawa A, Wakabayashi Y, Murakami Y (1999) Polarization-dependent resonant X-ray diffraction in charge- and orbital-ordering phase of $Nd_{1/2}Sr_{1/2}MnO_3$, *Phys Rev B*, **60**, 2425–2428.

16. Caciuffo R, Paixão JA, Detlefs C, et al. (2003) Multipolar ordering in NpO_2 below 25 K, *J Phys: Condens Matter*, **15**, S2287–S2296.

17. Mannix D, Tanaka Y, Carbone D, Bernhoeft N, Kunii S (2005) Order parameter segregation in $Ce_{0.7}La_{0.3}B_6$: 4f octopole and 5d dipole magnetic order, *Phys Rev Lett*, **95**, 117206.

18. Spaldin NA, Fiebig M, Mostovoy M (2008) The toroidal moment in condensed-matter physics and its relation to the magnetoelectric effect, *J Phys: Condens Matter*, **20**, 434203.

19. Staub U, Garcia-Fernandez M, Mulders AM, Bodenthin Y, Martinez-Lope MJ, Alonso JA (2007) Soft X-ray resonant magnetic powder diffraction on $PrNiO_3$, *J Phys: Condens Matter*, **19**, 092201.

20. Braicovich L, Ament LJP, Bisogni V, et al. (2009) Dispersion of magnetic excitations in the cuprate La_2CuO_4 and $CaCuO_2$ compounds measured using resonant X-ray scattering, *Phys Rev Lett*, **102**, 167401.

21. Wilkins SB, Forrest TR, Beale TAW, et al. (2009) Nature of the magnetic order and origin of induced ferroelectricity in $TbMnO_3$, *Phys Rev Lett*, **103**, 207602.

22. Beutier G, Marty A, Livet F, et al. (2009) Magnetic memory in discrete media observed by coherent soft X-ray resonant scattering, *New J Phys*, **11**, 113026.

23. Kim BJ, Ohsumi H, Komesu T, et al. (2009) Phase-sensitive observation of a spin-orbital mott state in Sr_2IrO_4, *Science*, **323**, 1329.

24. Blume M (1985) Magnetic scattering of X rays, *J Appl Phys*, **57**, 3615–3618.

25. Lovesey SW (1993) Photon scattering by magnetic solids, *Rep Prog Phys*, **56**, 257–326.

26. Paolasini L, Detlefs C, Mazzoli C, et al. (2007) ID20: a beamline for magnetic and resonant X-ray scattering investigations under extreme conditions, *J Synchrotron Rad*, **14**, 301–312.

27. Wilkins SB, Hatton PD, Roper MD, Prabhakaran D, Boothroyd AT (2003) Soft X-ray resonant magnetic diffraction, *Phys Rev Lett*, **90**, 187201.

28. McWhan DB, Vettier C, Isaacs ED, et al. (1990) Magnetic X-ray-scattering study of uranium arsenide, *Phys Rev B*, **42**, 6007–6017.

29. Bombardi A, Rodriguez-Carvajal J, Di Matteo S, et al. (2004) Direct determination of the magnetic ground state in the square lattice S = 1/2 antiferromagnet Li_2VOSiO_4, *Phys Rev Lett*, **93**, 027202.

30. Hill JP, McMorrow DF (1996) X-ray resonant exchange scattering: polarization dependence and correlation functions, *Acta Cryst A*, **52**, 236–244.

31. Collins SP, Bombardi A (2010) Resonant X-ray scattering and absorption, in *Magnetism and Synchrotron Radiation: New Trends* (Beaurepaire E, Bulou H, Scheurer F, Kappler JP, ed), Springer, New York, Heidelberg, p. 223.

32. Als-Nielsen J, McMorrow D (2001) *Elements of Modern X-ray Physics*, Wiley, Chichester.

33. Squires GL (1978) *Introduction to the Theory of Thermal Neutron Scattering*, Cambridge University Press, Cambridge.

34. Yakimov SS, Ozhogin VI, Gamlitskii VY, Cherepanov VM, Pudkov SD (1972) Magnetic field induced antiferromagnetism in FeBO, *Phys Lett*, **39**A, 421–423.

35. Collins SP, Bombardi A, Marshall AR, et al. (2009) Diamond beamline I16 (Materials & magnetism), *AIP Conf Proc*, **1234**, 303–306.

36. Brown SD, Bouchenoire L, Bowyer D, et al. (2001) The XMaS beamline at ESRF: instrumental developments and high-resolution diffraction studies, *J Synchrotron Rad*, **8**, 1172–1181.

37. Freeland JW, Lang JJ, Srajer G, Winarski R, Shu D, Mills DM (2002) A unique polarized X-ray facility at the advanced photon source, *Rev Sci Instrum*, **73**, 1408 1450.

38. Busing WR, Levy HA (1967) Angle calculations for 3- and 4- circle X-ray and neutron diffractometers, *Acta Cryst*, **22**, 457–464.

39. Nisbet G (2011) *Simulations of Multiple Scattering Peaks in I16*, Diamond Annual Review 2010/11.

40. Collins MF (1989) *Magnetic Critical Scattering*, Oxford University Press, New York.

41. Dmitrienko VE, Ovchinnikova EN, Kokubun J, Ishida K (2010) Dzyaloshinskii-Moriya interaction: how to measure its sign in weak ferromagnets? *JETP Lett*, **6**, 383–387.

42. Lang JC, Lee DR, Haskel D, Srajer G (2004) Imaging spiral domains in Ho metal using circularly polarized Bragg diffraction, *J Appl Phys*, **95**, 6537–6539.

43. Thole BT, Carra P, Sette F, van der Laan G (1992) X-ray circular dichroism as a probe of orbital magnetization, *Phys Rev Lett*, **68**, 1943–1946.

44. Carra P, Thole BT, Altarelli M, Wang X (1993) X-ray circular dichroism and local magnetic fields, *Phys Rev Lett*, **70**, 694–697.

45. Lovesey SW, Balcar E, Knight KS, Fernández Rodrígueze J (2005) Electronic properties of crystalline materials observed in X-ray diffraction, *Phys Rep*, **411**, 233–289.

46. Brouder C, Juhin A, Bordage A, Arrio M-A (2008) Site symmetry and crystal symmetry: a spherical tensor analysis, *J Phys: Condens Matter*, **20**, 455205.

47. Blume M (1994) Magnetic effects in anomalous dispersion, in *Resonant Anomalous X-Ray Scattering: Theory and Applications* (Materlik G, Sparks CS, Fischer K, ed), North-Holland, Amsterdam, p. 495-512.

48. Templeton DH, Templeton LK (1986) X-ray birefringence and forbidden reflections in sodium bromate, *Acta Cryst A,* **42**, 478–481.

49. Kuramoto Y, Kusunose H, Kiss A (2009) Multipole orders and fluctuations in strongly correlated electron systems, *J Phys Soc Jpn*, **78**, 072001.

50. Santini P, Carretta S, Magnani N, Amoretti G, Caciuffo R (2006) Hidden order and low-energy excitations in NpO_2, *Phys Rev Lett*, **97**, 207203.

51. Fernández-Rodríguez J, Scagnoli V, Mazzoli C, et al. (2010) Experimental evidence of anapolar moments in the antiferromagnetic insulating phase of V_2O_3 obtained from X-ray resonant Bragg diffraction, *Phys Rev B*, **81**, 085107.

52. Konings S, Schußler-Langeheine C, Ott H, et al. (2011) Magnetic domain fluctuations in an antiferromagnetic film observed with coherent resonant soft X-ray scattering, *Phys Rev Lett*, **106**, 077402.

53. Boeglin C, Beaurepaire E, Halté V, et al. (2010) Distinguishing the ultrafast dynamics of spin and orbital moments in solids, *Nature*, **465**, 458–462.

54. Gutt C, Streit-Nierobisch S, Stadler LM, et al. (2010) Single-pulse resonant magnetic scattering using a soft X-ray free-electron laser, *Phys Rev B*, **81**, 100401.

Chapter 8

Nuclear Resonant Scattering of Synchrotron Radiation: Applications in Magnetism

Ralf Röhlsberger
Deutsches Elektronen Synchrotron (DESY), Notkestraße 85, 22607 Hamburg, Germany
ralf.roehlsberger@desy.de

8.1 Introduction

Nuclear resonant scattering of synchrotron radiation has become an established field of condensed-matter research over the past 25 years. Synchrotron radiation with its outstanding brilliance, transverse coherence, and polarization has opened the field for many unique studies, for fundamental research in the field of light–matter interaction as well as for materials science. This applies, in particular, to the electronic and magnetic structure of very small sample volumes such as micro- and nanostructures and samples under extreme conditions of pressure and temperature. This chapter is devoted to the application of the technique to nanomagnetic systems such as thin films and multilayers. After a basic introduction

X-Ray Diffraction: Modern Experimental Techniques
Edited by Oliver H. Seeck and Bridget M. Murphy

ISBN 978-981-4303-59-0 (Hardcover), 978-981-4303-60-6 (eBook)
www.panstanford.com

into the method, a number of examples are presented to illustrate how magnetic spin structures within such layer systems can be revealed.

Atomic nuclei can be very sensitive probes of condensed matter properties. A number of spectroscopic techniques have been developed to probe condensed matter via nuclear hyperfine interactions, constituting the field of *nuclear condensed matter physics* [1,2]. A very prominent method in this field is the Mössbauer effect that probes internal magnetic and electric fields in the sample by analyzing the energetic hyperfine splitting of the nuclear levels via absorption of γ-radiation. Traditionally, this technique is performed in the energy domain where the radiation from a monochromatic radioactive source is Doppler tuned to measure the absorption spectrum around the nuclear resonance with nano-eV resolution. The hyperfine interaction of the nuclei in the sample typically lifts the degeneracy of the nuclear levels that manifests as a splitting and/or broadening of the absorption line(s).

Synchrotron radiation effectively allows one to perform the Fourier transform of Mössbauer spectroscopy from the energy domain into the time domain: The radiation is energetically broad (in contrast to the monoenergetic γ-radiation emitted by a radioactive source), and it comes in pulses with a duration of 50–100 ps (in contrast to the continuously emitting radioactive source). Thus, these pulses excite all hyperfine-split resonances in the sample at the same time. In the subsequent coherent decay, the interference of the emitted waves leads to a temporal beat pattern (as illustrated in in Fig. 8.1), very similar to the temporal acoustic beats of slightly detuned tuning forks. From such a beat pattern, the hyperfine interaction parameters of the nuclei in the sample can be deduced.

The brilliance of synchrotron radiation sources exceeds that of radioactive sources by several orders of magnitude. This was realized early by Ruby [3] and the first observation of the time-based analogue of Mössbauer spectroscopy as known as nuclear resonant scattering of synchrotron radiation was reported by Gerdau et al. [4] in Bragg scattering geometry and later by Hastings et al. [5] in forward scattering geometry. Since then, the method rapidly developed and is now available at many third-generation synchrotron sources across the world, covering a broad range in all fields of the natural sciences [6,2].

8.2 Basic Principles of Nuclear Resonant Scattering

Due to the narrow nuclear resonance width in the range of neV–μeV, the scattering process takes place on time scales ranging from microseconds to nanoseconds. This allows for a discrimination of the resonantly scattered radiation from the instantaneous nonresonant charge scattering that proceeds on time scales below 10^{-15} s. A synopsis of the method of time-resolved coherent nuclear resonant scattering is given in Fig. 8.1.

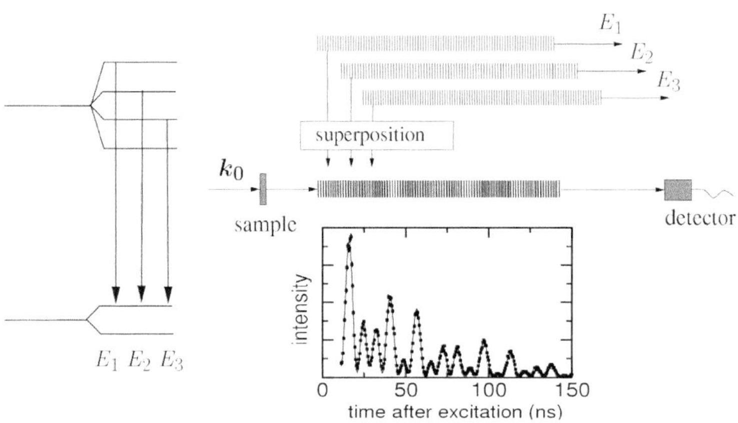

Figure 8.1 Principle of coherent nuclear resonant scattering in forward direction. Transitions between nuclear levels are excited simultaneously by synchrotron radiation pulses at $t = 0$. In this example, three selected waves with slightly different frequencies are emitted from the decaying nuclei in the sample. Their superposition leads to quantum beats in the temporal evolution of the decay. This is illustrated here by overlaying three wave trains with slightly different spatial frequencies, leading to a Moiré pattern. The intensity modulation of this pattern corresponds to the temporal quantum beats that are detected with a time-resolving detector like an avalanche photodiode or (APD).

The coherence and the collective nature of the scattering process lead to interesting effects in the temporal evolution of the scattering process:

(a) *Speed-up (acceleration)* of the decay compared to that of an isolated nucleus, resulting from superradiant emission in the coherent decay channels and

(b) *Quantum beats* resulting from the coherent superposition of waves emitted from nuclei at different frequencies.

The technique of time-resolved nuclear resonant scattering can be compared with diffraction from a set of slits, as illustrated in Fig. 8.2. In both cases, an array of objects (slits/resonances) is illuminated by a (spatially/energetically) broad beam of radiation.

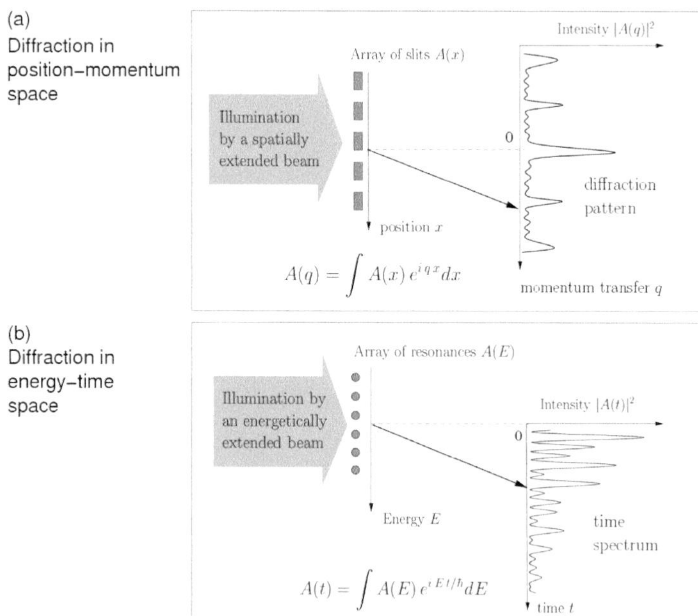

Figure 8.2 Correspondence between diffraction from an array of slits and temporal beats resulting from simultaneous excitation of an array of resonances. In both cases, these arrays are illuminated by a spatially viz. energetically extended (and coherent) incident beam. The diffraction pattern results from a Fourier transformation of the spatial/energetic arrangement of diffracting objects. The symmetry between both techniques is broken on the time scale, because causality has to be observed.

While conventional X-ray diffraction works in position–momentum space, time-resolved nuclear resonant scattering works in energy–time space. Recording a diffraction pattern in reciprocal space allows for obtaining structural information about the diffracting object. Likewise, recording a temporal quantum-beat pattern allows to determine the relative energetic positions and the shape of the resonance lines. In both methods, the field amplitudes in the complementary representations are related by Fourier transformation. The only difference is that the transformation to the time scale is restricted by causality. Besides that, both methods can be united into one space-time formulation of X-ray diffraction, leading to the quantum theory of crystal optics [7,8].

As a consequence, the spatial arrangement of resonant scatterers will not only determine the distribution of the scattered radiation in reciprocal space but also influence the temporal evolution of the decay. This fact finds its expression in the speed-up of the radiative decay of an ensemble of nuclei that are radiatively coupled through the standing wave that builds up in a Bragg reflection, for example. Likewise, the energetic arrangement of resonant scatterers may influence the distribution of radiation in reciprocal space. This manifests itself in the appearance of pure nuclear Bragg reflections from materials with a magnetic superstructure, for example. This discussion illustrates that the concept of X-ray diffraction naturally extends into the four-dimensional space–time continuum.

The transmission of X-rays through a slab of material that contains resonant nuclei can be described in a straightforward fashion via the energy-dependent index of refraction $n(\omega)$:

$$A(\mathrm{z}, \omega) = \exp[in(\omega)k_0 z]A_0,$$

which gives the amplitude of the radiation field in depth z of the sample. A_0 is the amplitude of the incident radiation field. The index of refraction n is related to the forward scattering length M via

$$n(\omega) = 1 + \frac{2\pi}{k_0^2}\sum_k \rho_k M_k(\omega)$$

with ρ_k being the number density of the kth atomic species in the sample. These equations can be generalized to a multitude of scattering problems, ranging from simple forward scattering to anisotropic optics, thin-film reflection and Bragg diffraction from crystals and gratings [9]. The number N of open scattering channels determines the dimension of M, that is, M is a scalar quantity for isotropic forward scattering, a 2 × 2 matrix for anisotropic (polarization-dependent) forward scattering, a 4 × 4 matrix for anisotropic (2-beam) Bragg diffraction that includes also nuclear resonant scattering in grazing incidence geometry.

Correspondingly, the quantities A_0 and $A(z)$ are vectors with N components that describe the amplitudes of the radiation field in the N scattering channels. This formalism naturally contains all multiple scattering processes within the sample, that is, the resulting radiation field $A(z)$ is a self-consistent solution to the scattering problem within the dynamical theory of X-ray scattering.

The specific interaction of the photons with the atoms in the material is contained in the scattering length M. It is well known that optical properties of a system change drastically if the photon energy approaches an atomic resonance. This is valid for the complete spectral range from the infrared into the hard X-ray regime. The remarkable features of X-ray scattering from inner-shell resonances have been discovered and exploited when high-brilliance synchrotron radiation became available [10,11]. This applies for X-ray scattering from nuclear resonances as well [4,6,12].

Figure 8.3 shows the real and imaginary parts of the forward scattering amplitude of ^{57}Fe over an energy range from 6 to 16 keV. Around 7 keV, one observes the electronic K-edge of the material as it results from excitation of an electron in the K-shell into the continuum. At a photon energy of 14.4 keV, one observes an extremely sharp feature that results from resonant excitation of the ^{57}Fe nucleus. The shape of this resonance becomes apparent only if one blows up the energy scale to the neV regime. The vertical axis is scaled in units of the classical electron radius r_0 which is the scattering length of a single electron. Thus, it appears that near the resonance, the scattering strength of the nucleus corresponds to an atom with $Z \approx 220$. This means that out of the small energy range around the resonance, a very strong scattering signal can arise that exceeds that of electronic resonances by far.

Near the resonance, the scattering amplitude of an atom with a single nuclear resonance can be written as follows:

$$M(\omega) = E(\omega) + N(\omega),$$

where E and N are the electronic and nuclear contributions to the scattering length, respectively:

$$[E(\omega)]_{\mu\nu} = (\boldsymbol{\varepsilon}_\mu \cdot \boldsymbol{\varepsilon}_\nu)[-Zr_0 + f_e' + i f_e''] = (\boldsymbol{\varepsilon}_\mu \cdot \boldsymbol{\varepsilon}_\nu)\left[-Zr_0 + f_e' + i\frac{k_0}{4\pi}\sigma(\omega)\right]$$

$$[N(\omega)]_{\mu\nu} = (\boldsymbol{\varepsilon}_\mu \cdot \boldsymbol{\varepsilon}_\nu)[f_n' + i f_n''] = (\boldsymbol{\varepsilon}_\mu \cdot \boldsymbol{\varepsilon}_\nu)\frac{f_0}{x+i}$$

$$= (\boldsymbol{\varepsilon}_\mu \cdot \boldsymbol{\varepsilon}_\nu)\left[\frac{f_0 x}{x^2+1} + i\frac{f_0}{x^2+1}\right],$$

where Z is the atomic number, r_0 the classical electron radius, and σ is the total absorption cross section. (ε_ν, ε_μ) are the vectors of the polarization basis. Their scalar product that appears in the expressions above indicates that the scattering process conserves the polarization, that is, the 2 × 2 matrices E and N are diagonal. This is in general not the case if the nucleus is subject to an electric or magnetic hyperfine interaction, as we shall see further.

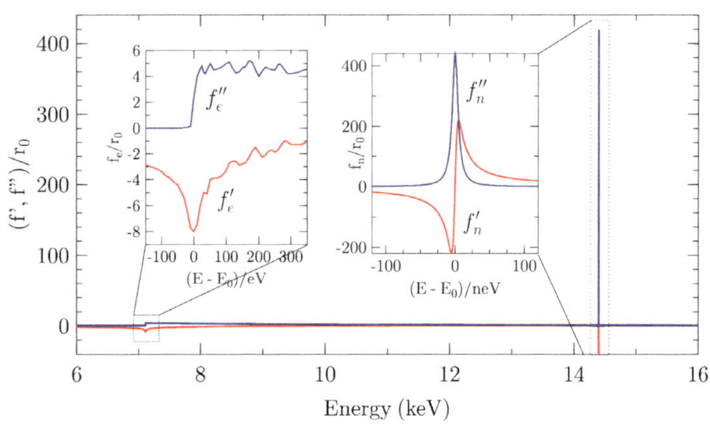

Figure 8.3 Anomalous scattering lengths for ^{57}Fe (given in units of r_0, the classical electron radius), including the K-edge around 7 keV and the nuclear resonance around 14.4 keV. While around the K-edge, the anomalous contributions amount to only a few electrons, around the nuclear resonance they correspond to >200 electrons!

The nuclear scattering length $N(\omega)$ is given here for a single, unsplit resonance line (i.e., no hyperfine interaction). $x = 2(E - E_0)/\Gamma_0$ denotes the deviation of the energy from the exact resonance energy E_0 measured in units of the natural linewidth Γ_0 of the transition. f_0 expresses the "oscillator strength" of the nuclear resonance:

$$f_0 = \frac{f_{LM}}{2k_0} \frac{2I_e + 1}{2I_g + 1} \frac{1}{1+\alpha}$$

where f_{LM} is the Lamb–Mössbauer factor that describes the relative fraction of photons that are elastically scattered (i.e., without recoil). I_e and I_g are the spins of the ground and excited nuclear state, respectively, and α is the coefficient of internal conversion.

For the 14.4 keV transition of ^{57}Fe, we have α = 8.21, I_g = 1/2, I_e = 3/2 and k_0 = 7.3·10^{10} m^{-1}. In α-Fe at room temperature, we have $f_{LM} \approx 0.8$ so that $f_0 \approx 1.2 \cdot 10^{-12}$ m. In the case of magnetic materials, a finite magnetic hyperfine field acts on the resonant nuclei. In ferromagnetic (FM) α-Fe, for example, the magnitude of this field is 33.3 T at room temperature. As a consequence, the degeneracy of nuclear levels with spin $I_N \geq 1/2$ is lifted, that is, these states split into $2I_N \geq 1$ magnetic sublevels between which electromagnetic transitions can occur according to the multipole selection rules. For an electric dipole transition, the nuclear scattering length reads:

$$[N(\omega)]_{\mu\nu} = \frac{3}{16\pi} \left\{ \begin{array}{l} (\boldsymbol{\varepsilon}_\mu \cdot \boldsymbol{\varepsilon}_\nu)[F_{+1} + F_{-1}] \\ -i(\boldsymbol{\varepsilon}_\mu \times \boldsymbol{\varepsilon}_\nu) \cdot \mathbf{m}[F_{+1} - F_{-1}] \\ +(\boldsymbol{\varepsilon}_\mu \cdot \mathbf{m})(\boldsymbol{\varepsilon}_\nu \cdot \mathbf{m})[2F_0 - F_{+1} - F_{-1}] \end{array} \right\}$$

The quantities $F_0 = F_0(\omega)$ and $F_{\pm 1} = F_{\pm 1}(\omega)$ are functions that describe the energy dependence of the scattering length for the transitions with change of magnetic quantum number Δm = 0, ± 1. In case of a magnetic dipole transition (as for the 14.4 keV transition of ^{57}Fe), the role of electric and magnetic fields of the radiation are interchanged. Then, one has to transform the polarization vectors in the expression above via $\varepsilon \rightarrow \varepsilon \times \mathbf{k}_0$.

The scalar and vector products involving the polarization-basis vectors ε_μ and ε_ν as well as the unit vector **m** in the direction of the magnetization indicate a strong polarization dependence of the scattering process. The three terms in the equation above represent different polarization dependencies. The first term is not sensitive to the sample magnetization. Its polarization dependence given by $\varepsilon_\mu \cdot \varepsilon_\nu$ is that of nonresonant charge scattering.

The second term describes circular dichroism because it depends on the difference between the resonant scattering amplitudes F_{+1} and F_{-1}. Since its polarization dependence is $\varepsilon_\mu \times \varepsilon_\nu$, it describes orthogonal scattering, for example, $\sigma \rightarrow \pi$ and $\pi \rightarrow \sigma$. The third term describes linear magnetic dichroism. Its polarization dependence allows for all scattering processes within the given polarization basis. Due to the very high sensitivity of the nuclear scattering length to the magnetization direction relative to the incident wave vector, nuclear resonant scattering constitutes a very sensitive probe of local spin structures in magnetic materials [13,14].

With this formalism, one can now calculate the index of refraction in the vicinity of the nuclear resonance. The index of refraction can then be used in any optical code, for example, Fresnel formulas for single interfaces or multibeam interference equations for calculating nuclear resonant reflectivities from thin films and multilayers [9].

Table 8.1 Parameters of Mössbauer isotopes that have been used in experiments with synchrotron radiation to date (October 2010). α is the natural abundance of the isotope, I_g and I_e are the spins of the ground and excited state, respectively, Γ_0 is the natural linewidth, and τ_0 is the lifetime

Isotope	$E\gamma$ [keV]	α [%]	Γ_0 [neV]	τ_0 [ns]	I_g	I_e	Multipolarity
^{181}Ta	6.23	99.9	0.067	9870	7/2	9/2	E1
^{169}Tm	8.41	100	114	5.8	1/2	3/2	M1
^{83}Kr	9.40	12.0	3.3	212	9/2	7/2	M1
^{57}Fe	14.41	2.1	4.7	141	1/2	3/2	M1
^{151}Eu	21.53	47.8	47.0	14.1	5/2	7/2	M1
^{149}Sm	22.49	13.8	64.1	10.3	7/2	4/2	M1
^{119}Sn	23.87	8.6	25.7	25.7	1/2	3/2	M1
^{161}Dy	26.65	18.9	16.2	40.8	5/2	5/2	E1
^{40}K	29.83	0.0117	160.5	4.13	4	3	M1
^{125}Te	35.49	7.0	310	2.14	1/2	3/2	M1
^{121}Sb	37.13	57.25	130.0	5.0	5/2	7/2	M1
^{61}Ni	67.40	1.25	88.3	7.5	3/2	5/2	M1

To illustrate the sensitivity of the method to the magnetization direction within the sample, we have assumed a 2 nm-thin Fe layer on a W substrate and calculated the time dependence of its reflectivity after resonant excitation. This is shown for selected magnetization geometries in Fig. 8.4.

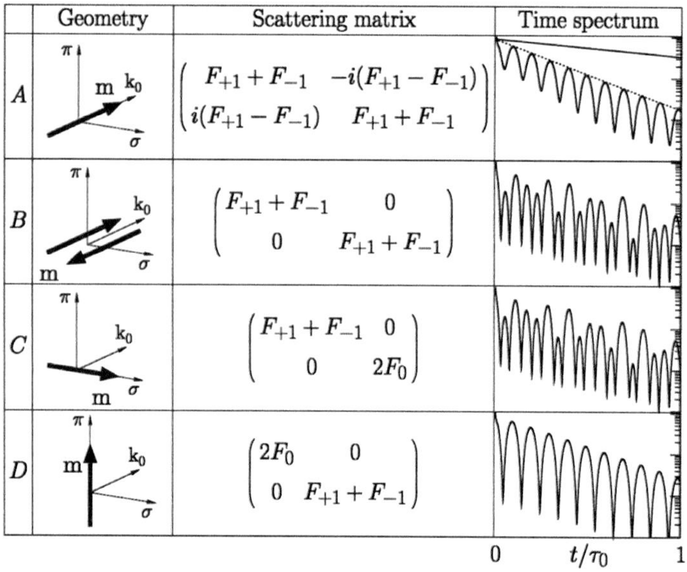

Figure 8.4 Time spectra of nuclear resonant scattering for selected orientations of the magnetic hyperfine field m relative to the incident wavevector k_0. The 2 × 2 matrix of the nuclear scattering length is given in a linear polarization basis (σ, π). The time spectra are calculated for a 2 nm thick, ferromagnetic α-^{57}Fe film on a tungsten substrate, assuming purely σ polarized incident radiation and unpolarized detection. This is the most frequently used scattering geometry in experiments with synchrotron radiation. A, C, and D display the results for a unidirectional magnetization of the sample. B results from the superposition of two magnetic sublattices in antiparallel alignment as in antiferromagnets, for example.

One clearly realizes that the time spectra of the nuclear resonant reflectivity can be used as fingerprints of the magnetic structure in the sample, although some care has to be taken since there are different magnetic configurations that lead to the same time spectrum. To rule out such ambiguities, it is recommended to record time spectra for at least two different orientations of the sample relative to the beam [15].

While most of the experiments to date have been done with ^{57}Fe, this is not the only isotope that has been employed in nuclear resonant scattering. Table 8.1 provides a list of these isotopes, whereby ^{121}Sb [16] and ^{125}Te [17] are the most recent additions.

8.3 Imaging the Magnetic Spin Structure of Exchange-Spring Magnetic Layers

The isotopic sensitivity of nuclear resonant scattering enables one to probe selected parts of a layer system by depositing ultrathin layers of a Mössbauer isotope. Enriching an element with its Mössbauer isotope (e.g., Fe by ^{57}Fe) does not disturb the chemical integrity of the layer system. Precise control over the isotopic probe layer thickness thus enables one to reach atomic resolution in the determination of magnetic properties.

This is particularly interesting for the study of magnetic heterostructures where the magnetic properties critically depend on the interfacial structure. Prominent examples are nanocomposites that consist of exchange-coupled soft- and hard-magnetic phases. Such materials are interesting candidates for new permanent magnets with giant magnetic energy products [18]. Thin bilayers consisting of a hard- and a soft-magnetic material are ideal model systems to investigate the fundamental properties of this coupling mechanism. As a characteristic property of such systems, the magnetization of the soft-magnetic film at the interface is pinned to the hard-magnetic film as a result of the exchange interaction. With increasing distance from the interface, the exchange coupling becomes weaker and the magnetization may rotate under the action of an external field. If, for example, the external field is applied orthogonal to the magnetization direction of the hard layer, the magnetic moments in the soft layer arrange in a spiral structure along the normal, as shown schematically in Fig. 8.5. Because of the reversible nature of this rotation, this is called the exchange-spring effect [19]. While a number of micromagnetical models have been developed to describe this behavior [20,21] direct measurements of the in-depth spin structure are scarce.

Here, we show how to apply nuclear resonant scattering from ultrathin probe layers of ^{57}Fe to directly measure the in-depth magnetic structure of exchange-spring bilayers [22]. The sample was a bilayer consisting of 11 nm soft-magnetic Fe on 30 nm hard-magnetic FePt. To minimize oxidation of the Fe layer, it was coated with a 3 nm thick Ag layer. All the layers were prepared by RF-magnetron sputtering in a high-vacuum system with a base pressure of $2{\cdot}10^{-7}$ mbar. The FePt layer was produced by cosputtering of Fe and Pt to obtain a composition close to $Fe_{55}Pt_{45}$. Subsequent annealing of this layer resulted in the formation of the hardmagnetic LI_0 phase with a coercivity of 0.95 T at room temperature. An ultrathin inclined probe layer of ^{57}Fe was embedded within the Fe layer, as shown in Fig. 8.5. The sample was mounted in a cryomagnet system and exposed to an external field of about 3.5 T to saturate the FePt and to introduce a remanent uniaxial magnetization along the direction of the incident wave vector $\mathbf{k}_0$. During the measurements, the sample was cooled to 4.2 K and subjected to variable external fields **H** perpendicular to the beam direction. To measure the depth dependence of the spin rotation, the sample was displaced transversely to the beam. Time spectra taken at various positions Δx are shown in Fig. 8.6 for an external field of H = 160 mT.

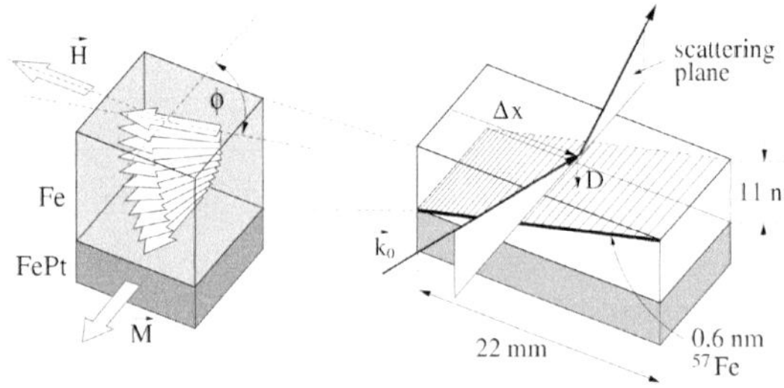

Figure 8.5 Left: Spin structure in an exchange-spring bilayer consisting of a soft-magnetic layer (Fe) on a hard-magnetic layer (FePt) with unidirectional anisotropy. An external field H is applied perpendicular to the remanent magnetization M of the hard layer. Right: To image the resulting magnetic spiral, an inclined ultrathin probe layer of ^{57}Fe is deposited within the soft layer. The magnetic properties in depth D of the film are probed by adjusting the transverse displacement Δx of the sample relative to the scattering plane.

The characteristic changes in the beat pattern reflect the rotation of the magnetization from an almost parallel alignment at Δx = 0 mm toward increasing perpendicular orientation with increasing Δx: This corresponds to a gradual transition from geometry A to geometry C in Fig. 8.4. From each of these spectra, the in-plane rotation angle φ of the magnetization was deduced by a fit of the data (solid lines). The resulting depth dependence of the rotation angle is shown in Fig. 8.6 for external fields of 160, 240 and 500 mT.

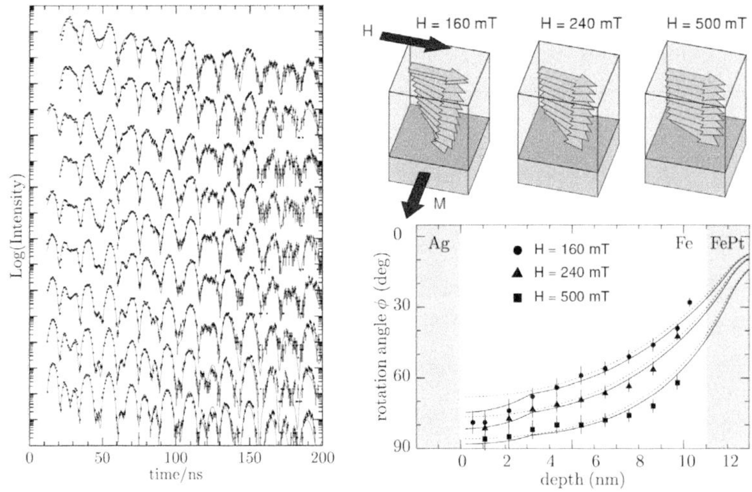

Figure 8.6 Measured time spectra of nuclear resonant grazing incidence reflection from the Fe/FePt exchange-spring system described in the text for various lateral positions starting from $\Delta x = 2$ mm (top) to $\Delta x = 18$ mm (bottom) in a perpendicular magnetic field of $B = 160$ mT. The solid lines are fits to the data with the rotation angle φ as the only free parameter. From such measurements, the depth dependence of the magnetization can be determined with very high spatial resolution, as shown in the right column for various external magnetic fields.

Dashed lines are simulations according to a one-dimensional micromagnetic model as outlined in Ref. [20], assuming the magnetic properties to be homogeneous throughout the Fe layer. Close to the Fe–Ag interface, however, the simulation clearly deviates from the measured data. This can be explained by a reduction of the exchange constant in the top 3 nm of the Fe layer to 1/3 of the value in the main body of the Fe layer. The solid lines are corresponding simulations that yield a very good description of the experimental data. This example demonstrates how ultrathin isotopic probe layers can be used to determine the depth-dependent magnetic spin structure in exchange-coupled thin films with subnanometer resolution.

In one of the next examples, we show how NRS from isotopic probe layers can contribute to understand the phenomenon of exchange bias (EB) that is observed when a ferromagnet is exchange-coupled to an antiferromagnet. The experiment relies on the use of an antiferromagnetically coupled Fe/Cr multilayer as the antiferromagnet of the EB system. As we shall see further, nuclear resonant scattering is a very powerful method to study the magnetic ordering in multilayers.

8.4 Antiferromagnetic Coupling in Fe/Cr Multilayers

Antiferromagnetic (AFM) multilayers can be used as model systems to study the behavior of finite magnetic spin systems. Here, we choose the archetypal Fe/Cr multilayer as an example. In such multilayers, the adjacent Fe layers are exchange coupled through the nonmagnetic Cr spacer in which standing waves of spin polarized conduction electrons mediate the coupling. The coupling is an oscillatory function of the Cr spacer layer thickness. At a Cr thickness of d = 1.3 nm, one observes the maximum of the AFM coupling strength. Due to the AF coupling of the Fe layers, the magnetic unit cell is twice as large as the chemical unit cell. Probing such a structure with a spin-sensitive scattering method leads to half-order Bragg reflections that are located between the Bragg peaks of the chemical structure [23]. Nuclear resonant scattering of synchrotron radiation exhibits a similar sensitivity to magnetic superstructures as neutron scattering does. The very high brilliance of modern synchrotron radiation sources, however, leads to particular advantages of the NRS technique such as high spatial resolution via microfocusing and short data acquisition times in the range of <1 hour for a reflectivity scan.

The Fe/Cr multilayer in this example consisted of 20 bilayers of (1.3 nm Fe/1.0 nm Cr) on a Si wafer, produced by sputter deposition in an Argon atmosphere from a Cr target and a Fe target enriched to 95 percent in ^{57}Fe. The angular dependence of the electronic and nuclear reflectivity is shown in Fig. 8.7. The chemical periodicity of the multilayer with a period of 2.3 nm results in the first-order electronic Bragg reflection at a grazing angle of 0.75°. The same peak is seen also in the nuclear reflectivity. However, in contrast to the electronic reflectivity, the nuclear reflectivity shows peaks also at the angular positions of the half-order reflections, corresponding to a spatial period of 4.6 nm. This is the period of a magnetic superstructure that results from the AFM coupling of the neighboring iron layers.

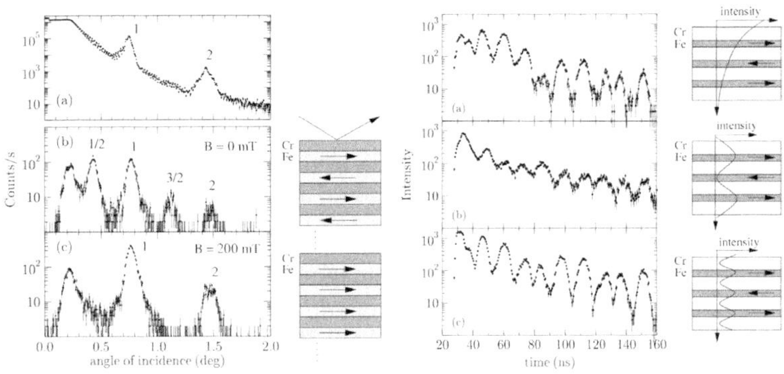

Figure 8.7 Left: Reflectivity of a multilayer consisting of 20 × (1.3 nm Fe/1.0 nm Cr) on Si. Numbers denote the order of the reflection. (a) Electronic charge scattering; (b) delayed nuclear resonant scattering in remanence after magnetization perpendicular to the incident wave vector. The Fe layers couple antiferromagnetically; and (c) delayed nuclear resonant scattering in an external field of 0.2 T with the moments of the Fe layers aligned parallel to each other. Right: Time spectra of nuclear resonant scattering from a Fe/Cr antiferromagnetic multilayer at three different angular positions: (a) the critical angle, (b) the half-order Bragg peak, and (c) the first-order Bragg peak. On the right-hand side, the intensity of the electric field as a function of depth in the layer system at these angular positions is shown.

Time spectra of the nuclear resonant reflectivity were recorded at the critical angle, the half-order Bragg reflection and the first-order Bragg reflection, as shown in Fig. 8.7. The shape of the spectra can be qualitatively explained by considering the distribution of the intensity of the radiation field inside the layer stack. This is illustrated on the right column of Fig. 8.7. While at the critical angle the intensity decays exponentially with increasing depth, at the Bragg reflections one observes a standing wave with the period of the diffracting lattice planes. Thus, at the critical angle and at the first-order peak the reflected signal is a coherent superposition of the contribution from all layers. Correspondingly, the time spectrum resembles to some degree the spectrum from two sites with opposite magnetization, as for the case displayed in Fig. 8.4b. For the half-order reflection, the time spectrum contains mostly the contribution from layers with the same magnetization direction, since the other

the magnetic ground state (see Fig. 8.6). The MOKE minor hysteresis loops of the coupled AFM/FM system show typical exchange bias characteristics (Fig. 8.8c). The hysteresis curve along the hard axis is symmetric and centered. However, the reversal parallel to the easy axis of the AFM shows a clear asymmetry, a strong increase of the coercivity field and a horizontal shift of the by a field of −1.3 mT. This is about a factor of 20 less than the value expected from the Meiklejohn–Bean model [28] for a rigid and uncompensated interface, and it coincides with the experimentally observed reduced bias fields in conventional systems.

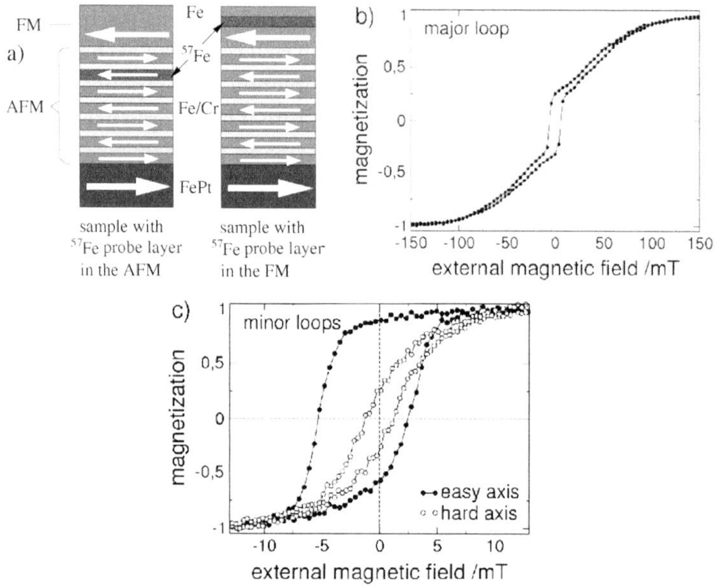

Figure 8.8 (a) Realization of an exchange bias model system. A hard-magnetic FePt buffer layer with remanent in-plane magnetization provides the unidirectional anisotropy, thus rendering the field cooling procedure as in conventional systems obsolete. The structure of the layer system leads to a domain-free AFM and thus to a fully uncompensated interface between the FM and the AFM where the magnetic structure is not affected by structural roughness. ^{57}Fe probe layers are employed to selectively address regions near the magnetic interfaces during the magnetic reversal process. (b) Major and (c) minor hysteresis loops of the layer system. The minor MOKE loops show a hysteresis that is typical for exchange bias samples. (b) and (c) from Ref. [15].

In order to clarify the origin of the observed exchange bias effect, we study the magnetic reversal via nuclear resonant scattering of synchrotron radiation from ^{57}Fe probe layers. Two chemically identical exchange bias systems were prepared with a 1.6 nm thick ^{57}Fe probe layer either in the central position of the FM or in the second Fe layer of the AFM (see Fig. 8.8a). The samples were illuminated in grazing incidence geometry close to the critical angle to maximize the reflected nuclear signal. From the measured time spectra, the orientation of the net magnetization $\mathbf{m}_{net}$ relative to the incoming photon wave vector can be determined, because the measured signal contains a contribution that is proportional to $\gamma|\mathbf{k}_0 \cdot \mathbf{m}|$, where $\mathbf{k}_0$ and $\mathbf{m}$ are unit vectors along the direction of the incoming wavevector and the direction of the net magnetization, respectively. $0 < \gamma < 1$ describes the relative degree of magnetization, that is, the fraction of uncompensated moments. Thus, a single time spectrum does not allow for a unique determination of the magnetization direction and the value of γ. Only the simultaneous evaluation of more than two time spectra recorded for different azimuthal orientations of the sample relative to the beam yields the relative angle φ between $\mathbf{m}$ and $\mathbf{k}_0$ as well as the fraction γ of uncompensated moments.

A detailed picture of the reversal mechanism in both sensor layers is obtained by plotting the vectorial state of their magnetization (represented by $\mathbf{m}$ and γ) in a polar graph, as shown in Fig. 8.9. The shape of the polar graphs allows one to distinguish between different reversal modes of the system: for a coherent rotation of the magnetization, the graph would be a half-circle connecting $\varphi = 0$ and $\varphi = \pi$. For a reversal via pure domain wall motion, the system would move along a straight line between these points. Here, we deduce a mixed reversal mode of the FM (red points) that is dominated by domain wall motion.

The polar graphs of the AFM probe layer magnetization clarify the origin of the asymmetric reversal of the FM. Due to the interlayer coupling to the FM, the individual moments of the AFM near the interface also take part in the first reversal. The important observation is that the AFM does not reverse completely. The incomplete reversal of the AFM near the interface at the end of the first reversal causes a partial compensation of the FM magnetization during the backward reversal. The backward reversal thus takes place at a reduced external magnetic field, which effectively results in a horizontal shift of the hysteresis loop.

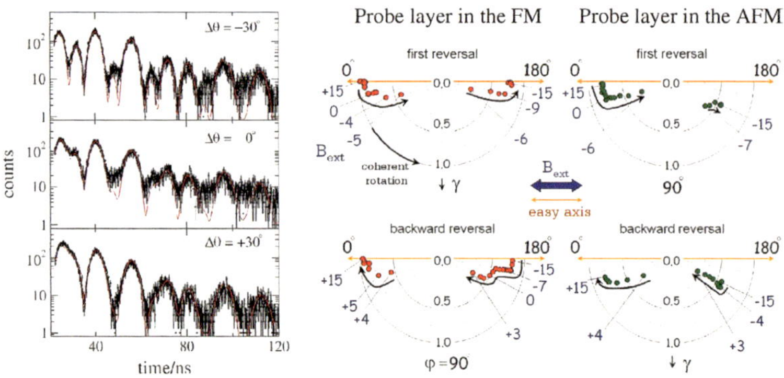

Figure 8.9 Left: Every magnetization state of the buried sensor layers was characterized by three NRS time spectra at different azimuthal orientations of sample + magnet relative to the photon beam. This procedure allows one to determine the spin direction and the fraction γ of uncompensated moments with high precision. Right: Polar graphs that display the magnetization direction and the value of γ for the probe layers in the FM and the AFM during the magnetic reversal along the easy axis. The numbers in blue display the magnetic field values (in mT) for selected points of the hysteresis. From Ref. [15].

Taking into account that the first Fe layer of the AFM in contact with the hardmagnetic FePt layer is fixed [13], two parts in antiphase within the AFM coexist after the first reversal. The coupling of both parts by a planar domain wall is energetically unfavorable due to the magnetocrystalline anisotropy energy. Only a lateral domain formation of the Fe layers can reduce the increased coupling energy inside the AFM, which occurs upon reversal of the FM moment direction.

These results identify the domain split interfacial spin structure of the AFM at the end of the first reversal to be the origin of the main archetypal exchange bias characteristics. The potential of NRS to obtain the fraction of compensated moments at the position of probe layers on both sides of the FM/AFM interface provides a qualitatively new picture of the reversal mechanism and the origin of the unidirectional exchange anisotropy.

8.6 Conclusion and Outlook

Nuclear resonant scattering of synchrotron radiation provides a powerful tool to study magnetic ordering in low-dimensional systems. The isotopic specificity of the method supports the use of probe layers to selectively address specific sites in the sample, in particular interfaces between materials with different magnetic properties such as hard/soft magnetic, FM/AFM, magnetic/superconducting layers. Future applications will focus on time-dependent phenomena as they occur during magnetic switching processes. Such processes critically rely on the depth dependence of the magnetic structure as it changes as function of time. With micro- and nanofocused beams the spatial resolution can be further increased, for example, by selecting individual magnetic layers in a cross-sectional transmission/reflection geometry. A very exciting perspective is the application of the method at future XFEL sources where it could help to understand the spatiotemporal evolution of ultrafast magnetic processes.

Acknowledgments

The experiments and studies described here have been supported by the enthusiasm and endurance of many coworkers during all the years. It is a pleasure to acknowledge the contributions of Heiko Thomas, Torsten Klein, Kai Schlage, Klaus Quast, Joachim Bansmann, Volkmar Senz, Karl-Heinz Meiwes-Broer, Eberhard Burkel, Thomas Diederich, Sebastien Couet, Jozef Korecki, Werner Keune, Balaram Sahoo, and others to the various projects in thin-film magnetism. Particular thanks goes to Rudolf Rüffer and his team at the beamlines ID18 and ID22N of the ESRF where all the nuclear resonant scattering experiments described here were performed.

References

1. Schatz G, Weidinger A (1996) *Nuclear Condensed Matter Physics: Nuclear Methods and Applications*, 2nd ed, Wiley, New York.
2. Röhlsberger R (2004) *Nuclear Condensed Matter Physics with Synchrotron Radiation: Basic Principles, Methodology and Applications*, Springer, Berlin; *Springer Tracts in Modern Physics*, vol. 208.

3. Ruby SL (1974) Mössbauer experiments without conventional sources, *J de Physique*, **35**, C6-209–C6-211.
4. Gerdau E, Rüffer R, Winkler H, Tolksdorf W, Klages CP, Hannon JP (1985) Nuclear Bragg diffraction of synchrotron radiation in yttrium iron garnet, *Phys Rev Lett*, **54**, 835–838.
5. Hastings JB, Siddons DP, van Bürck U, Hollatz R, Bergmann U (1991) Mössbauer spectroscopy using synchrotron radiation, *Phys Rev Lett*, **66**, 770–773.
6. Gerdau E, de Waard H (1999) Nuclear resonant scattering of synchrotron radiation, part A, *Hyp Interact*, **123/124**.
7. Hannon JP, Trammell GT (1968) Mössbauer diffraction. I. Quantum theory of gamma-ray and X-ray optics, *Phys Rev*, **169**, 315–329.
8. Hannon JP, Trammell GT (1969) Mössbauer diffraction. II. Dynamical theory of Mössbauer optics, *Phys Rev*, **186**, 306–325.
9. Röhlsberger R (1999) Theory of X-ray grazing incidence reflection in the presence of nuclear resonance excitation, *Hyp Interact*, **123/124**, 301–325.
10. Gibbs D, Harshman DR, Isaacs ED, McWhan DB, Mills D, Vettier C (1988) Polarization and resonance properties of magnetic X-ray scattering in holmium, *Phys Rev Lett*, **61**, 1241–1244.
11. Hannon JP, Trammell GT, Blume M, Gibbs D (1988) X-ray resonance exchange scattering, *Phys Rev Lett*, **61**, 1245–1248.
12. Gerdau E, de Waard H (2000) Nuclear resonant scattering of synchrotron radiation, part B, *Hyp Interact*, **125**.
13. Röhlsberger R, Bansmann J, Senz V, et al. (2003) Nanoscale magnetism probed by nuclear resonant scattering of synchrotron radiation, *Phys Rev B*, **67**, 245412.
14. L'abbé C, Meersschaut J, Sturhahn W, et al. (2004) Nuclear resonant magnetometry and its application to Fe/Cr multilayers, *Phys Rev Lett*, **93**, 037201.
15. Schlage K, Röhlsberger R, Klein T, Burkel E, Strohm C, Rüffer R (2009) Spatially resolved magnetic reversal in a multilayered exchange bias system, *New J Phys*, **11**, 013043.
16. Wille H-C, Shvyd'ko YV, Alp EE, et al. (2006) Nuclear resonant forward scattering of synchrotron radiation from ^{121}Sb at 37.13 keV, *Europhys Lett*, **74**, 170–176.
17. Wille H-C, Hermann RP, Sergueev I, et al. (2010) Nuclear forward and inelastic spectroscopy on ^{125}Te and $Sb_2{}^{125}Te_3$, *Europhys Lett*, **91**, 62001.

18. Skomski R, Coey JMD (1993) Giant energy product in nanostructured two-phase magnet, *Phys Rev B*, **48**, 15812–15816.

19. Kneller EF, Hawig R (1991) The exchange-spring magnet: a new material principle for permanent magnets, *IEEE Trans Magn*, **27**, 3588–3600.

20. Fullerton EE, Jiang JS, Grimsditch M, Sowers CH, Bader SD (1998) Exchange-spring behavior in epitaxial hard/soft magnetic bilayers, *Phys Rev B*, **58**, 12193–12200.

21. Jiang JS, Kaper HG, Leaf GK (2001) Hysteresis in layered spring magnets, *Discrete Contin Dynam Syst Ser B*, **1**, 219–232.

22. Röhlsberger R, Thomas H, Schlage K, Burkel E, Leupold O, Rüffer R (2002) Imaging the magnetic spin structure of exchange-coupled thin films, *Phys Rev Lett*, **89**, 237201.

23. Toellner TS, Sturhahn W, Röhlsberger R, Alp EE, Sowers CH, Fullerton EE (1995) Observation of pure nuclear diffraction from a Fe/Cr antiferromagnetic multilayer, *Phys Rev Lett*, **74**, 3475–3478.

24. Diederich Th, Couet S, Röhlsberger R (2007) Noncollinear coupling of iron layers through native iron oxide spacers, *Phys Rev B*, **76**, 054401.

25. Couet S, Diederich Th, Stankov S, et al. (2009) Probing the magnetic state of Fe/FeO/Fe trilayers by multiple isotopic sensor layers, *Appl Phys Lett*, **94**, 162501.

26. Nogues J, Schuller IK (1999) Exchange bias, *J Magn Magn Mater*, **192,** 203–232.

27. Berkowitz AE, Takano K (1999) Exchange anisotropy - a review, *J Magn Magn Mater*, **200,** 552.

28. Meiklejohn WH, Bean CP (1957) New magnetic anisotropy, *Phys Rev*, **105,** 904–913.

29. Malozemoff AP (1987) Random-field model of exchange anisotropy at rough ferromagnetic-antiferromagnetic interfaces, *Phys Rev B*, **35,** 3679.

30. Mauri D, Siegmann HC, Bagus PS, Kay E (1987) Simple model for thin ferromagnetic films exchange coupled to an antiferromagnetic substrate, *J Appl Phys*, **62**, 3047–3049.

Chapter 9

Reflectivity at Liquid Interfaces

Bridget M. Murphy
Ruprecht Haensel Laboratory/Institut für Experimentelle und Angewandte Physik, Christian-Albrechts-Universität zu Kiel, Leibnizstr. 19, 24118 Kiel, Germany
murphy@physik.uni-kiel.de

Liquids are in many ways similar to solids, but due to their nature and their lack of long-range order, there are some special considerations both experimentally and theoretically to be taken into account. Of special interest are liquid surfaces and inter faces, where often long-range ordering effects are observed. The reason is that a flat smooth surface is in effect in a one-dimensional (1D) ordered state and competes against the disorder in the liquid's bulk.

Interfacial properties and phenomena of liquid and in particular liquid/liquid interfaces are recognized as having great importance in the fields of organics [1], polymer science [2], bioscience [3], and electrochemistry [4]. Many of these systems have been neglected up to now due to demanding technical and instrumental requirements.

In this chapter, the principles of liquid interfaces will be introduced along with the X-ray scattering theory appropriate for liquids and liquid interfaces. Experimental considerations for X-ray scattering investigations at liquid interfaces will be discussed, and

X-Ray Diffraction: Modern Experimental Techniques
Edited by Oliver H. Seeck and Bridget M. Murphy

ISBN 978-981-4303-59-0 (Hardcover), 978-981-4303-60-6 (eBook)
www.panstanford.com

recent developments in this area will be considered. The theory and experimental details will be illustrated with examples at the liquid–vapor interface and liquid–liquid interface.

9.1 Introduction

Surfaces of soft matter materials or liquids are part of everybody's life. An obvious example is water. The surface of water is easily excited by external forces such as gravity, wind, or by direct contact and forms more or less macroscopic wave patterns. For viscous liquids, such as oil or honey, the behavior is different, excited surface modes damp out quickly and give rise to very smooth surfaces. Interestingly, the surface of mercury is very flat not due to high viscosity but due to high surface tension. Liquid surfaces can even appear at solid matter interfaces: Many surfaces of hard matter are frozen states of the molten material. Examples are glasses or polymers whose surfaces can almost be compared with a "snapshot" of a liquid surface.

The static and dynamic surface properties of liquids or soft matter materials are determined by parameters such as surface tension, viscoelasticity, gravitation, temperature and externally applied fields. For an introduction to liquid surfaces and interfaces, it is sufficient to use simple concepts of thermodynamics and hydrodynamics. Qualitatively speaking, the liquid surface arises from a balance between enthalpy and entropy: Whereas the enthalpy, determined by the interaction in the liquid bulk and external fields, has ordering effects and favors completely flat surfaces the entropy, connected with the thermal distortions, gives raise to disorder at the surface, called roughness (see Fig. 9.1).

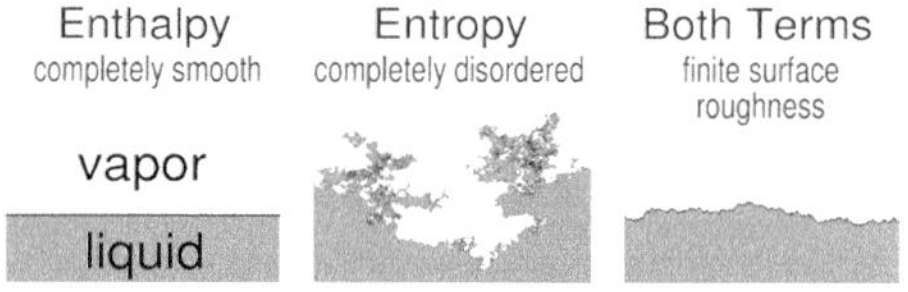

Figure 9.1 Schematic sketches of liquid surfaces shown some time after starting from a completely smooth interface [5]. Left (completely smooth): Molecular interactions exceed the entropic terms by many orders of magnitude. Center (completely disordered): Thermal entropy dominates completely. Right: Real liquid with both terms comparable. A fluctuating surfaces with finite roughness results.

9.2 X-Ray Reflectivity

The reader has been introduced to the basics of X-ray scattering in Chapter 1, however, only on a very general basis that is not easily applicable in the case of liquid surfaces. In this section, the Fresnel reflectivity of a solid layer system will be introduced as the result is part of the specular scattered signal from liquid surfaces. We will present the calculations and the results briefly. For detailed reviews of reflectivity both from solid state and liquid interfaces, the reader is referred to other articles and text books (see Refs. [6–9]).

There are three main approaches to considering reflectivity: (1) From classic optics, the classical Fresnel formulism by considering radiation incident on a planar dielectric interface or discontinuity; (2) the first-order Born approximation or kinetic approximation for which we consider weak interactions of the X-rays with matter; and (3) the distorted-wave Born approximation (DWBA), which is a compromise between the intuitive Born approximation that fails at small incident angles and the exact Fresnel formulism that cannot be extended to describe liquid surfaces. (1) and (2) will be introduced here. The DWBA, first developed by Vineyard in 1982 [10], is based on the kinetic approximations but combines some features of the dynamical treatment such as the transmission function for both the incoming and the scattered radiation. However, this book only gives insight into (1) and (2), and therefore the DWBA will be omitted. For further readings on the DWBA, we recommend more specialized publications [11–14].

9.3 Fresnel Reflectivity

The geometry for specular reflectivity $\alpha_i = \alpha_f$ is shown in Fig. 9.2. An incoming X-ray beam with angle α_f is reflected with an angle α_f. In this case, the wave vector transfer is perpendicular to the scattering surface.

Due to photoabsorption, the intensity of X-rays after travelling through a material of thickness l displays an exponential fall off proportional to the absorption coefficient of the material μ so that

$$I_0(l) = I_0 e^{-\mu l}. \qquad (9.1)$$

The refractive index for X-rays may be expressed as follows:

$$n = 1 - \delta + i\beta, \tag{9.2}$$

with dispersion $\delta = \lambda^2 r_e \rho/(2\pi)$ and absorption coefficient $\beta = \lambda\mu/(4\pi)$ where λ is the electromagnetic wavelength, r_e = 2.818·10^{-15}m the classical Thomson electron radius and ρ is the electron density. It follows that in the case of X-rays, the refractive index for almost all materials is lower than that of vacuum. Therefore, total external reflection is observed below a critical angle that may be determined from Snell's law. The critical angle of total external reflection α_c is very small, $1 - \alpha_c^2/2 \approx 1 - \delta$ so that we get the useful relation

$$\alpha_c = \sqrt{2\delta}\text{, with } \alpha_c \text{ given in radian.} \tag{9.3}$$

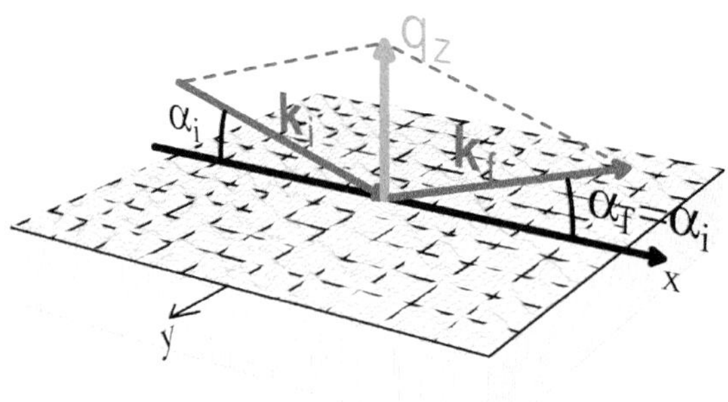

Figure 9.2 Scattering geometry for reflectivity. Shown are the incoming X-ray beam with angle α_i and the reflected beam α_f. The wave vectors and the wave vector transfer for specular reflectivity are also drawn.

For incoming X-rays with $\alpha_i < \alpha_c$, only an evanescent wave is transmitted in the propagating direction of the wave with a penetration depth of a few nanometers into the surface. The reflected and transmitted coefficients may be expressed in terms of the Fresnel equations well known from optics [15].

The reflected and transmitted coefficients F and T may be expressed as follows:

$$F \equiv \frac{a_R}{a_I} = \frac{k - k'}{k + k'} \text{ and } T \equiv \frac{a_T}{a_I} = \frac{2k}{k + k'}, \tag{9.4}$$

where α_I, α_R and α_T are the amplitudes of the incoming (I) reflected (R) and transmitted (T) X-ray components. Further, we use the definitions $k = k_0 \sin \alpha_i$ and $k' = k_0\sqrt{n^2 - \cos^2\alpha_i}$ with $k_0 = 2\pi/\lambda$. The reflected intensity of a single smooth surface is given by $R_F = |F|^2$. Well above the critical angle, the Fresnel reflectivity decays dramatically as a function of angle

$$R_F \cong \left(\frac{\alpha_c}{2\alpha_i}\right)^4, \tag{9.5}$$

so that reflectivity scans need to be collected over many orders of magnitude and are often presented on a logarithmic scale.

One very famous way to interpret reflectivity from multiple interfaces is to use the Parratt recursion algorithm from 1954 [16]. The total reflectivity R_F for N layers with thickness d_m and refractive index n_m may be calculated by considering the Fresnel reflectivity F and Snell's law at each interface giving:

$$R_{m-1,m} = \frac{R}{T} = A_{m-1}^{4}\left[\frac{R_{m,m+1} + F_{m-1,m}}{R_{m,m+1}F_{m-1,m} + 1}\right], \quad \text{where} \quad F_{m-1,m} = \frac{k_{m-1} - k_m}{k_{m-1} + k_m}. \tag{9.6}$$

Here, $Am = \exp(-ik_0\, k_m\, d_m/2)$ is an amplitude factor that depends on the thickness and $k_m = k_0\sqrt{n_m^2 - \cos^2\alpha_i}$ is the wave vector in medium m. To solve this recursion, the starting point is $R_{N,N+1} = 0$, which means that no reflection comes from the lower surface of the (infinite) bulk. Finally, $R_{1,2}$ is the reflected amplitude at the top surface and $R_F = |R_{1,2}|^2$ is the reflected intensity, which is measured by the detector.

The Fresnel formalism introduced above is valid for completely smooth layers. It can be extended to certain cases with rough interfaces when the diffuse scattering caused by this roughness is negligibly small compared to R_F. The Fresnel reflectivity R_F can then be multiplied with a so-called Névot–Croce factor to account for surface roughness [17]. For a sample with a single rough surface, this factor is given by $\exp(-q_z^2\,\sigma^2)$. Here, $q_z = 2k_0 \sin \alpha_i$ is the component of the wave vector transfer perpendicular to the surface, and σ is the root-mean square (rms) surface roughness parameter and is originated from the Gaussian statistics of the interface. More detailed reviews on scattering of rough surfaces are published elsewhere [11, 14,18–22].

9.4 Roughness at Liquid Surfaces

The information one can extract from liquid reflectivity experiments is the electron density profile in the direction perpendicular to the interface, the thickness and density of layers at the interface, and the interface roughness. When we study a liquid surface or interface, the measured reflectivity will differ from the theoretical Fresnel reflectivity. There are three main causes for this behavior. There will be an inherent roughness due to the finite size of the atoms. It is not possible to obtain an interface or surface flatter. The broadening of the profile is due to the following:

- Finite size of atoms
- Capillary waves
- Other roughness contributions

In order to determine the structure of our interface, we need to separate the capillary roughness from other sources of roughness that might be present, which contribute to the surface diffuse scattering.

Roughness is defined as a statistical deviation of the local surface from the mean surface. Therefore, statistical parameters and functions such as the root mean square deviation from the mean surface (the rms roughness), correlation lengths, correlation functions or power spectral densities can be introduced. For fluctuating surfaces, these values depend on space and time [5].

For a liquid in an equilibrium state, these statistical parameters exclusively depend on the physical properties of the liquid and its surroundings. In the simplest case, the surface tension, the viscosity, temperature, and gravity are sufficient to describe a liquid surface or interface. As liquid surfaces are thermally excited at temperatures T $\neq 0$, the local surface $z(x, y)$ is not static but fluctuating.

Energy density: $\varepsilon(\mathbf{r}, t)$ Temperature field: $T(\mathbf{r}, t)$

with the position vector $\mathbf{r}$ and the time t. In equilibrium, the sum of all forces is zero. In a simple approach, only the forces acting on volume elements $\rho\, \partial\mathbf{v}/\partial t$, resulting from pressure gradients $\nabla\mathbf{p}$ and friction $-\eta\Delta\mathbf{v}$ (with the viscosity η) are taken into account. From this follows

$$\rho\frac{\partial\mathbf{v}}{\partial t} = -\nabla p + \eta\Delta\mathbf{v}\,, \tag{9.7}$$

which is a simplified form of the Navier–Stokes equation. To solve Eq. (9.7) in the case of capillary waves, J. Jäckle used the fact that small deviations of the local surface from the mean surfaces can be approximated by a sum of plane waves. For incompressible and source-free liquids, the additional condition $\nabla \cdot \mathbf{v} = 0$ applies. This condition can be written as follows:

$$ik'v_x + \frac{\partial}{\partial z}v_z = 0 \qquad (9.8)$$

The whole problem can be decoupled in v_x and v_y. One boundary is simply given by $\mathbf{v} = 0$ deep in bulk liquid or at a solid–liquid interface. At the surface, the situation is more complicate and the stress tensor σ has to be included into the calculation with the surface tension γ, gravity g and the vertical surface displacement $u_z(x,y)$ (see Fig. 9.3).

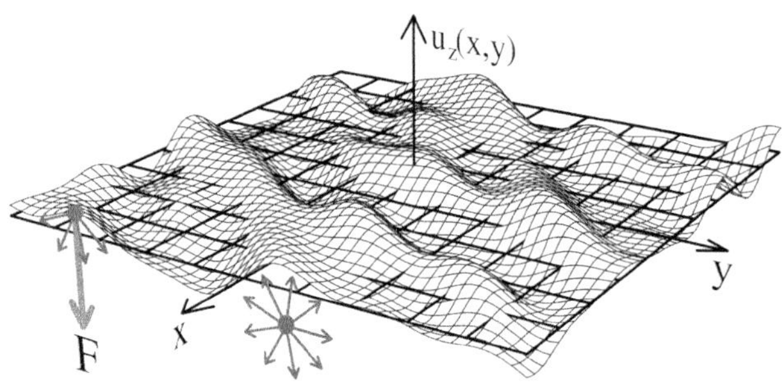

Figure 9.3 A liquid surface [5]. The close meshed grid is the real surface the wide the averaged mean surface. The local distance in **z**-direction between both surfaces is the vertical surface displacement $u_z(x, y)$. Also shown is the concept of surface tension: A volume element in the bulk is homogeneously surrounded by the liquid. All forces compensate each other. At the surface, the symmetry is broken and a resulting force points into the bulk.

A complete statistical picture of the surface structure in space and time can be derived from the dynamical power spectral density (PSD) $P(k', \omega)$, which is a measure of the number of surface modes in a particular k'-range as a function of time. The PSD can be calculated by using the following expression:

Also, an upper cut-off $Q_u = 2\pi/\kappa$ has to be introduced where κ is the minimum distance at which the liquid can still be accounted as homogeneous matter, that is, the atomic distance. In other words, at smaller distances, the liquid is not a continuum but governed by the properties of atoms or molecules. Typically, κ is on the order of a nanometer.

Coming back to Eq. (9.12), we now use the ACF of the liquid surface (see Eq. (9.11)) to continue:

$$I(\mathbf{q}) \propto \frac{1}{q_z^2} e^{-q_z^2\sigma^2} \iint \exp\left[-q_z^2 \frac{B}{2}\left(\ln(Q_c R/2) + \gamma_E\right)\right] \exp(iq_x x + iq_y y)dxdy$$

$$= \frac{1}{q_z^2} e^{-q_z^2\left[\sigma^2 + \frac{1}{2}B\gamma_E\right]} \iint \left(\frac{Q_c R}{2}\right)^{-\frac{B}{2}q_z^2} \exp(iq_x x + iq_y y)dxdy$$

$$= \frac{1}{q_z^2} e^{-q_z^2\left[\sigma^2 + \frac{1}{2}B\gamma_E\right]} \iint \left(\frac{Q_c^2}{4}(x^2 + y^2)\right)^{-\frac{B}{4}q_z^2} \exp(iq_x x + iq_y y)dxdy \qquad (9.15)$$

The integral can be solved by introducing a resolution function of the X-ray experiment (see [6,30] for the details). The resolution function takes into account the finite size of the coherently illuminated area L and as a consequence restricts the available Q-space to values larger than $Q_L = 2\pi/L$. It is very important to take care of the resolution function and be aware of it when treating the scattering data from liquid interfaces. Finally, after solving the integral for the specular condition ($\mathbf{Q}$ = 0), the result for the reflectivity of a liquid surface is the following:

$$I(\mathbf{q}) \propto \frac{1}{q_z^4} e^{-q_z^2\sigma_{eff}^2} \quad \text{with} \quad \sigma_{eff}^2 = \frac{B}{4}\ln\left(\frac{Q_u^2 + Q_c^2}{Q_c^2}\right) + \frac{1}{2}B\gamma_E - \frac{1}{2}B\ln\left(\frac{Q_L}{Q_c}\right). \qquad (9.16)$$

That means the reflectivity depends on the temperature and the surface tension (via B and Q_c, see Eq. (9.11)), on the lower limit at which the liquid properties break down (via Q_u which is connected with the molecular size), and on the resolution of the X-ray experiment (via Q_L).

The last term of σ_{eff} that is related to the resolution of the experiment needs more considerations. As in $Q_L = 2\pi/L$ the coherently illuminated length L depends on the incident angle (the projection of the beam on the sample surface is proportional to $1/\sin\alpha_i$), it is evident that Q_L also depends on $\sin\alpha_i$, and therefore on q_z. The exact dependence depends on the special realization of the experiment; however, the reflectivity of a liquid surface should be written as follows:

$$I(\mathbf{q}) \propto R_F \exp\left(-q_z^2 \sigma_{eff}(q_z)^2\right), \tag{9.17}$$

where R_F is the Fresnel reflectivity which replaces the factor $1/q_z^4$ from the Born approximation to achieve a more realistic picture at small incident angles close to the critical angle.

9.5.1 Experimental Considerations

Reforming the above expression, we see that reflectivity as measured contains the reflectivity (Fresnel) and also a contribution from the capillary waves [31]:

$$R(q_z) = R_F(q_z) \cdot |\Phi(q_z)|^2 \cdot CW(q, T, \gamma). \tag{9.18}$$

Here, R_F is the Fresnel reflectivity of a completely smooth surface, whereas $|\Phi|^2$ is a measure of the interface profile and has the meaning of a form factor. $CW(q, T, \gamma)$ is the capillary wave term and depends upon momentum transfer, temperature and the interface surface tension. Once we know the Fresnel reflectivity, we can extract the diffuse scattering component and thereby obtain the roughness of a liquid interface [11,12,30–32]. The value

$$\eta = \frac{1}{2} B q_z^2 = \frac{k_B T}{2\pi\gamma} q_z^2 \tag{9.19}$$

known as the capillary exponent, with the term η related to stiffness or surface tension, also includes a q_z^2 term and it is therefore not trivial to extract useful information. Additionally, one must consider all causes of surface roughness and not just capillary wave induced roughness. Ocko et al. [33] and references therein describe the useful expression for intrinsic roughness σ_i and capillary wave roughness σ_{cw} where the total roughness σ_T is given by the following equation:

$$\sigma_T^2 = \sigma_i^2 + \sigma_{cw}^2, \tag{9.20}$$

where

$$\sigma_T^2 = \sigma_i^2 + \frac{k_b T}{2\pi\gamma} \ln\left(\frac{k_{\max}}{k_{\min}}\right), \tag{9.21}$$

which is very similar to the findings of Eq. (9.16). The wave vector cut-offs are $k_{\max} = \pi/a$ where a is the molecular size and $k_{\min}$ the instrumental resolution. Looking at this expression, we can see even more clearly why the instrument resolution is so important for liquid scattering. For values of $\eta > 2$, it is not possible to distinguish surface scattering from bulk diffuse scattering as both exhibit correlations of the same length scale [34] and the exponent for bulk diffuse scattering is also 2.

9.5.2 Bulk Scattering

The penetration depth of the X-rays strongly depends on the incident and the exit angles. At incident angles smaller than α_c, an evanescent wave develops and typical penetrations are limited to <5 nm. At larger angles, the penetration depth increases rapidly to some micro meters. In this case, a significant amount of bulk liquid is illuminated by the X-rays and scatters strongly. This bulk scattering adds to the surface scattering. Fortunately, the specular surface scattering is confined to a very narrow angular range that is determined by the resolution and the condition $\alpha_i = \alpha_f$. In addition, close to the critical angle, the reflectivity is orders of magnitude higher than the bulk scattering which scatters basically into 4π. However, as the surface scattering drops as $1/q_z^4$ whereas, the bulk scattering increases with q_z due to the liquid structure factor. Thus, at larger incident angles, the bulk scattering eventually dominates the measured signal.

To differentiate the bulks scattering from reflected signal from the interface, the background scattering has to be measured very precisely, especially at large angles. Then, it has to be subtracted from the intensity at the specular condition. When investigating liquid–liquid interfaces, the background subtraction becomes even more delicate. As the X-rays always have to penetrate the top bulk liquid before being scattered at the liquid–liquid interface one has to consider the bulk scattering from the top liquid, additionally. This increases the amount of bulk scattering by at least a factor of two, which makes careful background subtraction even more important. In any cases, the background scattering is scattered over 4π, therefore it is easy to measure and subtract either by rotating the detector slightly away from the specular position ($\Delta q_x \approx 0.1$ Å^{-1}) or by using a two dimensional detector.

9.6 Instrumentation

9.6.1 Single-Crystal Liquid Diffractometer

The study of liquid–liquid interfaces requires special instrumental considerations. In the case of a liquid surface diffractometer, it is necessary that the instrument tilts the beam with respect to the sample. This geometry may be easily realized on a laboratory source by moving the anode [35], but often the investigations possible are limited due to a lack of intensity. Synchrotron X-ray sources can provide orders of magnitude of intensity more, and hence are very attractive for studying weakly scattering systems such as liquids. However, the experimental geometry is more complicated since it is necessary to deflect the beam down, which requires the use of a deflecting X-ray optic and some careful experimental considerations. Early designs by Als-Nielsen and Pershan [36] achieved this by using a setup similar to a triple-axis neutron spectrometer.

Many specialized designs for synchrotron instruments since then have used a single crystal to tilt the incoming X-ray beam down onto the sample (including those described in references [7,37-40]). In this arrangement, the beam moves on the surface of a cone and therefore both the sample and the detector have to be moved vertically and horizontally to follow the beam. These movements can involve up to six motors. This design works very well, but since the sample height has to be corrected for each new angle of incidence, a waiting time is required to allow vibrations at the liquid sample surface to decay before data collection can continue (see Fig. 9.4). Additionally, it is essential to check and correct for tracking errors after a careful alignment.

9.6.2 High-Energy Liquid Diffractometer

In a novel development suitable for liquid reflectivity at high energies, a double-crystal tilt stage has been implemented in an instrument at the ESRF [41]. Using a matching pair of deflecting crystals in transmission Laue geometry usually at 71 keV, it is possible to tilt the incident beam without moving the sample. The setup provides very accurate angular movement—a key requirement for working at high energies. Unfortunately, Laue geometries are not feasible at lower energies due to lack of transmission through the crystals.

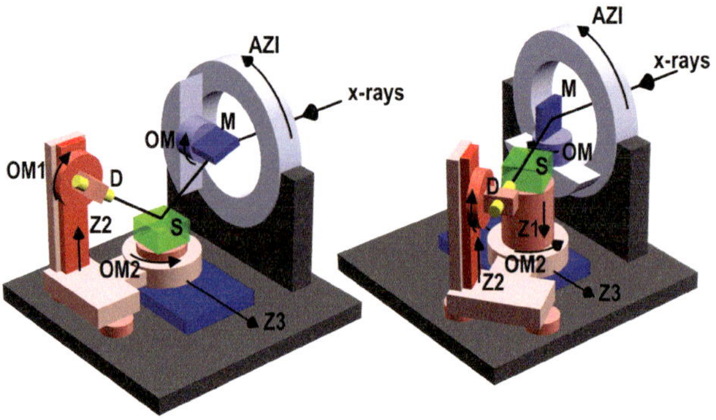

Figure 9.4 A sketch of a typical liquid diffractometer with one monochromator M. Left: Maximum incident angle on sample S. Right: Zero incident angle on S. To move the incident angle and follow with the detector, many motors have to be active and the sample has to be moved.

While there are extensive discussions as to the optimum energy for X-ray scattering experiments, recent studies have shown that in particular for investigations of biological systems energies in the range 24–34 keV are desirable to optimize the diffraction intensity while minimizing radiation damage [42–44]. In this region, the energy is sufficiently high to minimize photoelectric absorption while being low enough to optimize the interaction cross-section relative to the degree of damage due to Compton scattering, which can result in bond breaking and secondary beam damage due to free radicals [41]. On the contrary, investigations of liquid–liquid interfaces usually require transmission of some 5 cm bulk liquid, which is not possible at low photon energies where the absorption is high, so that a very useful energy window lies between 20 and 35 keV.

This energy range is too low for high energy optics in Laue geometry as the absorption at the reflection crystals would be too high. Furthermore, it may be interesting to investigate samples with typical biological active metals such as iron, copper, and so on at liquid–vapor interfaces. In these cases, the scattering contrast can be enhanced by anomalous scattering techniques. This means to choose photon energies at the absorption edges, which are ~8 keV. Again a high-energy Laue-based diffractometer would not be the appropriate choice.

9.6.3 Double-Crystal Liquid Diffractometer

As discussed in the previous section, it is desirable to optimize the photon energy with respect to penetration through liquid–liquid sample cell, the scattering strength and minimum radiation damage. The optimal range between 20 and 35 keV is outside the useful range of the Laue geometry as stated previously.

To overcome this problem, a double-crystal liquid diffractometer in reflection Bragg geometry has been developed (see Fig. 9.5 for a schematic sketch and Fig. 9.6 for the drawings): The Liquid interfaces scattering apparatus (LISA) at P08 at PETRA III at DESY [45] is a dedicated surface diffractometer operated by means of a tilting double crystal monochromator in Bragg geometry over the energy range 6.4–30 KeV allowing studies of liquid surfaces and liquid–liquid interfaces in an energy range optimized for absorption, the scattering cross-section and radiation damage. This modular spectrometer design is optimized to exploit the high brilliance of third-generation synchrotron sources and, in particular, the low horizontal divergence of 20 μrad × 10 μrad (horizontal × vertical) of the third generation synchrotron radiation source [46] A simular instrument is also in operation on I07 at the Diamond light source [47].

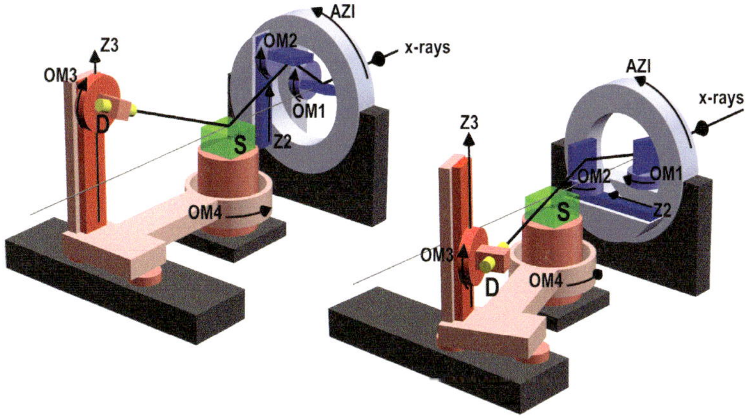

Figure 9.5 Schematic sketch of the liquid diffractometer LISA with two monochromators at OM1 and OM2. Left: Maximum incident angle on sample S. Right: Zero incident angle on S. To move the incident angle at fixed energy, only the motor AZI (the beamtilter) has to moved. The detector follows with OM3, OM4 and Z3. The sample does not move [45].

LISA provides access to a wide range of momentum space without moving the liquid sample to be investigated, which means that high-quality reflectivity data can be collected without the need for a waiting time as in the case of the standard liquid reflectivity design. This is achieved by using an asymmetric double-crystal setup mounted in the beam tilter consisting of a Si111 and Si220 crystal to bend the incoming X-ray beam down on to the sample [46]. The crystals may be rotated about the X-ray beam axis in order to change the angle of incidence (see Fig. 9.5). This means that in horizontal scattering geometry, one has to consider polarization effects. At 12 keV the loss due to polarization is about 35 percent; whereas at energies above 20 keV, it is less than 10 percent over a full reflectivity. Notably, the horizontal scattering geometry occurs at low AZI angles, that is, where the intensity loss due to polarization is at maximum. As the angle of incidence is increased the AZI rotation rotates toward the vertical scattering geometry and the available intensity increases being at maximum at highest angle of incidence. This means that the polarization effect is zero at high q_z where the reflected signal is weakest.

With this choice of crystals, it is possible to reach a vertical q_z-range of at least 2.5 Å^{-1} over the entire energy available. In the horizontal, the maximum **Q**-range is significantly larger and depends on the energy. The crystals share a rotation axis AZI so that motor movements are minimized during measurements. This design provides excellent stability and a high degree of accuracy for the beam position at the sample. In order to complete a reflectivity scan, the beamtilter is rotated about the primary beam axis as described earlier.

Using this setup, the liquid sample can remain stationary and the detector moves in height and angle to follow the specular reflection or any chosen offset off specular scan direction, therefore, leading to an enormous reduction in data collection time as will be illustrated later. Grazing incident scattering measurements are easily performed by rotating the detector arm horizontally. In this case, as above, one has to take care of polarization effects. To change the energy, it is necessary to select the appropriate Bragg angles for both crystals and translate the second crystal height perpendicular to the rotation axis.

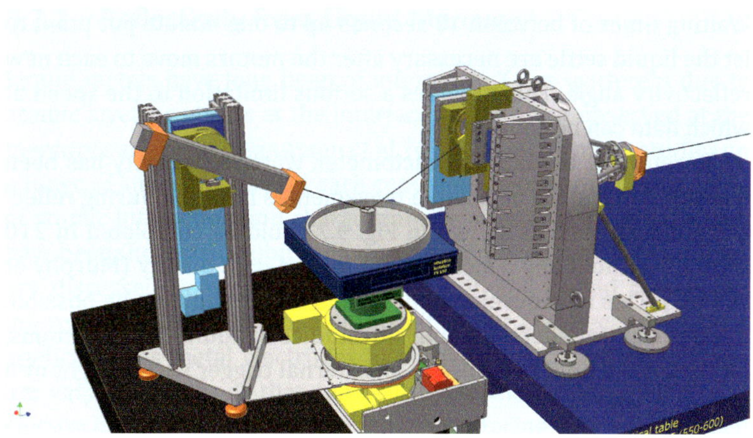

Figure 9.6 The LISA liquid spectrometer. From left to right: the detector stage, the sample stage and the beamtilter [48].

9.7 Examples

9.7.1 Reflectivity from Water

Water is the most common and most essential liquid on Earth, and because of this, it is a fascinating topic for scientists. Upon measuring the X-ray reflectivity from a water surface in 1985 Braslau et al. [31] observed that the reflectivity curve decayed faster than the Fresnel reflectivity implying that the surface was rough (see Fig. 9.7). They were the first group to measure the roughness of any liquid. By careful consideration, the roughness was found to be 3.2 Å. Where could this roughness come from? As we have discussed in this chapter, there are possible contributions from capillary waves, the size of the water atoms, and due to thermal disorder. Since the authors could determine the temperature and had prior knowledge of the size of the water molecules, in this remarkable experiment, they were able to confirm that the roughness was caused by the thermally excited capillary wave effect. In order to measure this interface, they had to wait for a long time at each new position on the reflectivity curve to allow the water to come to rest after the sample height correction once the other motors moved to the required angles. This waiting time is still required today when using a classical liquid spectrometer.

References

1. Kraack H, Ocko BM, Pershan PS, Sloutskin E, Deutsch M (2002) Structure of a Langmuir film on a liquid metal surface, *Science*, **298**, 1404–1407.
2. Fukuto M, Heilmann RK, Pershan PS, Yu SJM, Griffiths JA, Tirrell DA (1999) Structure of poly(gamma-benzyl-L-glutamate) monolayers at the gas-water interface: a Brewster angle microscopy and X-ray scattering study, *J Chem Phys*, **111**, 9761–9777.
3. Holt SA, Le Brun AP, Majkrzak CF, et al. (2009) An ion-channel-containing model membrane: structural determination by magnetic contrast neutron reflectometry, *Soft Matter*, **5**, 2576–2586.
4. Ocko BM, Wang J, Davenport A, Isaacs H (1990) Insitu X-ray reflectivity and diffraction studies of the Au(001) reconstruction in an electrochemical-cell, *Phys Rev Lett*, **65**, 1466–1469.
5. Seeck OH (2002) Capillary waves on polymer thin films, in *Soft Matter: Complex Materials on Mesoscopic Scale*, vol. 10 (Dhont JKG, Gompper G, Richter D, ed), Schriften des Forschungszentrums Jülich, Jülich, p. B7.1–B7.15.
6. Tolan M (1999) *X-Ray Scattering from Soft-Matter Thin Films - Materials Science and Basic Research*, Springer, Berlin.
7. Als-Nielsen J, Christensen F, Pershan PS (1982) Smectic-A order at the surface of a nematic liquid-crystal - synchrotron X-ray-diffraction, *Phys Rev Lett*, **48**, 1107–1110.
8. Als-Nielsen J, Jacquemain JD, Kjaer K, Leveiller F, Lahav M, Leiserowitz L (1994) Principles and applications of grazing-incidence X-ray and neutron-scattering from ordered molecular monolayers at the air-water-interface, *Phys Rep*, **246**, 252–313.
9. Als-Nielsen J, McMorrow D (2001) *Elements of Modern X-Ray Physics*, Wiley, New York.
10. Vineyard GH (1982) Grazing-incidence diffraction and the distorted-wave approximation for the study of surfaces, *Phys Rev B*, **26**, 4146–4159.
11. Sinha SK, Sirota EB, Garoff S, Stanley HB (1988) X-ray and neutron-scattering from rough surfaces, *Phys Rev B*, **38**, 2297–2311.
12. Sinha SK (1994) X-ray diffuse-scattering as a probe for thin-film and interface structure, *J Phys III*, **4**, 1543–1557.
13. Dietrich S, Haase A (1995) Scattering of X-rays and neutrons at interfaces, *Phys Rep*, **260**, 1–138.

14. Daillant J, Quinn K, Gourier C, Rieutord F (1996) Grazing incidence surface scattering of X-rays, *J Chem Soc Faraday T*, **92**, 505–513.
15. Born M, Wolf E (1959) *Principles of Optics: Electromagnetic Theory of Propagation, Interference and Diffraction of Light*, Pergoman Press, Oxofrd.
16. Parratt LG (1954) Surface studies of solids by total reflection of X-rays, *Phys Rev*, **95**, 359–369.
17. Nevot L, Croce P (1980) Characterization of surfaces by grazing X-ray reflection - application to study of polishing of some silicate-glasses, *Rev Phys Appl*, **15**, 761–779.
18. Sinha SK, Sanyal MK, Satija SK, et al. (1994) X-ray-scattering studies of surface-roughness of GaAs/ALAs multilayers, *Physica B*, **198**, 72–77.
19. Chou CH, Regan MJ, Pershan PS, Zhou XL (1997) Model-independent reconstruction of smooth electron density profiles from reflectivity data of liquid surfaces, *Phys Rev E*, **55**, 7212–7216.
20. Zhou XL, Chen SH (1995) Theoretical foundation of X-ray and neutron reflectometry, *Phys Rep*, **257**, 223–348.
21. Carbone D, Biermanns A, Ziberi B, et al. (2009) Ion-induced nanopatterns on semiconductor surfaces investigated by grazing incidence X-ray scattering techniques, *J Phys Condens Matter*, **21**, 224007.
22. Schlossman, ML. (2002) Liquid–liquid interfaces: studied by X-ray and neutron scattering, *Curr. Opin. Colloid Interface Sci.*, **7**, 235–243.
23. Jäckle J, Kawasaki K (1995) Intrinsic roughness of glass surfaces, *J Phys: Condens Matter*, **7**, 4351–4358.
24. Jäckle J (1998) The spectrum of surface waves on viscoelastic liquids of arbitrary depth, *J Phys: Condens Matter*, **10**, 7121–7131.
25. Tidswell IM, Rabedeau TA, Pershan PS, Kosowsky SD (1991) Complete wetting of a rough surface: an X-ray study, *Phys Rev Lett*, **66**, 2108–2111.
26. Fredrickson GH, Ajdari A, Leibner L, Carton J-P (1992) Surface modes and deformation energy of a molten polymer brush, *Macromolecules*, **25**, 2882–2889.
27. Tolan M, Press W (1998) X-ray and neutron reflectivity, *Z Kristallogr*, **213**, 319–336.
28. Chason E, Mayer TM (1997) Thin film and surface characterization by specular X-ray reflectivity, *Crit Rev Solid State*, **22**, 1–67.
29. Sanyal MK (1998) X-ray scattering studies of surfaces and interfaces, *Radiat Phys Chem*, **51**, 487–495.

30. Daillant J (2009) Recent developments and applications of grazing incidence scattering, *Curr Opin Colloid In*, **14**, 396–401.

31. Sanyal MK, Sinha SK, Huang KG, Ocko BM (1991) X-ray scattering study of capillary-wave fluctuations at a liquid surface, *Phys Rev Lett*, **66**, 628–631.

32. Braslau A, Pershan PS, Swislow G, Ocko BM, Als-Nielsen J (1988) Capillary waves on the surface of simple liquids measured by X-ray reflectivity, *Phys Rev A*, **38**, 2457–2470.

33. Braslau A, Deutsch M, Pershan PS, Weiss AH, Als-Nielsen J, Bohr J (1985) Surface-roughness of water measured by X-ray reflectivity, *Phys Rev Lett*, **54**, 114–117.

34. Ocko BM, Wu XZ, Sirota EB, Sinha SK, Deutsch M (1994) X-ray reflectivity study of thermal capillary waves on liquid surfaces, *Phys Rev Lett*, **72**, 242–245.

35. Tostmann H, DiMasi E, Pershan PS, Ocko BM, Shpyrko OG, Deutsch M (2000) Microscopic surface structure of liquid alkali metals, *Phys Rev B*, **61**, 7284–7287.

36. Weiss AH, Deutsch M, Braslau A, Ocko BM, Pershan PS (1986) Novel X-ray diffractometer for liquid surface studies, *Rev Sci Instrum*, **57**, 2554–2559.

37. Pershan PS, Braslau A, Weiss AH, Als-Nielsen J (1987) Smectic layering at the free-surface of liquid-crystals in the nematic phase - X-ray reflectivity, *Phys Rev A*, **35**, 4800–4813.

38. Pershan PS, Als-Nielsen J (1984) X-ray reflectivity from the surface of a liquid-crystal - surface-structure and absolute value of critical fluctuations, *Phys Rev Lett*, **52**, 759–762.

39. Als-Nielsen J (1984) X-ray studies of phase-transitions on surfaces, *Physica B&C*, **126**, 145–148.

40. Schlossman ML, Synal D, Guan YM, et al. (1997) A synchrotron X-ray liquid surface spectrometer, *Rev Sci Instrum*, **68**, 4372–4384.

41. Lin BH, Meron M, Gebhardt J, Graber T, Schlossman ML, Viccaro PJ (2003) The liquid surface/interface spectrometer at ChemMatCARS synchrotron facility at the advanced photon source, *Physca B: Condens Matter*, **336**, 75–80.

42. Honkimaki V, Reichert H, Okasinski JS, Dosch H (2006) X-ray optics for liquid surface/interface spectrometers, *J Synchr Rad*, **13**, 426–431.

43. Paithankar KS, Garman EF (2010) Know your dose: RADDOSE, *Acta Cryst D*, **66**, 381–388.

44. Nave C, Garman EF (2005) Towards an understanding of radiation damage in cryocooled macromolecular crystals, *J Synchr Rad*, **12**, 257–260.

45. Elsen A, Murphy BM, Ocko BM, et al. (2010) Surface layering at the mercury-electrolyte interface, *Phys Rev Lett*, **104**, 105501.

46. Seeck OH, Deiter C, Pflaum K, et al. (2012) The high-resolution diffraction beamline P08 at PETRA III, *J Synchrotron Rad*, **19**, 30–38.

47. Arnold T, Nicklin C, Rawle J, Sutter J, Bates T, Nutter B, McIntyre G, Burt M (2012) Implementation of a beam deflection system for studies of liquid interfaces on beamline I07 at Diamond, *J. Synchrotron Rad.*, **19**, 408–416.

48. Murphy BM, Greve M, Runge B, et al. (2010) A new diffractometer for studies of liquid - liquid interfaces, *AIP Conf Proc*, **1234**, 155–158.

49. Murphy BM, Greve M, Runge B, Koops CT, Elsen A, Stettner J, Seeck OH and Magnussen OM (2014) A novel X-ray diffractometer for studies of liquid-liquid interfaces, *J Synchrotron Rad*, **21**, 45–56.

50. Magnussen OM, Ocko BM, Regan MJ, et al. (1995) X-ray reflectivity measurements of surface layering in liquid mercury, *Phys Rev Lett*, **74**, 4444–4447.

51. Pershan PS (2011) X-ray scattering: liquid metal/vapor interfaces, *Eur Phys J D*, **196**, 109–120.

52. DiMasi E, Tostmann H, Shpyrko OG, Deutsch M, Pershan PS, Ocko BM (2000) Surface-induced order in liquid metals and binary alloys, *J Phys: Condens Matter*, **12**, A209–A214.

53. Regan MJ, Kawamoto EH, Lee S, et al. (1995) Surface layering in liquid gallium - an X-ray reflectivity study, *Phys Rev Lett*, **75**, 2498–2501.

54. Regan MJ, Magnussen OM, Kawamoto EH, et al. (1996) X-ray studies of atomic layering at liquid metal surfaces, *J Non-Cryst Solids*, **207**, 762–766.

55. Pershan PS, Regan MJ, Magnussen OM, et al. (1997) X-ray scattering studies of the liquid metal surface, *Abstr Pap Am Chem S*, **214**, 215-HYS.

56. Tostmann H, DiMasi E, Shpyrko OG, Pershan PS, Ocko BM, Deutsch M (1998) Surface phases in binary liquid metal alloys: an X-ray study, *Ber Bunsen Phys Chem*, **102**, 1136–1141.

57. Kraack H, Ocko BM, Pershan PS, Sloutskin E, Deutsch M (2002) Structure of a Langmuir film on a liquid metal surface, *Science*, **298**, 1404–1407.

very high pressures and temperatures encountered in field of warm condensed matter physics (Fig. 10.1).

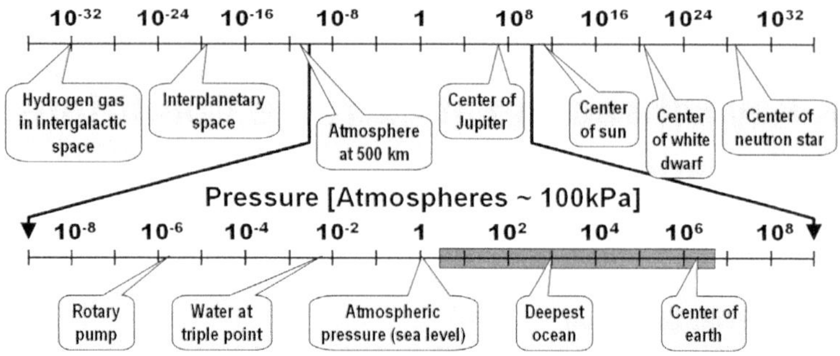

Figure 10.1 Extreme pressure conditions that encompass extreme conditions research. The gray bar marks the range that may be explored with a diamond anvil device and that is subject of this chapter. The numbers are given in atmospheres.

A description of the entire spectrum of experimental techniques that are employed in this wide region of pressure–temperature space is certainly beyond the scope of this chapter. Thus, it is convenient to limit the range of pressure and temperature conditions to those that are encountered in planetary geophysics and solid-state physics and that can be explored by means of static compression experiments within the DAC devices (Fig. 10.2). For a detailed description of the usage and different types of DAC devices, the reader might consult the reviews by Eremets [1] and Miletich et al. [2] and references within. This chapter also includes a small discussion on the use of DAC vs. Large Volume Press (LVP) devices, since there have been many new developments in both areas so that a reevaluation of the applicable pressure temperature range of both techniques seems justified. The latter should clarify for the reader when the respective techniques can be used to their maximum capability.

The reminder of the chapter will focus on examples of the different DAC techniques that are currently employed at third-generation synchrotron facilities to investigate the state of matter by means of powder and single crystal X-ray diffraction of crystalline and non-crystalline materials.

The chapter will conclude with an outlook on new developments that aim at extending the application of static compression techniques to simulate more dynamic environments. Such new experimental techniques may be used in future to conduct research in Pressure–Temperature–time (P-T-t) space that could be very successfully implemented at X-ray free-electron-laser (XFELs) such as the Linear Coherent Light Source (LCLS) at Stanford University, USA, and the European XFEL in Hamburg, Germany.

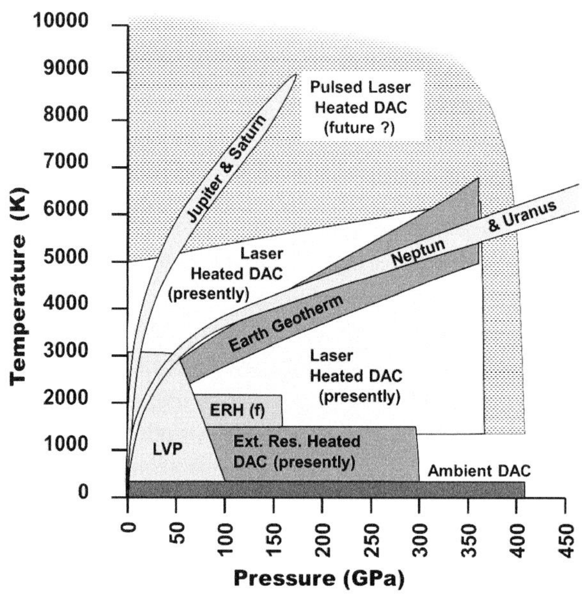

Figure 10.2 Pressure temperature regime that can be simulated by the different extreme conditions techniques used in geo-physics. "LVP" means "Large Volume Press" and "ERH (f)" means "Externally Resistive Heated DAC (future)". The lines labeled Neptun, Uranus, Jupiter and Saturn describe the proximate "geotherms" of these planets. The range of the Laser Heated DAC was recently extended by Tateno et al. [4].

10.1.1 Why X-Ray Diffraction at Extreme Conditions

The positions of Bragg X-ray diffraction peaks from powders and single crystals may be used to determine the size and symmetry of the unit cell, whereas the atomic arrangement in the unit cell may be inferred from the intensity of the diffraction peaks (see Chapter 2). By studying these atomic parameters as a function of "intensive" thermodynamic variables [3], such as pressure and temperature, we are able to deduce information on the compression and thermal expansion of the material, respectively. To describe the compression behavior of the unit cell, we use Equations of State (EOS) that, for example, model the change in volume as a function of pressure (PV-EOS). On the contrary, the intensity variation of diffraction peaks (both single crystals and powder) have been used to observe changes in bond lengths and angles between the atoms in the unit cell as a function of pressure and/or temperature, in order to learn more about the atomistic response of a structure to these intensive variables and investigate why a change from one structure to another may occur (e.g. Hazen et al. [5]). Both sets of information help us to systematically understand phase changes in pressure–temperature space and allow predictions on the possible behavior of compounds with similar chemical composition, as well as predict new structures in completely different chemical systems (Hrubiak et al. [6]). Finally, these data are used in geophysics to model the composition, structure, and dynamics of planetary bodies such as the Earth [7–9].

Although information on the EOS was initially the most important application of X-ray diffraction data obtained at extreme conditions, the field has broadened significantly in the last decades, and we are now able to collect even more information on materials of geophysical and scientific interest. Below, we will give some important examples that describe the actual use of diffraction data obtained at extreme conditions and the critical issues that one should be aware of when collecting such data.

10.1.1.1 Precise high-P and high-T equation of state studies

From powder and single crystal diffraction patterns, we are able to extract parameters of the unit cell and thus determine its volume as a function of pressure and temperature. Isothermal EOS are used to describe the pressure–volume (P-V) change. The concept of applying EOS to solids was originally used by Murnaghan [10,11] and further modified by Birch [12] and relates the volume/pressure slope to the "bulk modulus." The bulk modulus is the inverse of the compressibility and is defined as the second derivative of the lattice energy with the change in volume (e.g. [13]). Over the years, different types of EOS have been suggested [14–16] to fit both compression and thermal expansion data and relate the P-V-T EOS to physical and thermodynamic properties such as bulk modulus, Grüneisen parameter, thermal expansion, etc. (for a detailed review see Ref. [14,15,17–19]). Until today, these data are used to interpret, for example, seismic profiles such as in the case of the Preliminary Reference Earth Model (PREM) [7,20–22].

The most critical issue in the study of EOS is to collect quasi hydrostatic diffraction data as well as the accurate determination of pressure and temperature. Piermarini et al. [23] found that a mixture of methanol : ethanol (4:1 by volume) is a rather effective pressure medium for compression experiments in the DAC that creates as close to hydrostatic conditions as possible up to pressures of ~10 GPa. At higher pressures, the methanol–ethanol mixture freezes and stress conditions are no longer hydrostatic. A good indicator of the presence of non-hydrostatic conditions is the shift of the diffraction spots and peak broadening in single crystals such as quartz [24,25]. In powders, on the contrary, macro-stresses produce strains that cause the diffraction lines to shift, whereas the strains produced by micro-stresses result in diffraction line broadening [26]. In powders of materials with cubic crystal structure, one can infer the macro-stresses by calculating the uniaxial stress component [27–29]. In single crystals, one can analyze the evolution of the measured Eulerian strain tensor to determine whether the hydrostatic character of the experiment is changing (e.g., [18,30,31]) or look directly at the deviatory stress (e.g., [24,32]). Many other pressure media have been suggested for the pressure range up to 10 GPa, but none seems to be as effective as the methanol–ethanol mixtures [25,33]. To create quasi hydrostatic conditions above 10–12 GPa, liquid media such as compressed Ar, Ne, and He have been used, from which the latter two seem to be most effective until they freeze at pressures of 4.8 and 12.1 GPa, respectively [33]. In general, any medium that is "soft" in comparison to the sample may be used as a pressure-transmitting medium [34]; for example, for "hard" samples, we have used solid pressure media that are rather soft in comparison to the sample, such as NaCl, CsCl, and more recently Al [34]. In addition, one can anneal the sample within a resistive or a laser-heated DAC (e.g., [35]) to reduce stresses in the sample. A final point of concern is the interaction of the sample with the pressure medium that can significantly change the compressibility behavior of the sample.

This kind of interaction is commonly found in microporous and mesoporous materials when compressed in "penetrating" liquid pressure media, even at moderate pressure [36,37].

The determination of pressure and temperature in the DAC is an ongoing subject of debate. The most widely used pressure scale is based on the shift of the ruby fluorescence lines [38–43] as well as change of unit cell volume of metals determined through shift of the X-ray diffraction peaks as described by the EOS of the metals [14,40,44–48]. However, there are ongoing discussions about the accuracy of these calibrations, especially at very high pressure and simultaneous high pressure and high temperatures. We will come back to this subject later in Section 10.2.3. For good examples of carefully determined high-pressure and temperature EOS, see publications by Fei et al. [49], Dorogokupets and Dewaele [50], Dewaele et al. [46,47], and Komabayashi and Fei [51].

10.1.1.2 Studies on crystallographic properties

Crystallographic studies at extreme conditions have many facets, but the most prominent ones are (a) identifying the crystal structure of a new phase formed during a phase transition or synthesized at extreme conditions and (b) the systematic study of the change of structural parameters as functions of pressure at high and low temperatures. In both cases, one uses the intensities of the powder or single crystal diffraction peaks to determine the position of the atoms in the unit cell and their vibrational properties. This type of full characterization of the crystal structure of new phases is a challenging undertaking and is particularly demanding when analyzing powder diffraction data of low symmetry compounds. Thus, the latter has only been used when no adequate single crystals are available (e.g., in the case of the post-perovskite phase [52– 54]).

Powder diffraction is used more often for systematic studies of crystallographic properties of phases with known structure, aimed to determine the bond lengths and angles as a function of pressure (e.g., [55,56]). On the contrary, single crystal diffraction is the choice of technique (assuming single crystals are available), because it offers more unambiguous information on the crystal structure, especially in the case of low symmetry compounds. Recent examples of single crystal studies of new "synthesized compounds" are thus on molecular solids of, for example, oxygen or hydrogen [57], on elemental metals [58] such as sodium [59] or noble gas compounds [60].

Nevertheless, even when using single crystal data for structure refinements, it can become rather cumbersome to determine, unambiguously, the crystal structure of a new phase because our access to reciprocal space is restricted by the openings in the DAC. For example, in the case of a "symmetric piston cylinder DAC," the access can be limited to a field of view of $\pm 20^\circ$ [61]. Fortunately, in recent years, we have seen significant improvements in the design of the DAC (e.g., [62,63]), so that it is now possible to use openings of $\pm 45^\circ$ and $\pm 55^\circ$ to explore more reciprocal space. A more fundamental challenge that has not been solved adequately is the structural refinement of new phases after a reconstructive phase transition. In most cases, the single crystals shatter into smaller crystallites resulting in a powder pattern. A good summary of some new ideas to overcome this challenge can be found in Ice et al. [64] or Dera [65] and some of the ideas have been recently implemented [66].

Good examples for systematic single crystal studies are on rock-forming silicates ([67–70]). This work shows clearly how powerful systematic single crystal studies are and how they result in a deep understanding of the crystal chemistry of, for example, silicates. Unfortunately, many of these single crystal diffraction studies were not done at synchrotron facilities but at laboratory sources and thus are limited to pressures below 10–12 GPa and room temperature [18]. There are two reasons for this: (a) at ~10 GPa the quasi hydrostatic pressure medium methanol–ethanol freezes and (b) at higher pressures, the single crystals have to be so small that they can no longer provide sufficient diffraction intensities when using a laboratory X-ray source. There has been a concerted effort in the last few years to extend the pressure range by using compressed Ne or He gas as a pressure medium and to implement fast data-collection techniques at synchrotron facilities [61,71,72].

Despite this progress, single crystal experiments are still conducted at room temperature [31,61,73–76], and only recently there have been efforts to conduct single crystal diffraction experiments at high pressure and simultaneous high/low temperatures in the DAC [2,77,78]. We will come back to these efforts in Section 10.2.

earth behave anisotropically during the propagation of seismic waves. This behavior is best understood for upper mantle anisotropies that are attributed to preferred orientation of olivine crystals, attained during convection. Anisotropy is also present in the crust, where macroscopic and microscopic fractures have a significant influence as well as in the lower mantle (particularly the lowermost D" zone) and in the inner core. Although there is general agreement that seismic anisotropy is caused by deformation, the detailed mechanisms in the deep earth are poorly understood. Thus, there has been an entire branch of high-pressure research dedicated to the study of powder diffraction patterns obtained perpendicular to the compression axis without a pressure transmitting medium (Fig. 10.3) to determine elastic-plastic properties.

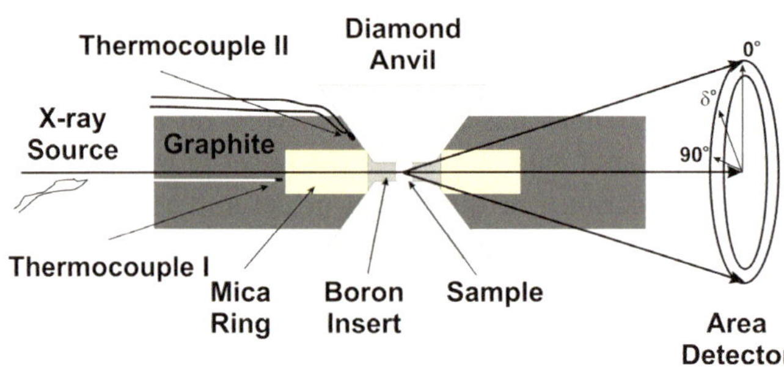

Figure 10.3 Setup of a resistive heating radial X-ray diffraction DAC (RH-RXD-DAC) experiment. The polycrystalline sample is confined under nonhydrostatic stress between the two diamond anvils. A monochromatic X-ray beam is sent through the gasket, perpendicular to the compression direction. Microscopic properties of the sample are studied with X-ray diffraction by analyzing the variation in the diffraction pattern as a function of the azimuthal angle δ. Reproduced from Liermann et al. [102].

Because most oxides and silicates become ductile at 15–20 GPa and deform under application of stress through the anvils, they have been able to infer likely deformation mechanisms by interpreting preferred orientation patterns (texture). The technique called radial diffraction allows the study of texture development during the compression experiment via Rietveld refinement and the extraction of the orientation distribution function (with specialized software such as Materials Analysis Using Diffraction (MAUD)). Diffraction data from deformed polycrystals, combined with a polycrystal plasticity model, emerge as a powerful tool to infer microscopic deformation mechanisms such as active slip systems. Deformation experiments at upper mantle conditions are best performed in a multi-anvil apparatus called the deformational DIA (D-DIA) as in the case of olivine [94,95], ringwoodite [96], perovskite [35,97,98], and periclase [99– 101].

However, at ultrahigh pressure and temperature, one has to use the DAC to produce high enough pressure and stress in order to observe changes in the preferred orientation pattern in situ through texture development. Critical for these experiments was the development of gaskets that consist of materials that are "transparent" to the incident and diffracted X-rays, such as beryllium [103,104], amorphous boron-epoxy [100,105–107], and c-BN. An extensive review on the subject can be found in Wenk et al. [97]. Recent advancements of the experimental technique focus on extending radial diffraction texture measurements to simultaneous high pressure and temperature [102,108]. This technique is now used to study the variations of orientation patterns, and thus deformation mechanisms, as a function of pressure and temperature on principal mantle phases such as perovskite, post-perovskite, magnesiowuestite, and ringwoodite to infer the mechanisms that produce anisotropy in the deep mantle and improve geodynamic models. Relevant examples of radial diffraction studies at lower mantle conditions in the DAC are those on post-perovskite at room temperatures by Merkel et al. [109,110] and more recently by Miyagi et al. [111]. Examples of radial diffraction experiments conducted at simultaneous pressure and temperature in the DAC are thus on iron by Miyagi et al. [112] performed in the laser-heated DAC and Liermann et al. [102] performed in the resistive-heated DAC. Note that these two studies also show the strength and weaknesses of the two heating methods when combined with radial X-ray diffraction, that is, that radial diffraction in the laser-heated DAC is significantly complicated by the large temperature gradients experienced by the sample. In contrast, the radial diffraction experiments in the resistive-heated DAC offer much better temperature stability, but cannot reach temperatures much higher than 2,000 K, still far from the temperature conditions that are present in the lower mantle. Thus, a combination of the two heating techniques in the future will allow better control on pressure and temperature at lower mantle conditions.

effort to use single crystal Laue diffraction [64], which does not require sample rotation as in the case of monochromatic single crystal diffraction. At the end of this chapter, we discuss what type of technical developments are necessary to further advance this technique and make it applicable to single crystal work in the DAC. Additional challenges have to be overcome in research in extreme conditions emerge from the fatigue of the pressure-creating anvils, that is, the breaking of diamond anvils at very high pressure and simultaneous temperature. Thus, the extreme conditions research community is starting to think about new ways to conduct very high pressure and high/low temperature experiments by employing pump and probe techniques, such as pulsed laser heating and the use of the dynamic DAC [118,119]. Such new techniques only slowly find their way to third-generation sources, because they require the usage of the bunch structure of these pulsed sources and gated detectors that are efficient at high energies. However, by studying P-T-t space, we may also open up new areas of structural research that have not been accessible before. For example, we can more accurately pinpoint solid–solid and solid–liquid phase transitions [93] by conducting single exposure diffraction experiments in the Hz region. Alternatively, we can explore different points during a phase transition through pump and probe experiments. Finally, pump and probe techniques enable structural studies of materials under nonequilibrium conditions combining static extreme conditions research with more dynamic sample environments.

In the following we will present examples for standard X-ray diffraction experiments currently conducted at third-generation synchrotrons. The final section of this chapter will describe new avenues in X-ray diffraction at extreme conditions that emerge with the development of the fourth-generation X-ray free-electron lasers such as the European XFEL or the LSLC at Stanford.

10.2 Standard X-Ray Diffraction Techniques and Sample Environments Used at Extreme Conditions

Sample environments for X-ray diffraction at extreme conditions in the DAC are very versatile. In fact, one could fill up a lifetime of research just by conducting experiments at simultaneous high pressure and temperature using simple elastic scattering techniques. Below, we will give some examples for the different sample environments commonly used at the third-generation synchrotron in conjunction with elastic scattering on powders and single crystals of crystalline materials. Finally, we will look at the emerging field of scattering from non-crystalline materials at extreme conditions.

10.2.1 Powder Diffraction at Simultaneous High Pressure and Temperature in the DAC

In general, there are two methods used to produce high pressure and simultaneous high temperatures in the DAC: (1) internal heating and (2) external heating of the DAC. The former is further subdivided into laser-heated and resistive-heated DAC. However, the laser-heated DAC is the more widely used technique because it covers the largest range of conditions in pressure–temperature space (Fig. 10.2). The group of externally heated DAC consists mostly of resistive-heated DAC with a few exceptions such as the inductive-heated DAC [120] that will not be further discussed here.

10.2.1.1 Laser-heated DAC

Laser heating is accomplished by coupling a powerful laser beam with the sample in the DAC while the temperature is measured optically through pyrometry. Lasers commonly employed for heating materials in the DAC have a wavelength in the infrared range, that is, 1,064 nm (Nd:YAG, Nd:YLF, Yb-fiber lasers) and 10,600 nm (CO_2 laser). Lasers with a wavelength of 1,064 nm are more frequently used at synchrotron facilities since they are more stable, easier to visualize on the sample, thus easier to handle/align, and consequently less dangerous. Nevertheless, lasers with a wavelength of 10,600 nm have the ability to heat samples that are optically transparent. This is due to the high lattice absorption of the sample in the range of 10^{-3}–10^{-4}. In contrast, lasers with a wavelength of 1,064 nm can only effectively couple to black or gray materials. Since heating is localized at the sample in the gasket hole, temperatures cannot be measured by means of thermocouples but have to be estimated through multi-wavelength spectroscopic pyrometry.

In pyrometry, the visible portion of the light spectrum (600–900 nm) emitted by the laser-heated sample is fitted with a Planck's function, assuming that the emissivity of the sample does not vary with the wavelength (gray body approximation). This approximation results in a rather large uncertainty for the pyrometric temperature measurements (see *Introduction* by Eremets [1] and review by Magunov [121]). In the following, we will discuss the two laser heating setups that are most frequently implemented at third-generation synchrotron and some new advances in the area of homogenous heating and more accurate temperature measurements made over the last decade.

In contrast to laser-heated DAC used at home laboratories, experiments at synchrotron radiation sources have an added challenge, that is, not only the lasers and the pyrometry beam path have to enter the DAC through the diamonds but also the X-ray diffraction beam. There are two principal ways to guide all three beam paths to the sample (see Fig. 10.4):

(a) Using an amorphous carbon or thin glassy mirror (CM) which is coated with silver or other reflecting materials that guides the laser and pyrometric beam at an angle of 90° to the sample, parallel to the compression axis. The X-ray beam is aligned collinear to the other two beams through the X-ray transparent mirrors. This setup is used at HPCAT [122] and GSECARS [123,124] at the APS and beamline 12.2.2, ES-2 at the Advanced Light Source [125].

(b) Introducing the laser beam from the side and the pyrometric beam path on axis via a mirror with a hole that allows the passage of the X-ray beam. In such a case, one needs reflective optics such as Schwarzfield-type objectives that collect light indirectly from the sample area so that one can have a hole in the mirror close to the sample through which the X-ray beam passes. This setup is favored at ID27 [126] at the ESRF.

Both laser-heating setups have their advantages and disadvantages. When using an X-ray transparent mirror it adds background to the diffraction image. This is particularly true for the downstream side mirror where the incident and diffracted X-ray beam interact with the mirror [127]. Nevertheless, the alignment of the system is "straight forward" [123]. In contrast, introducing the laser from the side allows optimization of the laser beam path as well as the pyrometric beam path for temperature measurements. However, such a setup makes the operation of the laser during the experiment more challenging because the laser beam travels through the diamond at an angle and requires to continuously adjust the focus. Nevertheless, it also offers the advantage to adjust the temperature by focusing and defocusing the laser beam [127].

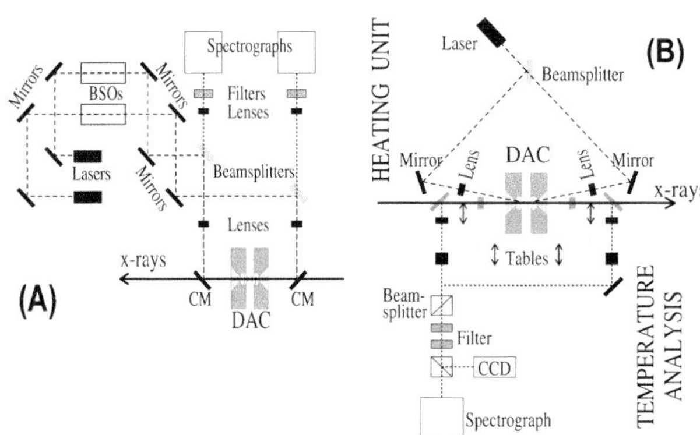

Figure 10.4 Two different optical layouts for Yb fiber laser heating systems in the DAC. (a) X-ray illuminating, heating and temperature sensing is possible simultaneously with the use of carbon mirrors CM [123]. (b) X-rays, heating laser and temperature sensing take different paths. Sensing and X-rays measurements are not simultaneous (see vertical arrows to move analysis module out) [127].

Finally, there is the issue of chromatic aberration that can cause significant errors in temperature measurements. Standard achromatic lenses used in case (a) might create temperature errors in the range of 1,000 K. In contrast, the custom made Schwarzschild-type telescope lenses often lack good surface quality and result in poor optical imaging resolution. An interesting alternative method to this is to use achromatic lenses in conjunction with an iris right in front of the DAC opening [127]. Unfortunately, the observation setup has to be removed when taking an X-ray diffraction image, which means that one will not be able to measure temperature while collecting X-ray data. It seems that there is no perfect solution at the moment, and one has to choose which setup will work better for the experiments that one wants to perform. Beamline BL10XU at

SPring8 favor a combination of both extremes introducing the laser beam from the side downstream and on axis upstream [128]. Here, temperature measurements are done from one side using fibers so that one can obtain temperatures from multiple locations on the sample. At beamline P02.2 at PETRA III, we will also implement a hybrid setup combining features of the design by Prakapenka et al. [123] and by Boehler et al. [127].

One of the biggest challenges in the early days of laser-heating experiments in the DAC, was the laser power and its pointing stability. The fact that the laser spot has a Gaussian intensity distribution and heats the sample very inhomogeneously creating a large temperature gradient across the sample, has been of further concern [1]. In addition, the huge thermal conductivity of the diamond along the compression axis of the DAC creates additional temperature gradients that, unless properly addressed, will make accurate measurements of melting, EOS, etc., very difficult. There are various approaches to overcome these difficulties. Just a few years ago, lasers did not have stabilities in the 0.1–0.2 percent ranges, so that the systems were equipped with polarizers and electronic feedback loops. In this technique, the laser is powered at much higher values than required for heating purposes and an electronic feedback loop is used to correct any instability [1,124–126].

Today, all synchrotron facilities use a fiber-pumped near-infrared laser such as Yb-fiber lasers [123,127]. These lasers offer high-power stability in the order of 0.2 percent and can be very finely tuned using the DC power supply of the laser, even at low powers. The intensity profile of the lasers are nearly perfect Gaussian (with a beam quality factor of $M^2 < 1.1$) and thus well suited for tuning the temperature gradient in the sample by changing the focusing on the sample. In addition, the change in laser beam size for relatively low pressures (up to 50 GPa) and high pressure (above 1 Mbar range) can be adjusted by using different focusing lenses [126,127].

Alternatively, the beam shape can be modified by using so-called π-shapers that are able to change the Gaussian shape of the laser in a flat top shape [123]. Although this new technology limits the maximum power of the laser, it reduces the spatial temperature gradient and offers new opportunities for laser-heated diffraction experiments. Temperature gradients along the compression axes of the diamonds are nearly impossible to access and can only be suppressed by insulating the samples from the diamonds with an insulating pressure medium (Ar, Ne, He) and/or a disk of an insulating material (Al_2O_3, MgO, BN, NaCl, KCl, etc., [1]). Nevertheless, at very high power and small focal spot of the laser-heating beam, the pressure medium seems to react with the diamonds. Eremets [1] explained that the deformed structure of, for example, NaCl pressure medium melts at lower temperatures and reacts with the diamonds resulting in dark areas on the diamond culet surface.

In the last few years, we have also seen more and more evidence that the diamonds themselves are being "overheated" and damaged [129] during laser heating. Thus, the mobilized carbon from the diamonds seems to diffuse through or around the pressure medium and reacts with the sample forming, for example, carbides in the case of a metal sample. The latter can significantly lower the melting temperature of the sample as was shown for lead [93] and more recently for tantalum [92]. These observations have stimulated a vigorous discussion about the validity of the results of many laser-heated DAC experiments (melting, solid–solid transition, EOS, etc.), because even when the diamonds are coated with alumina (Al_2O_3), this phenomena seems to persist. Time-resolved (seconds range) diffraction experiments can reduce the duration of the experiment and minimize the time carbon has to travel and react with the sample. Furthermore, continuous exposure of the sample allows determining cross contaminations very quickly as indicated by the formation of carbides in the sample, so that the data can be discarded as shown by Dewaele et al. [92].

It might be noted that the mobility of carbon is also a challenge that resistive-heated DACs have to face in the future, as they will reach higher temperatures. In the near-future, technical developments, such as pulsed laser heating setups operating in a "pump and probe" mode, can reduce the time that the diamond anvils and the sample are experiencing high temperatures. Thus, one might be able to limit the diffusion of carbon in the sample and get more accurate data on, for example, melting curves of elements and compounds.

Additional opportunities might arise from using the fourth-generation light sources such as the European XFEL (see Section 10.3).

Finally, there has been much effort to make the laser-heated DAC system portable and thus available to more than the beamlines optimized for high pressure. In general, laser-heating systems are dedicated to one beamline and thus permanently installed and not movable to other beamlines that offer other synchrotron techniques, for example, inelastic scattering and spectroscopy (XANES, EXAFS, and fluorescence). The development of fiber-pumped near-infrared lasers such as Yb-fiber lasers has promoted the implementation of a variety of portable laser-heating systems. Thus, their beam paths mimic the stationary systems described above, that is, using either amorphous X-ray transparent mirrors combining all beam paths [130,131] or using optimized beam paths for each component, laser, pyrometry and X-ray [127]. Which of the system will become standard at the third- and fourth-generation light sources will be seen in the coming years.

10.2.1.2 Resistive-heated DAC

In a resistive-heated DAC, there are two basic ways to heat the sample contained in a gasket in between two diamonds (Figs. 10.5–10.7): (a) Through thin wires that offer a large electrical resistance, requiring low current but large voltage or (b) heaters that consist of various forms of graphite that require large current with relatively small voltage. All resistive heating techniques successfully implemented at third-generation synchrotrons can be reduced to variations of one of these two techniques. Table 10.1 lists some of the different resistive-heated DACs and the pressure–temperature range at which they are currently employed at synchrotrons. The list starts with techniques for heating the entire DAC (Fig. 10.5), followed by those that only heat the inner portions of the DAC or the diamond tip and the gasket (Fig. 10.6), and concludes with relatively new internally heated techniques that only heat the inside of the gasket or the top of the culet surface (Fig. 10.7). Each of the techniques has its advantages and disadvantages, in terms of temperature stability and the pressure/temperature range. Thus, "entirely heated" DAC can offer high-temperature stability and very accurate temperature control, whereas the pressure range is limited due to mechanical stability of the different components of the DAC body, the gasket materials, and the anvils themselves. Of particular interest in the last decade has been the so-called "hydrothermal DAC" originally described by Bassett et al. [133]. It has been widely used to conduct larger volume low-pressure experiments at very high temperature to study hydrothermal systems in conjunction with fluorescence measurements on solutions to determine in situ fractionations between solids and liquids as well as melts and liquids (e.g. [134,135]).

Table 10.1 Table of resistive heater devices

Heater type	p_{max} [GPa]	T_{max} [K]	X-ray technique	Reference
Wire heating* (Nichrome, Pt, Mo)	100	1,073 (Nichrome) 1,372 (Pt) 1,573 (Mo)	Powder diffraction	[132]
Wire heating* (hydrothermal DAC)	2.5	1,473 (Mo)	Powder diffraction fluorescence	[133–136]
Wire heating* (3 pin DAC)	30	673 (Pt)	Powder and single crystal diffraction	[77,137]
Wire heating* (membrane DAC)	100	1,073 (Pt)	Powder diffraction	[138]
Graphite*	300	1,573	Powder diffraction Diffuse scattering Inelastic scattering	[102,139–142]
Metal wire§	43	2,373 (Fe)	Powder diffraction	[143]
Strip heater§	80	2,200 (Fe) 2,200 (Re)	Powder diffraction	[144–147]
Micro wire heating§ (designer diamonds)	20	2,000	Powder diffraction	[148,149]

Notes: *External resistive heater DAC, §Internally resistive heater DAC.

In contrast, internally heated DACs can reach much higher temperature and pressure but suffer from the large thermal conductivity of the diamonds that transport heat very efficiently away from the gasket/sample area. In addition, temperature can only be measured by pyrometry, limiting measurements to above 800 K, with an accuracy of about ± 50 K. The range of pyrometry can be extended to as low as 400–500 K using infrared-sensitive spectrometers, but with large uncertainties because of the lack of emissivity data at these low temperatures.

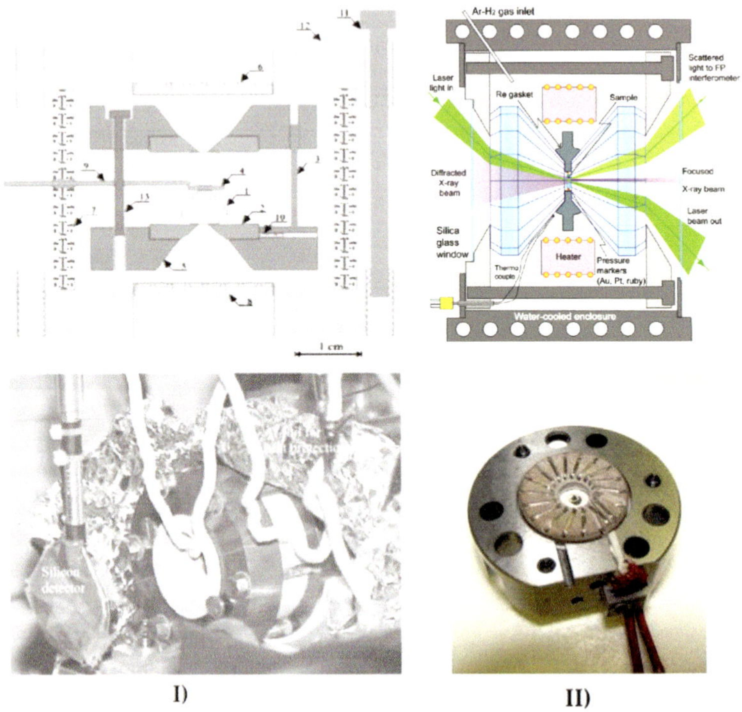

Figure 10.5 Two examples of wire resistive heated DACs used for powder and single crystal X-ray diffraction experiments at synchrotron facilities. Left column from Dubrovinsky and Dubrovinskaia [132] and right column from Sinogeikin et al. [77].

Finally, graphite resistive heated DACs represent a kind of hybrid between the two extreme "wire heating" techniques, that is, the graphite only heats a portion of the tip of the diamonds as well as the entire gasket [141]. This results in relative homogenous heating of the sample and allows accurate temperature measurements through thermocouples placed next to the gasket and at the tip of the diamond anvil (± 30 K, e.g., [102]). It provides access to very high pressures and relatively high temperatures (i.e., up to 300 GPa and 1,500 K [142]).

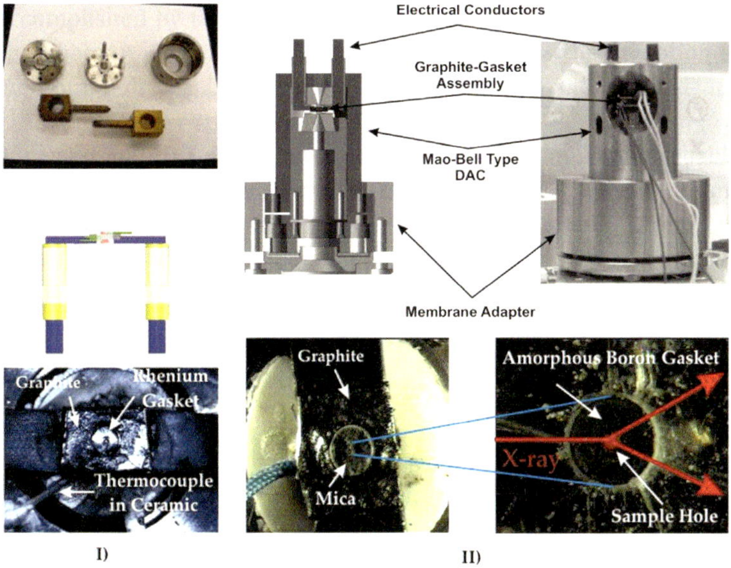

Figure 10.6 Examples of graphite resistive heated DACs used for powder X-ray diffraction experiments at synchrotron facilities. Column (I) shows a four pin DAC used by Shen et al. [139] and column (II) displays a modified Mao Bell Type DAC used by Liermann et al. [102].

The pressure–temperature limit of any of the resistive-heated techniques originates from the instability of the single crystal diamonds and probably more important, the failing of the heat-softened metal gasket. Thus, there is space for improvement by using either ceramic gasket materials [102,106] and/or the use of polycrystalline diamond (so-called nano diamonds) that show extremely high hardness in comparison to single crystal diamond at high temperatures [150,151].

One interesting question in this context concerns the implementation of hydrostatic conditions in the resistive-heated DAC. As mentioned, the application of the resistive-heated cells at much higher pressure and temperature conditions can only be accomplished through the use of ceramic gaskets. The latter excludes the use of pressure media such as methanol, ethanol or gases such as

medium. At this point, one needs to choose a different standard or use one of the spectroscopic standards.

Table 10.2 Overview of the most prominent examples of pressure scales developed over the last decades. The table includes the applicable pressure–temperature range and the reference with the most recent summaries/calibrations for the particular standard. The pressure is displayed at the highest temperature, that is, at lower temperature the pressure range can be larger. The first five entries list spectroscopic standards (fluorescence and Raman). P-V-T EOS standards based on the unit cell volume determined through X-ray diffraction measurements follow.

Standard	p [GPa]	T [K]	Extrapolation	Technique	Reference
Ruby	95	900		Fluorescence	[43,158*,159]
SrB_4O_7: Sm^{2+}	53	900		Fluorescence	[159*,162]
Sm:YAG	95	900		Fluorescence	[158*,163, 164]
Diamond (^{12}C)	33	700	100 GPa, 2,473 K	Raman	[158*#,159, 165,166]
c-BN	100	973	100 GPa, 2,473 K	Raman	[159*#, 167–169]
bcc-Fe †				EOS	[51*,170]
hcp-Fe	20	1,500	360 GPa, 7,000 K	EOS	[45,47#,85*]
fcc-Fe †	68	2,410		EOS	[51*,137]
MgO	203	3,663		EOS	[160,171*, 172]
NaCl (B2)	100	1,000	150 GPa, 3,000 K	EOS	[49*,146,173]
Au	140	2,330		EOS	[49,174*, 175]
Pt	80	1,900	300 GPa, 3,000 K	EOS	[49#,146*, 176]
Diamond	55	900	80 GPa, 900 K	EOS	[46*#]
Ne	55	900	125 GPa, 2,000 K	EOS	[46*,49#]

Notes: †Pressure temperature limited by the phase stability field (no extrapolation). *Reference from which pressure–temperature range was taken (#reference from which extrapolation was taken).

The latter require a rather accurate online spectrometric system (both fluorescence and Raman) to determine the pressure at simultaneous high temperatures. Most third-generation synchrotron beamlines that are dedicated to extreme conditions research nowadays provide such online systems. Unfortunately, diffraction beamlines that are not dedicated to this research usually have only limited access to such equipment. There are efforts under way to make such "standard" equipment available to all beamlines at third-generation synchrotrons but progress is rather slow. Furthermore, the use of online spectroscopic systems requires the movement of the optical beam path in and out of the X-ray beam, and hence many researchers rather use an X-ray diffraction standard. This way it is also possible to determine pressure in the sample and the standard in one shot to avoid any unambiguity caused by potential pressure or temperature gradients.

10.2.4 Diffraction on Nano-Crystalline Powders, Amorphous Solid and Liquids: Use of the Total Scattering Function in the DAC

Using the total scattering intensity function $S(Q)$ to study the local structure of amorphous compounds and melts has been a fast growing field of research for X-ray diffraction techniques at third-generation synchrotrons (see Chapter 2 and review by Fisher et al. [177]). In particular, the possibility to study these weakly scattering materials at high energies to collect the normalized total scattering intensity $S(Q)$, also called the total scattering function, as a function of large ranges of scattering vector $Q = 4\pi\lambda^{-1} \sin\theta$ (which is also called momentum transfer) has made high-energy diffraction beamlines at synchrotrons an ideal tool. More recently, the application of the total scattering intensity $S(Q)$ to nano-crystalline materials and the analysis of its pair distribution function (PDF) to interpret the diffuse scattering has significantly advanced the structural analysis of nano-crystalline materials (e.g. [178]).

The study of the total scattering function $S(Q)$ at simultaneous high pressure and temperatures is a rather new subject in the field of extreme conditions research but has become of significant importance for understanding the dynamics of the interior of the

Earth and other planetary bodies (see reviews by Duffy [87,179]). In particular, the study of the structure of glasses as analogues for melts or the study of melts themselves at high pressure can help us to understand the chemical differentiation of planetary bodies (such as the Earth) over time (e.g. [180,181]). We gain information on how Earth-like planetary bodies evolved from a potentially homogenous molten magma ocean to the segregated structure that we see today (crust, mantle, and core). Of particular interest for these studies are the changes of the glass/melt structure as a function of pressure and temperature and the correlated changes in basic thermodynamic and transport properties such as density and viscosity, respectively. Structural changes of the glasses and melts can be studied by looking at the First Strong Diffraction Peak (FSDP) that is related to the inter-polyhedral linkages, as well as the long range oscillations correlated to the nearest-neighbor (tetrahedral) packing (see review by Wilding et al. [182]). Change in density has been studied both in the LVP (e.g. in the Paris Edinburgh press by Sanloup et al. [183]) and within the DAC [184–186] while viscosities are mostly studied in a LVP such as the multi anvil device through radiography of falling spheres [187,188]. However, since the pressure/ temperature range of the LVP is still limited to the upper portion of the lower mantle (approx. 140 GPa, 1,300 K) the resistive- and laser-heated DAC has been used as an alternative technique to explore the structure and density of glasses and melts at higher pressures and temperatures [139,185,186,189–192].

Measurements of the total scattering function $S(Q)$ of nano-crystalline materials at higher pressures and the analysis of its Fourier Transform, the PDF, is an even newer subject in extreme conditions research (see *introduction* by Parise et al. [193]). It aims at complementing structural analysis of Bragg diffraction patterns by looking at the change in atomic arrangement in a nano-crystalline material that leads to diffuse scattering and thus significant broadening of the diffraction peaks [194]. The ultimate goal is to understand why nano-materials behave so differently at high pressure compared to bulk materials, for example, the phase transition in the nano-crystalline material occurs at much higher pressure in comparison to its bulk crystalline counterpart (e.g. [194,195]). Such studies, although still in their infancy, may become important because the compression of materials, especially under non-hydrostatic conditions, leads to grain size reduction down to nano-crystalline regime (e.g., [34,196]). This is a very scarcely explored subject of extreme conditions research in the DAC. A good example for a total scattering function/PDF study is the work on iron sulfide (FeS) summarized in the work by Ehm et al. [194].

In the early days, the diffraction patterns of amorphous and nano-crystalline materials were collected via the energy-dispersive diffraction technique, which discriminates the weakly scattering signal of the sample from the Compton and parasitic scattering of the diamonds [189–191]. Because the scattering is so weak, larger sample size can significantly decrease the collection times. Hence, many of the early diffraction experiments were conducted in LVPs such as the Paris Edinburgh cell [197] or a large-volume moissanite DAC [198,199]. Still, the collection times for one data point were significant (8 hr according to Parise et al. [193]), and of course, pressure range was limited to 20–30 GPa. The implementation of new detection strategies (see review by Parise et al. [193]) involving online image plates and fast amorphous silicon detectors up to a range of 30 $Å^{-1}$ momentum transfer [200–202], as well as focusing at high energies ($E > 60$ keV), has significantly reduced the collection time for the diffraction patterns of amorphous compounds in LVP to the tens of minutes and in the DAC to hours [193]. Furthermore, the use of perforated diamond anvils on the downstream side of the DAC has reduced significantly the Compton and parasitic scattering of the diamonds [186,200,201], so that the scattering of amorphous compounds in the DAC is becoming a more widely used technique. One may note that the addition of temperature in conjunction with high pressure might open new opportunities for exciting research (e.g., [114,139,191]).

10.3 New Directions in Extreme Conditions Research at the Third- and Fourth-Generation Light Sources

With the advent of very brilliant XFELs (such as the LCLS at Stanford, USA) which are also called fourth-generation light sources, X-ray diffraction at Extreme Conditions Research will enter a new dimension. The amount of peak and average brilliance emanating from these sources is unprecedented and will increase by a factor of 104 in comparison to the best third-generation light sources (see Fig. 10.9) as predicted, for example, for the European XFEL. This amount of photons will tremendously reduce collection times for static extreme conditions research, so that one will be able to collect diffraction patterns in a μs or shorter time scale (one-shot experiments) and resolve time-dependent processes down to the femto-second (fs) time regime using pump and probe techniques. Thus, the XFELs will allow us to conduct dynamic extreme conditions experiments in devices such as pulsed laser-heated/ dynamic DAC [119,203]. Furthermore, the short exposure times for one-shot experiments will enable detailed investigations of shockwave experiments created by powder guns or shock-driving lasers [204]. Thus, one can already "dream" about exploring physical and geophysical properties of matter at extreme conditions at a completely new level, and it seems justified to spend a few words on exploring the scientific and experimental possibilities.

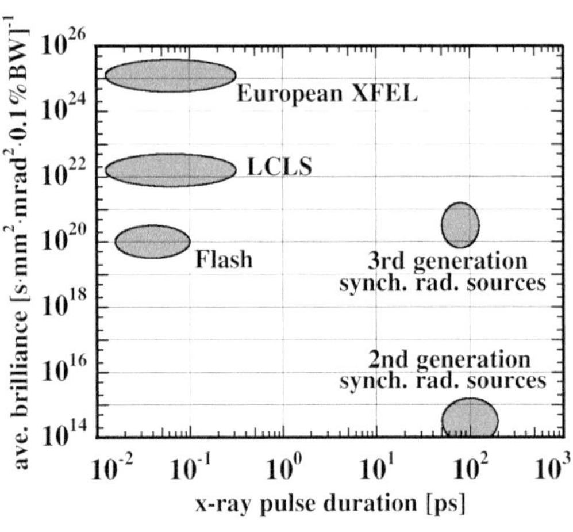

Figure 10.9 Average brilliance as a function of X-ray pulse duration for second and third generation light sources and the XFELs to illustrate that the average brilliance of a source is decisive for the duration of one shot experiments and that the pulse duration is controlling the time resolution of pump and probe experiments.

Recent workshops on the subject of dynamic experiments at extreme conditions have indicated two major thrust directions: (a) time-resolved investigations and (b) accessing very high pressure, temperature, and electromagnetic fields. The list below gives a small overview of the current and near future research focus areas in this relative new field of dynamic extreme condition research.

Time-resolved investigations

Studying kinetics of reactions (sec—μs)

- o Diffusion of impurities (carbon) and melts in the heated DAC
- o Trace element partitioning and diffusion (spatial and time domain) in the heated DAC
- o Development of stress as a function of compression rate, heating rate, etc., in the DAC

Observation of transient processes (ms—μs)

- o Meta-stable phases created during super heating/cooling and super compression
- o Grain growth and nucleation in transient phases

Investigation of nucleation processes, phase transitions, crystal structure (ps—fs)

- o Atomistic mechanics (time and length scale) involved in the nucleation of polymorphs and the propagation of phase transitions (liquid–solid, solid–solid)
- o Observing electronic transitions during dynamic compression, for example, spin transition of Fe in Earth sciences, spin density correlation (using, e.g., Mößbauer, EXAFS, Compton Scattering, phonon scattering)
- o Change in crystal structure on an atomistic size scale, that is, within one unit cell

Reaching regimes of pressure, temperature, and electromagnetic field not yet accessible

Studying shockwave processes (μs—fs)

- o Determine EoS and density at P-T conditions of the earth's core or other planetary bodies (e.g. Jupiter)
- o New phases, their crystallographic properties (using pink beam Laue diffraction), crystallization, domain growth, melting above 4 Mbar (not reachable through DAC), for example, at 10 Mbar.

o Dynamics of meteorites and asteroid impacts and formation of impact craters, for example, development of lattice strain

Physics in the warm dense matter (WDM) regime and beyond (ps—fs)

o Understand transition from "normal" solids to plasmas (correlations, depression of ionization potential, energy-level shifts, etc.)

o Fundamental understanding of X-ray plasma creation and its dynamics (astrophysics, inertial fusion)

Especially, the possibility to probe shockwave and particular isentropic compression experiments with X-rays will allow us to explore pressure and temperature regimes that are outside those reachable with static high-pressure and high-temperature experiments described in Fig. 10.2. Although a detailed discussion of the subject is beyond the scope of this chapter, Fig. 10.10 from Lee et al. [205] illustrates well the placement of the static high-pressure regimes (CM = Cold Matter) in comparison to the very extreme regions of WDM and hot dense matter (HDM) in density–temperature space (1 eV = 11,605 K).

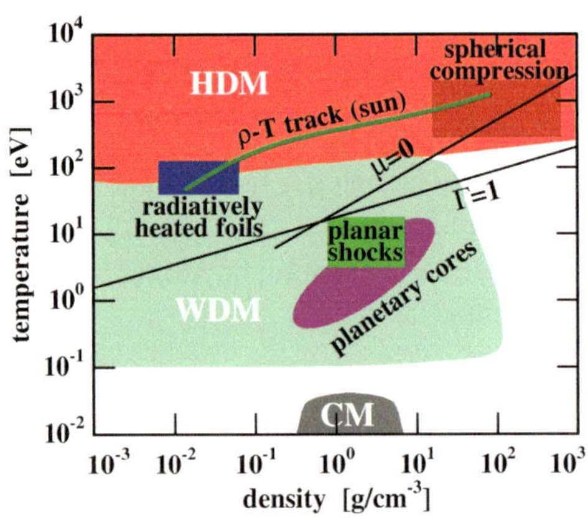

Figure 10.10 This image, reproduced from Lee et al. [205], illustrates the temperature density phase diagram of hydrogen and the placement of the static high-pressure regimes (CM = Cold Matter) in comparison to the very extreme regions of Warm Dense Matter (WDM) and Hot Dense Matter (HDM) in density-pressure space (note that 1 eV = 11,605 K). In the figure, μ is the chemical potential and Γ is the coupling.

Although the XFEL will offer new opportunities to conduct cutting-edge extreme and very extreme conditions experiments, it also produces significant experimental challenges. For example, the amount of photons deposited in the sample can be so significant (especially when using the first harmonic of the undulator) that the samples portion interacting with the X-ray beam might be destroyed. Thus, one needs to start thinking about new ways of conducting extreme conditions experiments by, for example, refreshing the sample after every interaction with the X-ray beam through, for example, rotation or other movements on the sample. Such new scientific and technical developments will take time to mature, as do the XFELs so that one could start the development of time-resolved experiments at the third-generation sources that are themselves pulsed. In fact, many of the experiments with dynamic DAC environments should be conducted at the new bright third-generation sources such as PETRA III, NSLS II, etc., and only when either the brilliance or the pulse length of the storage rings is insufficient one should consider conducting experiments at the XFELs. Hence, the third-generation sources seem to be an ideal place to start the development of dynamic X-ray diffraction experiments.

To illustrate the differences, capabilities and limitations of the third-generation sources in comparison to the XFELs, we compare below the time structure of a typical third generation light source (PETRA III) and the time structure of an XFEL, in particular the European XFEL, that is currently been built in Hamburg, Germany. Finally, we give an overview of the types of time-resolved X-ray diffraction experiments that are currently possible, we give few examples of X-ray diffraction experiments that have already been conducted at the third generations to probe extreme conditions and discuss possible future X-ray diffraction experiments at the XFELs.

10.3.1 Types of Dynamic Experiments to Be Conducted at the Third- and Fourth-Generation Sources

There are in principle two types of dynamic experiments that can be conducted to explore phenomena at extreme conditions, single-exposure (S-E) and pump and probe (P-P) experiments (see Table 10.3). Single-exposure experiments consist of a series of single diffraction exposures that contain the entire diffraction/spectroscopic information. The minimum duration of the individual exposures at extreme conditions (using high energies) is mostly dependent on the amount of photons necessary to collect a diffraction or spectroscopic pattern. On the basis of the current brilliance of the third- and fourth-generation light sources and the detector capabilities, the fastest phenomena that single-exposure experiments can study are in the ms and in the μs range, respectively. Figure 10.9 provides a number of average brilliance and pulse durations of synchrotron and XFEL sources. One can see that XFELs provide higher brilliance as well as much shorter pulse durations, which might be their most important feature.

Table 10.3 List of possible types of phenomena that can be studied at extreme conditions, the time scale of these experiments and the types of dynamic X-ray diffraction experiments that can be conducted to investigate these phenomena

Mechanism	Time scale	Experiment
Dislocation nucleation	10^{-12} s = 1 ps	P–P at 4th
Phase transformations	10^{-12} s = 1 ps	P–P at 4th
Spallation	10^{-8} s = 10 ns	P–P at 4th and 3rd
Transformation propagation	10^{-7} s = 100 ns	P–P at 4th and 3rd
Strain hardening 3D	10^{-7} s = 100 ns	P–P at 4th and 3rd
Damage evolution (voids/cracks)	10^{-6} s = 1 μs	P–P at 4th and 3rd, S-E at 4th
Dynamic recrystallization	10^{-6} s = 1 μs	P–P at 4th and 3rd, S-E at 4th
Adiabatic shear localization	10^{-5} s = 10 μs	P–P at 4th and 3rd, S-E at 4th
Fracture and diffusion (melting?)	10^{-3} s = 1 ms	P–P at 4th and 3rd, S-E at 4th and 3rd

Notes: S-E: Single exposure, P-P: Pump and probe.

The second type of experiments that one can conduct are pump and probe experiments, that is, repeated initiation of an excited state and probing with X-rays at or after the excitation (with different delays) until enough photons are collected on the detector to obtain a good diffraction or spectroscopic pattern.

Processes to be studied by pump and probe experiments have to be reversible and reproducible. The time resolution of these types of experiments is dependent on the pulse duration/structure of the source, that is, the shortest pulses at the synchrotrons will be in the tens of ps range (see Chapter 3 for an overview of pulse lengths). However, if one wants to study the phase transitions on an "atomistic level time scale" and see the nucleation initiation of polymorph as an indication of the phase transition, including the breaking of bonds that take place in the 1–10 ps range, the 100 fs long or shorter X-ray pulses of XFELs are needed for such experiments.

10.3.2 Possible Single-Exposure and Pump and Probe Experiments Using the Time Structure of PETRA III (ECB) and the European XFEL (HED) for Dynamic Experiments at Extreme Conditions

In this section some examples of single exposure and pump and probe experiments are briefly introduced that have already been conducted, are possible at third generation synchrotron sources or that could be implemented at XFELs such as the LCLS or the European XFEL.

10.3.2.1 Single exposure experiments at third generation synchrotron

One single pulse from a synchrotron radiation source is in most cases too weak for an experiment such as taking a diffraction image. A "Single Exposure" experiment in this context means the integration over many pulses. This leads to a minimum exposure time for a good diffraction pattern in the ms range (assuming materials consisting of heavy *Z* elements and using the first harmonic of the undulator). As indicated in Table 10.3, there are several slow phenomena, in particular diffusion controlled processes taking place in the ms regime.

A good example of such a process is the release and diffusion of carbon from the diamond anvil surface into metal sample during laser-driven melting experiments within the DAC. In these types of melting experiments one tries to reduce the time during which the sample is at high temperatures by quickly heating the sample while continuously collecting diffraction patterns to pinpoint the onset of melting through the disappearance of diffraction peaks and the appearance of diffuse scattering (e.g. [92,93]). So far, these types of experiments have been limited to a few seconds, because of the absence of detectors that have frame rates faster then 2–3 seconds. New detectors (PerkinElmer XRD 1621) optimized to high energy and brighter sources (PETRA III, NSLS II) will enable us to conduct these experiments in the sub-second regime. In fact, if one exposes only part of the detector using a very fast shutter one can gain diffraction images in only 35 μs as recently shown by Yoo et al. [206]. One can use this type new developments to conduct diffraction experiments that measure the hysteresis of phase transitions or to study chemical reactions at various compression and heating rates.

One recent example of such an experiment is illustrated in Fig. 10.11. Here, *bcc*-type iron was compressed in a symmetric DAC that was adapted with a membrane and compressed at a rate of 4 bar/min by pressurizing helium in the driving membrane (measured by the group of Liermann at PETRA III). The transition from *bcc*- to *hcp*-type iron as function of time can be seen very nicely. Similar studies are ideally suited for investigating the effect of varying compression rate on the development of stress state of the sample and the location of the transition pressure.

10.3.2.2 Pump and probe experiments at third generation synchrotron

Pump and probe experiments in the pulsed laser-heated DAC have been recently conducted at GSECARS at the APS by Goncharov et al. [203] (see Fig. 10.12). In this experiment, the pulsed heating laser was used to create an excited state of high temperatures in platinum powder loaded in a DAC. X-ray diffraction is used to probe the atomic structure of matter in this case to determine the unit cell volume of platinum. The authors collected 2 μs diffraction patterns at different delays after the arrival of the laser pulse and on a gated Pilatus detector repeatedly over 10 sec (in this scenario a 2 μs interval sees a total of 13 bunches assuming a 24 bunch mode of the APS storage ring). While this experiment has a relatively short exposure time, it does not make effective use of the short pulse structure of the APS storage ring. However, since the publication of their results, Goncharov et al. have moved on to collect single bunch diffraction patterns using the gated Pilatus detector.

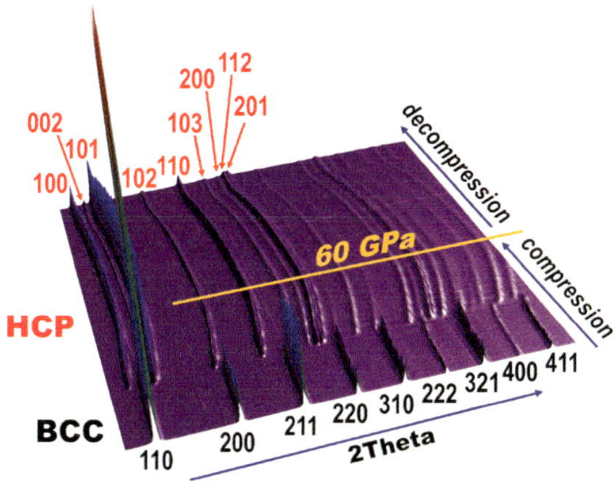

Figure 10.11 3D plot of a diffraction experiment on the fast compression of iron bcc to hcp structure. The data were collected on a Perkin Elmer XRD 1621 with a sampling rate of 5 frames/sec in the membrane driven DAC with a membrane compression rate of 4 bar/min. You can see clearly the transition from *bcc* iron to *hcp* iron. After decompression, there are small signatures of recovering bcc iron.

Similar experiments, using a pulsed pumping device (pulsed-laser heating, dynamic DAC, pulsed magnetic field, etc.) in conjunction with the pulsed structure of the third generation storage rings and either gated detectors or choppers with non-gated detectors, can be used to study reversible mechanisms at extreme conditions all the way down to the hundreds of ps range. While this technique cannot give an insight into the processes on an "atomistic level time scale", for example, the initiation of nucleation of a new polymorph due to a phase transition, it does allow us to observe the propagation of phase transitions within the sample. It also enables the exploration of pressure and temperature conditions that we have not been able to reach in the past because of the tremendous stress created on the diamond anvils at static high pressure and temperatures. These classes of experiments promise to make it possible to reach temperatures and pressures of planetary interiors such as those present in the earth's core.

10.3.2.3 Single exposure experiments at fourth generation XFEL

As the average brilliance of XFELs is a factor 10^4 higher than at any of the third generation synchrotron sources, one will be able to collect Single-Exposure" diffraction patterns in a much shorter time. We like to illustrate the potential for conducting diffraction experiments at an XFEL source such as the European XFEL using the example of Goncharov et al. [203]. Initial estimations indicate that an XFEL pulse might contain about 10^{10} ph/pulse at 30 keV. With this intensity and at this energy, it is expected that the sample survives several X-ray flashes. Assuming that we collect for 2 μs duration (see Ref. [203]), the sample would see 10 × 100 fs long pulses each containing 10^{10} ph/pulse at 30 keV. This should result in a good diffraction pattern by all estimates today, as we need 10^{10} ph/sec to create a good diffraction pattern on a PerkinElmer XRD 1621 detector. This means we no longer need to repeat these diffraction exposures but can get all information within a 2 μs illumination period. Furthermore, since each bunch train is about 666 μs long, one will be able to collect diffraction data at 2 μs intervals over the entire laser heating pulse in one bunch train. Detectors suitable to cope with these timing requirements (repetition rate) are under development and are expected to be available at the start of the European XFEL facility.

10.3.2.4 Pump and probe experiments at the fourth-generation XFEL

Pump and probe experiments at XFELs will be similar to those at the third generation, except that the pulse duration is a factor 1,000 shorter and the pulses are significantly more powerful. Thus, one will be able to really look at the "atomistic time scale" phenomena (e.g., nucleation phenomena due to a phase transition, see Table 10.3).

However, there is an entire new area of transient phase transition physics (what some people would call "disequilibrium physics") that can be explored with pump and probe techniques at the XFELs. As an example, we like to take another look at the experiment by Goncharov et al. [203]. The initial state of the laser-heating pulse arrival indicates a sharp rise of intensity and temperature (see at time 0 in Fig. 10.12). The authors specifically avoided this time regime of the pulsed heating experiments, because it most likely displays highly disequilibrium/transient conditions. Although this choice is understandable when studying equilibrium thermodynamics, studying this initial state will give us an insight into nonequilibrium behaviors and specifically what might occur in a "shockwave". Covering this pulse period of about 2 ... 4 ps, with a dense grid of 100 fs pulses in a pump and probe experiment, one could get a very detailed view of the physics that occurs during a shock laser pulse or a shock front propagation through the sample.

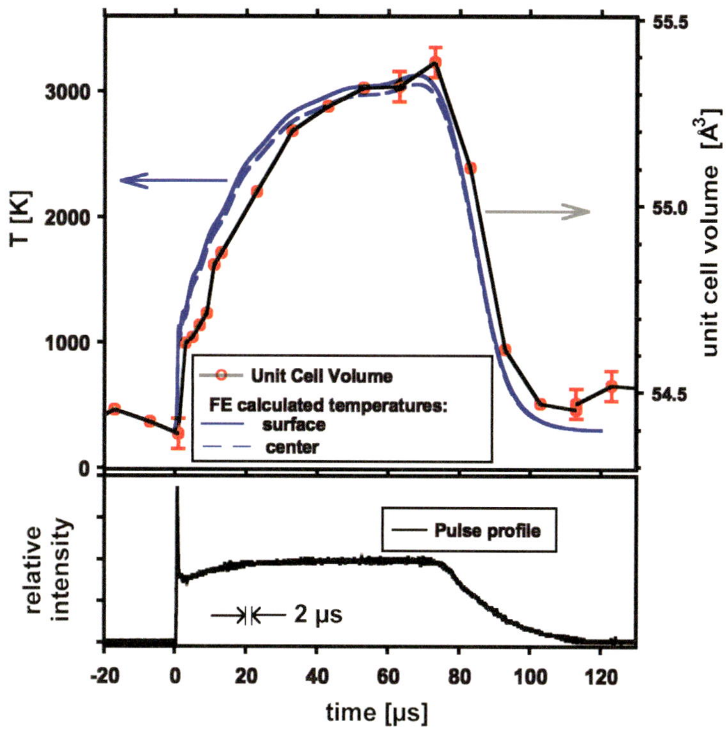

Figure 10.12 Unit cell volume of platinum powder and modeled temperature as a function of time (from Goncharov et al. [203]). The insert at the bottom shows the pulse profile of the heating laser as determined from radiometric temperature measurements. Shown on the bottom is the duration of an individual exposure of 2 μs. To achieve good statistics it has been stroboscopically collected over 10 sec.

10.4 Summary

This chapter shows that extreme conditions research has flourished over the last decades and continues to grow at third-generation synchrotron facilities. This is particularly true as we are able to collect diffraction images in less than one second due to decreased emittance of the new third-generation storage rings and improved focusing devices. With the advent of the even more powerful XFELs, in the near future, time-resolved experiments will take a more prominent role in extreme conditions research and allow us to explore dynamic regimes of extreme conditions so that we can probe matter at ever higher pressures and temperatures.

References

1. Eremets MI (1996) *High Pressures Experimental Methods*, Oxford University Press, Oxford, New York, Tokio, p. 390.
2. Miletich R, Allan DR, Kuhs W (2000) High-Pressure Single-Crystal Thechniques, in *High-Pressure Single-Crystal Techniques. Reviews in Mineralogy and Geochemistry* (Hazen RM, Downs RT, ed), Mineralogical Society of America, Washington, DC, p 445.
3. Ganguly J (2009) *Thermodynamics in Earth and Planetary Sciences*, 1st ed, Springer, Berlin.
4. Tateno S, Hirose K, Ohishi Y, Tatsumi Y (2010) The structure of iron in earth's inner core, *Science*, **330**, 359.
5. Hazen RM, Downs RT, Prewitt CT (2000) Principles of comparative crystal chemistry. High-temperature and high-pressure crystal chemistry, *Rev Min Geo*, **41**, 1–33.
6. Hrubiak R, George L, Saxena SK, Rajan K (2009) A materials database for exploring material properties, *JOM*, **61**, 59–62.
7. Dziewonski AM, Anderson DL (1981) Preliminary reference Earth model, *Phys Earth Planet Int*, **25**, 297–356.
8. Tirone M, Ganguly J, Morgan JP (2009) Modeling petrological geodynamics in the Earth's mantle, *Geochem Geophys Geosys*, **10**, Q04012.
9. Connolly JAD (2009) The geodynamic equation of state: what and how, *Geochem Geophys Geosys*, **10**, Q10014.
10. Murnaghan FD (1937) A theory of elasticity, *Phys Rev*, **51**, 593.

11. Murnaghan FD (1944) The compressibility of media under extreme pressures, *Proc Natl Acad Sci USA*, **30**, 244–247.

12. Birch F (1947) Finite elastic strain of cubic crystals, *Phys Rev*, **71**, 809–824.

13. Glob RJ, Scott DS (1985) The relationship between mineral hardness and compressibility (or bulk modulus), *Can Min*, **23**, 273–285.

14. Anderson OL (1995) *Equation of State of Solids for Geophysics and Ceramic Science*, Oxford University Press, New York, pp. 243–274.

15. Fei Y (1995) Thermal expansion, in *Handbook of Physical Constants* (Ahrens TJ, ed), American Geophysical Union, Washington, DC, pp. 29–44.

16. Angel RJ, Downs RT, Finger LW (2000) High-temperature high-pressure diffractometry, in *High-Temperature and High-Pressure Crystal Chemistry* (Hazen RM, Downs RT eds), Reviews in Mineralogy and Geochemistry 41, Mineralogical Society of America, Washington DC.

17. Holzapfel WB (1998) Equation of state of solids under strong compression, *High Press Res*, **16**, 81–126.

18. Hazen RM, Downs RT (2000) *High-Temperature and High-Pressure Crystal Chemistry*, Reviews in Mineralogy and Geochemistry 41, Mineralogical Society of America, Washington DC.

19. Stacey FD, Davis PM (2004) High pressure equations of state with applications to the lower mantle and core, *Phys Earth Planet Inter*, **142**, 137–184.

20. Saxena SK, Chatterjee N, Fei Y, Shen G (1994) *Thermodynamic Data on Oxides and Silicates*, Springer, Heidelberg.

21. Saxena SK (1996) Earth mineralogical model: Gibbs free energy minimization computation in the system MgO-FeO-SiO2, *Geochim Cosmochim Acta*, **60**, 2379–2395.

22. Fabrichnaya O, Aldinger F (2004) Assessment of thermodynamic parameters in the system ZrO_2-Y_2O_3-Al_2O_3, *Zeit Metall*, **95**, 27–39.

23. Piermarini GJ, Block S, Barnett JD (1976) Hydrostatic limits in liquids and solids to 100 kbar, *J Appl Phys*, **44**, 5377–5382.

24. Downs RT, Singh A (2006) Analysis of deviatoric stress from nonhydrostatic pressure on a single crystal in a diamond anvil cell: the case of monoclinic aegirine, $NaFeSi_2O_6$, *J Phys Chem Solids*, **67**, 1995–2000.

25. Angel RJ, Bujak M, Zhao J, Gatta GD, Jacobsen SD (2007) Effective hydrostatic limits of pressure media for high-pressure crystallographic studies, *J Appl Cryst*, 40, 26–32.
26. Singh AK, Liermann HP, Saxena SK, Mao HK, Devi SU (2006) Nonhydrostatic compression of gold powder to 60 GPa in a diamond anvil cell: estimation of compressive strength from X-ray diffraction data, *J Phys Cond Matter*, **18**, S969–S978.
27. Singh AK (1993) The lattice strains in a specimen (cubic system) compressed nonhydrostatically in an opposed anvil device, *J Appl Phys*, **73**, 4278–4286.
28. Singh AK, Mao H-K, Shu J, Hemley RJ (1998) Estimation of single-crystal elastic moduli from polycrystalline X-ray diffraction at high pressure: application to FeO and iron, *Phys Rev Lett*, **80**, 2157–2160.
29. Singh AK, Balasingh C, Mao H-K, Hemley RJ, Shu J (1998) Analysis of lattice strains measured under nonhydrostatic pressure, *J Appl Phys*, **83**, 7567–7575.
30. Gatta D, Kantor I, Boffa Ballaran T, Dubrovinsky L, McCammon C (2007) Effect of non-hydrostatic conditions on the elastic behaviour of magnetite: an in situ single-crystal X-ray diffraction study, *Phys Chem Minerals*, **34**, 627–635.
31. Jacobsen SD, Holl HM, Adams KA, et al. (2008) Compression of single-crystal magnesium oxide to 118 GPa and a ruby pressure gauge for helium pressure media, *Am Min*, **93**, 1823–1828.
32. Zhao J, Angel RJ, Ross NL (2010) Effects of deviatoric stresses in the diamond-anvil pressure cell on single-crystal samples, *J Appl Cryst*, **43**, 743–751.
33. Klotz S, Chervin J-C, Munsch P, Le Marchand G (2009) Hydrostatic limits of 11 pressure transmitting media, *J Phys D: Appl Phys*, **42**, 075413.
34. Singh AK, Liermann H-P, Akahama Y, Saxena SK, Menéndez-Proupin E (2008) Strength of polycrystalline coarse-grained platinum to 330 GPa and of nanocrystalline platinum to 70 GPa from high-pressure X-ray diffraction data, *J Appl Phys*, **103**, 063524.
35. Wenk HR, Lonardell I, Pehl J, et al. (2004) In situ observation of texture development in olivine, ringwoodite, magnesiowüstite and silicate perovskite at high pressure, *Earth Planet Sci Lett*, **226**, 507–519.
36. Gatta GD (2008) Does porous mean soft? On the elastic behaviour and structural evolution of zeolites under pressure, *Z Kristallogr*, **223**, 160–170.

37. Gatta GD (2010) Extreme deformation mechanisms in open-framework silicates at high-pressure: evidence of anomalous inter-tetrahedral angles, *Micropor Mesopor Mater*, **128**, 78–84.

38. Barnett JD, Block S, Piermarini GJ (1973) An optical fluorescence system for quantitative pressure measurement in the diamond-anvil cell, *Rev Sci Instrum*, **44**, 1.

39. Piermarini GJ, Block S, Barnett JD, Forman RA (1975) Calibration of the pressure dependence of the R1 ruby fluorescence line to 195 kbar, *J Appl Phys*, **46**, 2774–2780.

40. Mao H-K, Bell PM, Shaner JW, Steinberg DJ (1978) Specific volume measurements of Cu, Mo, Pd and Ag and calibration of the Ruby R1 fluorescence pressure gauge from 0.06 to 1 Mbar, *J Appl Phys*, **49**, 3276–3283.

41. Mao H-K, Xu J, Bell PM (1986) Calibration of the Ruby pressure gauge to 800-kbar under quasi-hydrostatic conditions, *J Geophys Res-Solid*, **91**, 4673–4676.

42. Holzapfel WB (2003) Refinement of the ruby luminescence pressure scale, *J Appl Phys*, **93**, 3.

43. Syassen K (2008) Ruby under pressure, *High Press Res*, **28**, 75–126.

44. Fei Y, Li J, Hirose K, et al. (2004) A critical evaluation of pressure scales at high temperatures by in situ X-ray diffraction measurements, *Phys Earth Planet Inter*, **143/144**, 516–526.

45. Dewaele A, Loubeyre P, Occelli F, Mezouar M, Dorogokupets PI, Torrent M (2006) Quasihydrostatic equation of state of iron above 2 Mbar, *Phys Rev Lett*, **97**, 215504.

46. Dewaele A, Datchi F, Loubeyre P, Mezouar M (2008) High pressure-high temperature equations of state of neon and diamond, *Phys Rev B*, **77**, 094106.

47. Dewaele A, Torrent M, Loubeyre P, Mezouar M (2008) Compression curves of transition metals in the Mbar range: experiments and projector augmented-wave calculations, *Phys Rev B*, **78**, 104102.

48. Dewaele A, Loubeyre P, Mezouar M (2004) Equations of state of six metals above 94 GPa, *Phys Rev B*, **70**, 094112.

49. Fei Y, Ricolleau A, Frank M, Mibe K, Shen G, Prakapenka V (2007) Toward an internally consistent pressure scale, *Proc Nation Acad Sci*, **104**, 9182–9186.

50. Dorogokupets PI, Dewaele A (2007) Equations of state of MgO, Au, Pt, NaCl-B1, and NaCl-B2: internally consistent high-temperature pressure scales, *High Press Res*, **27**, 431–446.

51. Komabayashi T, Fei Y (2010) Internally consistent thermodynamic database for iron to the Earth's core conditions, *J Geophys Res*, **115**, B03202.
52. Shim S-H, Duffy TS, Jeanloz R, Shen G (2004) Stability and crystal structure of $MgSiO_3$ perovskite to the core-mantle boundary, *Geophys Res Lett*, **31**, L10603.
53. Shim S-H, Catalli K, Hustoft J, et al. (2008) Crystal structure and thermoelastic properties of $(Mg_{0.91}Fe_{0.09})SiO_3$ postperovskite up to 135 GPa and 2,700 K, *Proc Natl Acad Sci USA (PNAS)*, **105**, 7382–7386.
54. Kubo A, Kiefer B, Shim S-H, Shen G, Prakapenka VB, Duffy TS (2008) Rietveld structure refinement of $MgGeO_3$ post-perovskite phase to 1 Mbar, *Am Min*, **93**, 965–976.
55. Hinrichsen B, Dinnebier RE, Liu H, Jansen M (2008) The high pressure crystal structures of tin sulphate: a case study for maximal information recovery from 2D powder diffraction data, *Zeit Kristall*, 223, 195–203.
56. Locherer T, Halasz I, Dinnebier R, Jansen M, Lundegaard LF (2010) Isothermal compressibility and anisotropic structural response of As_2O_5 under pressure, *Solid State Com*, **150**, 201–204.
57. Lundegaard LF, Gregoryanz E, McMahon MI, Guillaume C, Loa I, Nelmes RJ (2009) Single-crystal studies of incommensurate Na to 1.5 Mbar, *Phys Rev B*, **79**, 064105.
58. McMahon MI, Nelmes RJ (2006) High-pressure structures and phase transformations in elemental metals, *Chem Soc Rev*, **35**, 943–963.
59. Lundegaard LF, Guillaume C, McMahon MI, Gregoryanz E, Merlini M (2009) On the structure of high-pressure high-temperature η-O_2, *J Chem Phys*, **130**, 164516.
60. Somayazulu M, Dera P, Goncharov AF, et al. (2010) Pressure-induced bonding and compound formation in xenon–hydrogen solids, *Nat Chem*, **2**, 50–53.
61. Dera P, Lavina B, Borkowski LA, et al. (2008) High-pressure polymorphism of Fe_2P and its implications for meteorites and Earth's core, *Geophys Res Lett*, **35**, L10301.
62. Boehler R, De Hantsetters K (2004) New anvil designs in diamond cells, *High Press Res*, **24**, 391–396.
63. Zaug JM, Soper AK, Clark SM (2008) Pressure-dependent structures of amorphous red phosphorus and the origin of the first sharp diffraction peaks, *Nat Mat*, **7**, 890–899.
64. Ice GE, Dera P, Liu W, Mao H-K (2005) Adapting polychromatic X-ray microdiffraction techniques to high-pressure research: energy scan approach, *J Synchrotron Rad*, **12**, 608–617.

65. Dera P (2010) All different flavors of synchrotron single crystal X-ray diffraction experiments, in *High-Pressure Crystallography: From Fundamental Phenomena to Technological, Applications* (Boldyreva E, Dera P, ed), Springer, Heidelberg, pp. 11–22.

66. Wang L, Ding Y, Yang W, et al. (2010) Nanoprobe measurements of materials at megabar pressures, *Proc Nat Acad Sci*, **107**(14), 6140–6145.

67. Thompson RM, Downs RT (2003) Model pyroxenes I: ideal pyroxene topologies, *Am Min*, **88**, 653–666.

68. Thompson RM, Downs RT (2004) Model pyroxenes II: structural variation as a function of tetrahedral rotation, *Am Min*, **89**, 614–628.

69. Thompson RM, Downs RT, Redhammer GJ (2005) Model pyroxenes III: volume of C2/c pyroxenes at mantle P, T, and x, *Am Min*, **90**, 1840–1851.

70. Thompson RM, Downs RT (2010) Packing systematics of the silica polymorphs: the role played by O-O nonbonded interactions in the compression of quartz, *Am Min*, **95**, 104–111.

71. Yamanaka T, Fukuda T, Hattori T, Sumiya H (2001) New diamond anvil cell for single-crystal analysis, *Rev Sci Inst*, **72**, 1458–1462.

72. Dera P (2007) *GSE-ADA Data Analysis Program for Monochromatic Single Crystal Diffraction with Area Detector*, GeoSoilEnviroCARS, Argonne, IL.

73. Jacobsen SD, Lin J-L, Angel RJ, et al. (2005) Single-crystal synchrotron X-ray diffraction study of wüstite and magnesiowüstite at lower-mantle pressures, *J Synchrotron Rad*, **12**, 577–583.

74. Yamanaka T, Uchida A, Nakamoto Y (2008) Structural transition of post-spinel phases $CaMn_2O_4$, $CaFe_2O_4$, and $CaTi_2O_4$ under high pressures up to 80 GPa, *Am Min*, **93**, 1874–1881.

75. Lavina B, Dera P, Downs RT, et al. (2009) Siderite at lower mantle conditions and effect of the pressure-induced spin-pairing transition, *Geophys Res Lett*, **36**, L23306.

76. Lavina B, Dera P, Downs RT, et al. (2010) Structure of siderite $FeCO_3$ to 56 GPa and hysteresis of its spin-pairing transition, *Phys Rev B*, **82**, 064110.

77. Sinogeikin S, Bass J, Prakapenka V, et al. (2006) Brillouin spectrometer interfaced with synchrotron radiation for simultaneous X-ray density and acoustic velocity measurements, *Rev Sci Inst*, **77**, 103905.

78. Dubrovinsky L, Boffa-Ballaran T, Glazyrin K, et al. (2011) Single-crystal X-ray diffraction at megabar pressures and temperatures of thousands of degrees, *High Press Res*, **30**, 620–633.

79. Holloway JR, Wood JB (1991) *Simulating the Earth: Experimental Geochemistry*, Springer Verlag, New York, Berlin, p. 196.
80. Cohen LH, Klement W, Kennedy GC (1966) Investigation of phase transformations at elevated temperatures and pressures by differential thermal analysis in piston-cylinder apparatus, *J Phys Chem Solids*, **27**, 179–168.
81. Bose K, Ganguly J (1995) Quartz-coesit transition revisited: reversed experimental determination at 500-1200°C and retrieved thermochemical properties, *Am Min*, **80**, 231–238.
82. Weir CE, Lippincott ER, Van Valkenburg A, Bunting EN (1959) Infrared studies in the 1-micron to 15-micron region to 30,000 atmospheres, *J Res Nat Bur Stand Sec A-Phys Chem*, **63**, 55–62.
83. Weinstein BA, Piermarini GJ (1975) Raman-scattering and phonon dispersion in Si and GPA at very high-pressure, *Phys Rev B*, **12**, 1172–1186.
84. Piermarini GJ, Weir CE (1962) A diamond cell for X-ray diffraction studies at high pressures, *J Res Nat Bur Stan Sec A – Phys Chem*, **66**, 325–331.
85. Uchida T, Wang Y, Rivers ML, Sutton SR (2001) Stability field and thermal equation of state of ε-iron determined by synchrotron X-ray diffraction in a multianvil apparatus, *J Geophys Res*, **106**, 21799–21810.
86. Catalli K, Shim S-H, Prakapenka V (2009) Thickness and Clapeyron slope of the post-perovskite boundary, *Nature*, **462**, 782–785.
87. Duffy TS (2005) Synchrotron facilities and the study of the Earth's deep interior, *Rep Prog Phys*, **68**, 1811–1859.
88. Hemely RJ, Mao H-K, Struzhkin VV (2005) Synchrotron radiation and high pressure: new light on materials under extreme conditions, *J Synchrtron Rad*, **12**, 135–154.
89. Kavick A, Mills DM, Ohta T (2005) Structure determination by single-crystal X-ray diffraction at megabar pressures, *J Synch Rad*, **12**, 547–710.
90. Murakami M, Hirose K, Kawamura K, Sata N, Ohishi Y (2004) Post-perovskite phase transition in $MgSiO_3$, *Science*, **304**, 855–858.
91. Hirose K, Kawamura K, Ohishi Y, Tateno S, Sata N (2005) Stability and equation of state of $MgGeO_3$ post-perovskite phase, *Am Min*, **90**, 262–265.
92. Dewaele A, Mezouar M, Guignot N, Loubeyre P (2010) High melting points of tantalum in a laser-heated diamond anvil cell, *Phys Rev Lett*, **104**, 255701.

93. Dewaele A, Mezouar M, Guignot N, Loubeyre P (2007) Melting of lead under high pressure studied using second-scale time-resolved X-ray diffraction, *Phys Rev B*, **76**, 144106.

94. Katayama I, Nakashima S, Yurimoto H (2006) Water content in natural eclogite and implication for water transport into the deep upper mantle, *Lithos*, **86**, 245–259.

95. Jung H, Katayama I, Jiang Z, Hiraga T, Karato S (2006) Effect of water and stress on the lattice-preferred orientation of olivine, *Tectonophysics*, **421**, 1–22.

96. Wenk H-R, Ischia G, Nishiyama N, Wang Y, Uchida T (2005) Texture development and deformation mechanisms in ringwoodite, *Phys Earth Planet Inter*, **152**, 191–199.

97. Wenk H-R, Lonardelli I, Merkel S, et al. (2006) Deformation textures produced in diamond anvil experiments analysed in radial diffraction geometry, *J Phys: Condens Matter*, **18**, S933–S947.

98. Karato S, Wang ZC, Liu B, Fujino K (1995) Plastic deformation of garnets: systematics and implications for the rheology of the mantle transition zone, *Earth Planet Sci Lett*, **130**, 13–30.

99. Yamazaki D, Karato S (2002) Fabric development in (Mg,Fe)O during large strain, shear deformation: implications for seismic anisotropy in Earth's lower mantle, *Phys Earth Planet Int*, **131**, 251–267.

100. Merkel S, Wenk H-R, Shu J, et al. (2002) Deformation of polycrystalline MgO at pressures of the lower mantle, *J Geophys Res*, **107**, 2271.

101. Heidelbach F, Stretton I, Langenhorst F, Mackwell S (2003) Fabric evolution during high shear strain deformation of magnesiowustite ($Mg_{0.8}Fe_{0.2}O$), *J Geophys Res-Solid Earth*, **108**, 2154.

102. Liermann H-P, Merkel S, Miyagi L, et al. (2009) Experimental method for in situ determination of material textures at simultaneous high pressure and high temperature by means of radial diffraction in the diamond anvil cell, *Rev Sci Inst*, **80**, 104501.

103. Hemley RJ, Mao HK, Shen GY, et al. (1997) X-ray imaging of stress and strain of diamond, iron, and tungsten at megabar pressures, *Science*, **276**, 1242–1245.

104. Mao HK, Shu J, Shen G, Hemley RJ, Li B, Singh AK (1999) Elasticity and rheology of iron above 220 GPa and the nature of the Earth's inner core, *Nature*, **399**, 280.

105. Mao H-K, Shu J, Fei Y, Hu J, Hemley R (1996) The wustite enigma, *J Phys Earth Planet Inter*, **96**, 135–145.

106. Lin JF, Shu J, Mao HK, Hemley RJ, Shen G (2003) Amorphous boron gasket in diamond anvil cell research, *Rev Sci Inst*, **74**, 4732.

107. Merkel S, Yagi T (2005) X-ray transparent gasket for diamond anvil cell high pressure experiments, *Rev Sci Instrum*, **76**, 046109.

108. Kunz M, Caldwell WA, Miyagi L, Wenk H-R (2007) In situ laser heating and radial synchrotron X-ray diffraction in a diamond anvil cell, *Rev Sci Inst*, **78**, 063907.

109. Merkel S, Kubo A, Miyagi L, et al. (2006) Plastic deformation of $MgGeO_3$ post-perovskite at lower mantle pressures, *Science*, **311**, 644–646.

110. Merkel S, McNamara AK, Kubo A, et al. (2007) Deformation of (Mg,Fe) SiO_3 post-perovskite and D anisotropy, *Science*, **316**, 1729–1732.

111. Miyagi L, Kanitpanyacharoen W, Kaercher P, Lee KKM, Wenk H-R (2010) Slip systems in $MgSiO_3$ post-perovskite: implications for D" anisotropy, *Science*, **329**, 1639–1641.

112. Miyagi L, Kunz M, Knight J, Nasiatka J, Voltolini M, Wenk H-R (2008) In situ phase transformation and deformation of iron at high pressure and temperature, *J Appl Phys*, **104**, 103510.

113. Ito E, Yamazaki D, Yoshino T, et al. (2010) Pressure generation and investigation of the post-perovskite transformation in $MgGeO_3$ by squeezing the Kawai-cell equipped with sintered diamond anvils, *Earth Planet Sci Lett*, **293**, 84–89.

114. Morard G, Mezouar M, Rey N, et al. (2007) Optimization of Paris-Edinburgh press cell assemblies for in situ monochromatic X-ray diffraction and X-ray absorption, *High Press Res*, **27**, 223–233.

115. Liermann H-P, Jain A, Singh AK, Saxena SK (2010) Compression of silver in a diamond anvil cell: pressure dependences of strength and grain size from X-ray diffraction data, *J Phys Chem Soli*, **71**, 1088–1093.

116. McMahon MI (2005) Structures from powders and poor-quality single crystals at high-pressure, *J Synchrotron Rad*, **12**, 549–553.

117. Nisr C, Ribarik G, Ungar T, Vaughan GBM, Cordier P, Merkel S (2012) High resolution three-dimensional X-ray diffraction study of dislocations in grains of MgGeO3 post perovskite at 90 GPa, *J Geophys Res*, 117, B03201.

118. Goncharov AF, Beck P, Struzhkin VV, Hemley RJ, Crowhurst JC (2008) Laser-heating diamond anvil cell studies of simple molecular systems journal of at high pressures and temperatures, *Phys Chem Sol*, **69**, 2217–2222

119. Evans WJ, Yoo C-S, Lee G-W, Cynn H, Lipp MJ, Visbeck K (2007) Dynamic diamond anvil cell (dDAC): a novel device for studying the dynamic-pressure properties of materials, *Rev Sci Inst*, **78**, 073904.

120. Shinoda K, Noguchi N (2008) An induction heating diamond anvil cell for high pressure and temperature micro-Raman spectroscopic measurements, *Rev Sci Inst*, **79**, 015101.

121. Magunov AN (2009) Spectral pyrometry, *Pribory i Tekhnika Eksperimenta*, **4**, 5–28.

122. Meng Y, Shen G, Mao HK (2006) Double-sided laser heating system at HPCAT for in situ X-ray diffraction at high pressures and high temperatures, *J Phys Condens Matter*, **18**, S1097–S1103.

123. Prakapenka VB, Kubo A, Kuznetsov A, et al. (2008) Advanced flat top laser heating system for high pressure research at GSECARS: application to the melting behavior of germanium, *High Press Res*, **28**, 225–235.

124. Shen G, Heinz DL (1998) High pressure melting of deep mantle and core materials, in *Ultra High Pressure Mineralogy: Physics and Chemistry of the Earth's Deep Interior* (Hemley RJ, ed), Mineralogical Society of America, Washington DC, pp. 369–396.

125. Caldwell WA, Kunz M, Celestre RS, et al. (2007) Laser-heated diamond anvil cell at the advanced light source beamline 12.2.2, *Nucl Inst & Meth Phys Res Sec A*, **582**, 221–225.

126. Schultz E, Mezouar M, Crichton W, et al. (2005) Double-sided laser heating system for in situ high pressure-high temperature monochromatic X-ray diffraction at the ESRF, *High Press Res*, **25**, 71–83.

127. Boehler R, Musshoff HG, Ditz R, Aquilanti G, Trapananti A (2009) Portable laser-heating stand for synchrotron applications, *Rev Sci Inst*, **80**, 045103.

128. Watanuki T, Shimomura O, Yagi T, Kondo T, Isshiki M (2001) Construction of laser-heated diamond anvil cell system for in situ X-ray diffraction study at SPring-8, *Rev Sci Inst*, **72**, 1289–1292.

129. Prakapenka VB, Shen G, Dubrovinsky L (2003) Carbon transport in diamond anvil cells, *High Temp-High Press*, **35/36**, 237–249.

130. Dubrovinsky L, Glazyrin K, McCammon C, et al. (2009) Portable laser-heating system for diamond anvil cells, *J Synchrotron Rad*, **16**, 737–741.

131. Shen G, Wang L, Ferry R, Mao HK, Hemley RJ (2010) A portable laser heating microscope for high pressure research, *J Phys: Conf Series*, **215**, 012191.

132. Dubrovinskaia N, Dubrovinsky L (2003) Melting curve of water studied in externally heated diamond-anvil cell, *High Press Res*, **23**, 307–311.

133. Bassett WA, Shen AH, Bucknum M, Chou I-M (1993) A new diamond anvil cell for hydrothermal studies to 2.5 GPa and from -190 to 1200°C, *Rev Sci Instrum*, **64**, 2340–2345.

134. Schmidt C, Rickers K (2003) In situ determination of mineral solubilities in fluids using a hydrothermal diamond-anvil cell and SR-XRF: solubility of AgCl in water, *Am Min*, **88**, 288–292.

135. Petitgirard S, Daniel I, Dabin Y, Cardon H, Tucoulou R, Susini J (2009) A diamond anvil cell for X-ray fluorescence measurements of trace elements in fluids at high pressure and high temperature, *Rev Sci Inst*, **80**, 033906.

136. Bassett WA (2003) High pressure-temperature aqueous systems in the hydrothermal diamond anvil cell (HDAC), *Eur J Mineral*, **15**, 773–780.

137. Sinogeikin SV, Bass JD (2000) Single-crystal elasticity of pyrope and MgO to 20 GPa by Brillouin scattering in the diamond anvil cell, *Phys Earth Planet Inter*, **120**, 43–62.

138. Iota V, Yoo C-S, Klepeis J-H, Jenei Z, Evans W, Cynn H (2007) Six-fold coordinated carbon dioxide VI, *Nat Mat*, **6**, 34–38.

139. Shen G, Liermann H-P, Sinogeikin S, et al. (2007) Distinct thermal behavior of GeO_2 glass in tetrahedral, intermediate, and octahedral forms, *Proc Natl Acad Sci USA (PNAS)*, **104**, 14576–14579.

140. Rekhi S, Dubrovinsky LS, Saxena SK (1999) Temperature-induced ruby fluorescence shifts up to a pressure of 15 GPa in an externally heated diamond anvil cell, *High Temp-High Press*, **31**, 299–305.

141. Dubrovinsky LS, Saxena SK, Tutti F, Rekhi S, Le Bihan T (1999) In situ X-ray study of the thermal expansion of iron at multimegabar pressure, *High Temp-High Press*, **31**, 553–559.

142. Dubrovinsky LS, Saxena SK, Dubrovinskaia NA, Rekhi S, Le Bihan T (2000) Grüneisen parameter of epsilon-iron up to 300 GPa from in situ X-ray study, *Am Mineral*, **85**, 386–389.

143. Boehler R (1986) The phase-diagram of iron to 430-kbar, *Geoph Res Lett*, **13**, 1153–1156.

144. Dubrovinsky LS, Saxena SK, Lazor P (1997) X-ray study of iron with in situ heating at ultra-high pressures, *Geophys Res Lett*, **24**, 1835–1838.

145. Zha CS, Bassett WA (2003) Internal resistive heating in diamond anvil cell for in situ X-ray diffraction and Raman scattering, *Rev Sci Inst*, **74**, 1255–1262.

146. Zha CS, Mibe K, Bassett WA, Tschauner O, Mao HK, Hemley RJ (2008) The P-V-T equation of state of platinum to 80 GPa and 1900 K using

new internal resistive heating diamond-anvil cell techniques, *J Appl Phys*, **103**, 054908.

147. Komabayashi T, Fei Y, Meng Y, Prakapenka V (2009) In situ X-ray diffraction measurements of the γ-ε transition boundary of iron in an internally-heated diamond anvil cell, *Earth Planet Sci Lett*, **282**, 252–257.

148. Weir ST, Jackson DD, Falabella S, Samudrala G, Vohra YK (2009) An electrical microheater technique for high-pressure and high-temperature diamond anvil cell experiments, *Rev Sci Inst*, **80**, 013905.

149. Zha CS, Krasnicki S, Meng Y-F, et al. (2009) Composite chemical vapor deposition diamond anvils for high-pressure/high-temperature experiments, *High Press Res*, **29**, 317–324.

150. Sumiya H, Irifune T (2005) Synthesis of high-purity nano-polycrystalline diamond and its characterization, *Sei Tech Rev*, **59**, 52–59.

151. Nakamoto Y, Sumiya H, Matsuoka T, Shimizu K, Irifune T, Ohishi Y (2007) Generation of multi-megabar pressure using nano-polycrystalline diamond anvils, *Jpn J Appl Phys 2*, **46**, L640–L641.

152. Daniels WB, Ryschkevich MG (1983) Simple double diaphragm press for diamond anvil cells at low-temperatures, *Rev Sci Inst*, **54**, 115–116.

153. LeToullec R, Loubeyre P, Pinceaux JP, Mao H-K, Hu J (1992) A system for doing low temperature-high pressure single crystal X-ray diffraction with a synchrotron source, *High Press Res*, **6**, 379–388.

154. Hazen RM, Finger LW (1980) High-temperature diamond-anvil pressure cell for single-crystal studies, *Rev Sci Instrum*, **52**, 75–79.

155. Ogata K, Takeuchi Y, Kudoh Y (1987) Structure of a-quartz as a function of temperature and pressure, *Z Kristallogr*, **179**, 403–413.

156. Hackwell TP, Angel RJ (1995) Reversed brackets for the $P1 \leftrightarrow I1$ transition in anorthite at high pressures and temperatures, *Phys Rev Lett*, **93**, 115.

157. Dubrovinsky L, Boffa-Ballaran T, Glazyrin K, et al. (2010) Single-crystal X-ray diffraction at megabar pressures and temperatures of thousands of degrees, *High Press Res*, **30**, 620–633.

158. Goncharov AF, Zaug JM, Crowhurst JC, Gregoryanz E (2005) Optical calibration of pressure sensors for high pressures and temperatures, *J Appl Phys*, **97**, 094917.

159. Datchi F, Dewaele A, Loubeyre P, Letoullec R, Le Godec Y, Canny B (2007) Optical pressure sensors for high-pressure-high-temperature studies in a diamond anvil cell, *High Press Res*, **27**, 447–463.

160. Zha CS, Mao H-K, Hemley RJ (2000) Elasticity of MgO and a primary pressure scale to 55 GPa, *Proc Natl Acad Sci USA (PNAS)*, **97**, 13494–12499.

161. Sinogeikin S, Bass J, Prakapenka V, et al. (2006) Brillouin spectrometer interfaced with synchrotron radiation for simultaneous X-ray density and acoustic velocity measurements, *Rev Sci Inst*, **77**, 103905.

162. Datchi F, LeToullec R, Loubeyre P (1997) Improved calibration of the $SrB_4O_7:Sm^{2+}$ optical pressure gauge: advantages at very high pressures and high temperatures, *J Appl Phys*, **81,** 3333–3339.

163. Zhao Y, Barvosa-Carter W, Theiss SD, Mitha S, Aziz MJ, Schiferl D (1998) Pressure measurement at high temperature using ten Sm:YAG fluorescence peaks, *J Appl Phys*, **84**, 4049–4059.

164. Sanchez-Valle C, Daniel I, Reynard B, Abraham R, Goutaudier C (2002) Optimization of Sm^{3+} fluorescence in Sm-doped yttrium aluminum garnet: application to pressure calibration in diamond-anvil cell at high temperature, *J Appl Phys*, **92**, 4349–4353.

165. Hanfland M, Syassen K, Fahy S, Louie SG, Cohen ML (1985) Pressure-dependence of the 1st-order Raman mode in diamond, *Phys Rev B*, **31**, 6896–6899.

166. Aleksandrov IV, Goncharov AF, Zisman AN, Stishov SM (1987) Diamond at high-pressurs - Raman-scattering, equation of state and high-pressure scale, *Zh Eksp Teor Fiz+*, **93**, 680–691.

167. Datchi F, Canny B (2004) Raman spectrum of cubic boron nitride at high pressure and temperature, *Phys Rev B*, **69**, 144106.

168. Datchi F, Dewaele A, Le Godec Y, Loubeyre P (2007) Equation of state of cubic boron nitride at high pressures and temperatures, *Phys Rev*, **75**, 214104.

169. Goncharov AF, Crowhurst JC, Dewhurst JK, et al. (2007) Thermal equation of state of cubic boron nitride: implications for a high-temperature pressure scale, *Phys Rev B*, **75**, 224114.

170. Boehler R, Ramakrishnan J (1980) Experimental results on the pressure-dependence of Grüneisen-parameter – a review, *J Geophys Res*, **85**, 6996–7002.

171. Speziale S, Zha C-Z, Duffy TS, Hemley RJ, Mao H-K (2001) Quasi-hydrostatic compression of magnesium oxide to 52 GPa: implication for the pressure-volume-temperature equation of state, *J Geophys Res*, **106**, 512–528.

172. Tange Y, Nishihara Y, Tsuchiya T (2009) Unified analyses for P-V-T equation of state of MgO: a solution for pressure-scale problems in high P-T experiments, *J Geophys Res*, **114**, B03208.

173. Ueda Y, Matsui M, Yokoyama A, Tange Y, Funakoshi K-I (2008) Temperature-pressure-volume equation of state of the B2 phase of sodium chloride, *J Appl Phys*, **103**, 113513.

174. Hirose K, Sata N, Komabayashi T, Ohishi Y (2008) Simultaneous volume measurements of Au and MgO to 140 GPa and thermal equation of state of Au based on the MgO pressure scale, *Phys Earth Plant Inter*, **167**, 149–154.

175. Hirose K, Sinmyo R, Sata N, Ohishi Y (2006) Determination of post-perovskite phase transition boundary in $MgSiO_3$ using Au and MgO pressure standards, *Geophy Res Lett*, **33**, L01310.

176. Matsui M, Ito E, Katsura T, et al. (2009) The temperature-pressure-volume equation of state of platinum, *J Appl Phys*, **105**, 013505.

177. Fisher HE, Barnes AC, Salmon PS (2006) Neutron and X-ray diffraction studies of liquids and glasses, *Rep Prog Phys*, **69**, 233–299.

178. Egami T, Billinge SJL (2003) *Underneath the Bragg Peaks: Structural Analysis of Complex Materials*. Pergamon Materials Series, vol. 7 (Cahn RW, ed), Elsevier, Kidlington, p. 316.

179. Duffy TS (2008) Some recent advances in understanding the mineralogy of Earth's deep mantle, *Phil Trans R Soc A*, **366**, 4273–4293.

180. Labrosse S, Hernlund JW, Coltice N (2007) A crystallizing dense magma ocean at the base of the Earth's mantle, *Nature*, **450**, 866–869.

181. Karki BB, Stixrude L (2010) First-principles study of enhancement of transport properties of silica melt by water, *Phys Rev Lett*, **104**, 215901.

182. Wilding CA, Wilson M, McMillan PF (2006) Structural studies and polymorphism in amorphous solids and liquids at high pressure, *Chem Soc Rev*, **35**, 964–986.

183. Sanloup C, Guyot F, Gillet P, Fiquet G, Mezouar M, Martinez I (2000) Density measurements of liquid Fe-S alloys at high-pressure, *Geophys Res Lett*, **27**, 811–814.

184. Shen G, Sata N, Rivers ML, Sutton SR (2002) Molar volumes of molten indium at high pressures measured in a diamond anvil cell, *Appl Phys Lett*, **81**, 1411–1413.

185. Hong X, Shen G, Prakapenka VB, Rivers ML, Sutton SR (2007) Density measurements of noncrystalline materials at high pressure with diamond anvil cell, *Rev Sci Inst*, **78**, 103905.

186. Mei Q, Sinogeikin S, Shen G, Amin S, Benmore CJ, Ding K (2010) High-pressure X-ray diffraction measurements on vitreous GeO_2 under hydrostatic conditions, *Phys Rev B*, **81**, 174113.

187. Kanzaki M, Kurita K, Fujii T, Kato T, Shimomura O, Akimoto S (1987) New technique to measure viscosity and density of silicate melts at high pressure, in *High-Pressure Research in Mineral Physics* (Manghnani MH, Syono Y, ed), American Geophysical Union Geophysical Monograph, Washington, DC, pp. 195–200.

188. Terasaki H, Kato T, Urakawa S, et al. (2002) Viscosity change and structural transition of Molten Fe at 5 GPa, *Geophy Res Lett*, **29**, 1227.

189. Meade C, Hemley RJ, Mao H-K (1992) High-pressure X-ray diffraction of SiO_2 glass, *Phys Rev Lett*, **69**, 1387–1390.

190. Katayama Y, Mizutani T, Utsumi W, Shimomura O, Yamakata M, Funakoshi K (2000) A first-order liquid-liquid phase transition in phosphorus, *Nature*, **403**, 170–173.

191. Shen G, Prakapenka VB, Rivers ML, Sutton SR (2003) Structural investigation of amorphous materials at high pressures using the diamond anvil cell, *Rev Sci Inst*, **74**, 3021–3025.

192. Shen G, Prakapenka VB, Rivers ML, Sutton SR (2004) Structure of liquid iron at pressures up to 58 GPa, *Phys Rev Lett*, **92**, 185701.

193. Parise JB, Ehm L, Michel FM (2010) Analysis of the total scattering using the quantitive high pressure pair distribution function: practical considerations, in *High-Pressure Crystallography: From Fundamental Phenomena to Technological Application* (Boldyreva E, Dera P, ed), Springer Science + Business Media B. V., New York, pp. 513–522.

194. Ehm L, Michel FM, Parise JB (2010) Analysis of the total scattering using the quantitative high pressure pair distribution function: case study, in *High-Pressure Crystallography: From Fundamental Phenomena to Technological Application* (Boldyreva E, Dera P, ed), Springer Science + Business Media B. V., New York, pp. 523–531.

195. Variano BF, Schlotter NE, Hwang DM, Sandroff CJ (1988) Investigation of finite size effects in a 1st order phase-transition – high-pressure Raman study of CDS microrystallites, *J Chem Phys*, **88**, 2848–2850.

196. Singh AK (2004) X-ray diffraction from solids under nonhydrostatic compression—some recent studies, *J Phys Chem Solids*, **65**, 1589–1596.

197. Chapman KW, Chupas PJ, Kurtz CA, Locke DR, Parise JB, Hriljac JA (2007) Hydrostatic low-range pressure applications of the Paris-Edinburgh cell utilizing polymer gaskets for diffuse X-ray scattering measurements, *J Appl Cryst*, **40**, 196–198.

198. Xu J-A, Mao H-K, Hemley RJ, Hines E (2004) Large volume high-pressure cell with supported moissanite anvils, *Rev Sci Inst*, **75**, 1034–1038.

199. Guthrie M, Tulk CA, Benmore CJ, et al. (2004) Formation and structure of a dense octahedral glass, *Phys Rev Lett*, **93**, 115.

200. Martin CD, Antao SM, Chupas PJ, Lee PL, Shastri SD, Parise JB (2005) Quantitative high-pressure pair distribution function analysis of nanocrystaline gold, *App Phys Lett*, **86**, 061910/1–3.

201. Chupas PJ, Qui XY, Hanson JC, Lee PL, Grey CP, Billinge SJL (2003) Rapid-acquisition pair distribution function (RA-PDF) analysis, *J App Cryst*, **36**, 1342–1347.

202. Chupas PJ, Chapman KW, Lee PL (2007) Applications of an amorphous silicon-based area detector for high-resolution, high-sensitivity and fast time-resolved pair distribution function measurements, *J Appl Cryst*, **40**, 463–470.

203. Goncharov AF, Prakapenka VB, Struzhkin VV, Kantor I, Rivers ML, Dalton DA (2010) X-ray diffraction in the pulsed laser heated diamond anvil cell, *Rev Sci Inst*, **81**, 113902.

204. Turneaure SJ, Gupta YM (2009) Real time synchrotron X-ray diffraction measurements to determine material strength of shocked single crystals following compression and release, *J Appl Phys*, **106**, 033513.

205. Lee RW, Baldis HA, Cauble RC, et al. (2002) Plasma-based studies with intense X-ray and particle beam sources, *Laser Part Beams*, **20**, 527–536.

206. Yoo C-S, Wei H, Chen J-Y, Shen G, Chow P, Xiao Y (2011) Time- and angle-resolved X-ray diffraction to probe structural and chemical evolution during Al-Ni intermetallic reactions, *Rev Sci Instrum*, **82**, 113901–113906.

Chapter 11

Synchrotron Tomography

Astrid Haibel
Helmholtz-Zentrum Geesthacht, Max Planck-Str 1, 21502 Geesthacht, Germany
Beuth Hochschule für Technik Berlin, Luxemburger Str. 10, 13353 Berlin, Germany
ahaibel@beuth-hochschule.de

The developmentof new third-generation synchrotron sources combined with the enormous advancement of detector technology, data storage, and processing capabilities has led to a strong rise of high-end synchrotron tomography facilities [1–6]. Tomography is well known from medicine and biology, but it is also used in materials science, geophysics, archeology, and other sciences. The wide field of applications in biology and medicine ranges from the visualization and quantitative analysis of tissues, teeth, structures of bones, implants, plants, small insects down to single cells in the submicrometer and even in the nanometer range of spatial resolution. In materials science and neighboring disciplines, the application field encompasses the investigation of defects, porous structures, granular structures, cracks, voids, fibers, precipitations or inclusions, small components, and inner structures of stones and soil. In this chapter, different tomographic measurement techniques,

X-Ray Diffraction: Modern Experimental Techniques
Edited by Oliver H. Seeck and Bridget M. Murphy

ISBN 978-981-4303-59-0 (Hardcover), 978-981-4303-60-6 (eBook)
www.panstanford.com

the data reconstruction as well as an example for the quantitative analysis of tomographic data sets will be described.

11.1 Measurement Principle of Synchrotron Tomography

Synchrotron tomography is based on the detection of radiographic images collected by illuminating a measurement object from many different angular directions. Thereby, X-rays pass through the object mounted on a rotation stage (see Fig. 11.1). The attenuated X-rays are recorded by a detector system mostly composed of a microscope optic connected on a CCD camera. In front of the microscope optics, a scintillator screen converts the X-ray light into visible light. The digital radiographic images are read out into a computer where they are collected and processed.

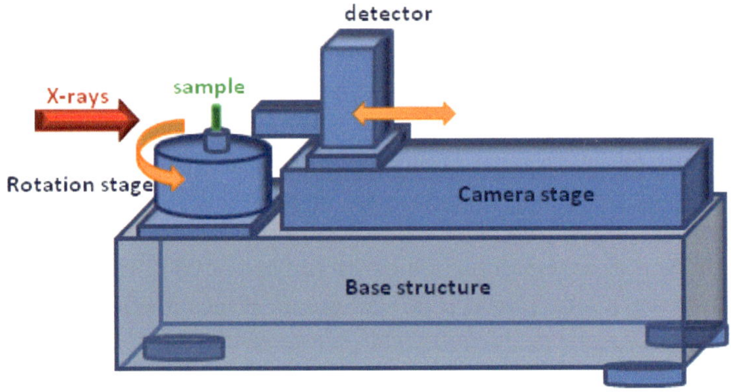

Figure 11.1 Sketch of the measurement principle of synchrotron tomography.

The decision for the choice of the scintillator material depends strongly on the efficiency of the conversion of X-ray light into visible light, that is, on the high X-ray stopping power and high light yield, on the short decay time or the low afterglow of the scintillator material and on the desired spatial resolution for the images. Ceramic powder scintillator materials (e.g., Gd_2O_2S, the dominant powder scintillator in medical imaging for decades) show a high light yield and a low afterglow, but for spatial resolutions <5 μm, single-crystal scintillators (e.g., CWO, BGO, YAG(Ce), LSO, GGG) are required [7]. Finally, which crystal will be used depends on the X-ray energy and on the quantum efficiency against the visible wavelength of the CCD sensor behind the scintillator.

Synchrotron tomography instruments are mostly composed of three parts, the detector stage, the sample stage and a base structure. Beside a high-quality detector system, the precondition for successful synchrotron tomography measurements is the high precision and stability of all mechanical components, and a high X-ray beam stability.

To achieve spatial resolutions near the physical limit, it is necessary to assure high mechanical stability for the components of the setup. Therefore, the detector stage as well as the sample rotation and translation axes are mostly high precision air-bearing stages, often mounted on granite substructures, adjustable in pitch, yaw, roll, height, and horizontal to the beam. In particular, for the achievement of spatial resolutions down to the nanometer range a high stability of the laboratory temperature is also very important.

11.1.1 Monochromatization

The first step before measuring is the preparation of an adequate X-ray beam. Basically, the X-rays provided by a synchrotron can be used for tomography without any further treatment. The polychromatic beam (also called white beam) provided by an insertion device (mostly an undulator or wiggler, see Chapter 2) is very intense, and is therefore used for fast tomography experiments [8]. But, the use is limited to a few cases due to imaging artifacts that can occur (such as beam hardening artifacts, known from X-ray tubes). Therefore, usually monochromatic X-rays are preferred.

Three types of monochromators are typically used, single-crystal monochromators, multilayer monochromators, and Laue monochromators. All types operate according to Bragg's law:

$$n\lambda = 2D \sin\theta.$$

The incoming beam hits the crystal or multilayer surface (with period length D) under a selected angle θ and only a certain wavelength band with center λ that interferes constructively in reflection (or in transmission for the Laue geometry) will be used for the measurement. Although, single crystals select a very small wavelength or energy band in the order of $\Delta E/E = 10^{-4}$ due to the mosaicity of the crystal, multilayer structures accept wavelength or energy bands in the order of $\Delta E/E = 10^{-2}$.

The reflected (or transmitted) wavelength is tunable by changing the inclination of the monochromator crystals relative to the incoming beam. Mostly, two crystals or multilayer structures are combined as double-crystal monochromators (DCMs) or double multilayer monochromators (DMMs). This concept preserves the beam direction with a parallel offset of the X-ray beam.

Which type of monochromator is used depends on the required energy range and band width. Due to the high X-ray absorption in the crystals for energies < E ~30–40 keV in the lower energy range, the Bragg geometry in reflection is used; for higher energies, the Laue geometry is preferred. By bending the Laue crystals, monochromatic X-rays with a bandwidth of $\Delta E/E = 10^{-4}$ is delivered.

The trade-off for the lower energy resolution of multilayer structures is by a factor of 100 higher flux, which is often a clear benefit and for the most applications an energy resolution of $\Delta E/E = 10^{-2}$ is completely sufficient. Therefore, to avoid an excessive loss of flux, in the most tomographic applications, multilayer monochromators are used.

11.2 Absorption Tomography

For absorption tomography, a monochromatic X-ray beam traverses an object of thickness d characterized by the spatial distribution of the linear X-ray attenuation coefficients $\mu(x, y)$ as shown in Fig. 11.2.

The transmitted intensity $I(x, y)$ is given by Lambert–Beer law as follows:

$$I(x,y) = I_0 \exp\left(-\int_0^d \mu(x,y)dy\right),$$

where I_0 is the original intensity of the beam and d the thickness of the object. The values of the material-dependent absorption coefficients $\mu(x, y)$ are dependent on the absorption and scattering of the incident radiation caused by several different mechanisms such as photoelectric absorption, Compton scattering (incoherent scattering), and Rayleigh scattering (coherent scattering). Using the mass density ρ, atomic number Z, and photon energy E, the absorption coefficient μ (beside in the vicinity of absorption edges) follows the relation

$$\mu \propto \frac{Z^4}{E^3}\rho .$$

The object is scanned under equidistant angles and a stack of 2D radiographic images, the so-called projections P_θ, are measured. Because synchrotron radiation is quasiparallel, a rotation from 0 –180° is sufficient for a complete data set. From these projections, 3D images can be reconstructed.

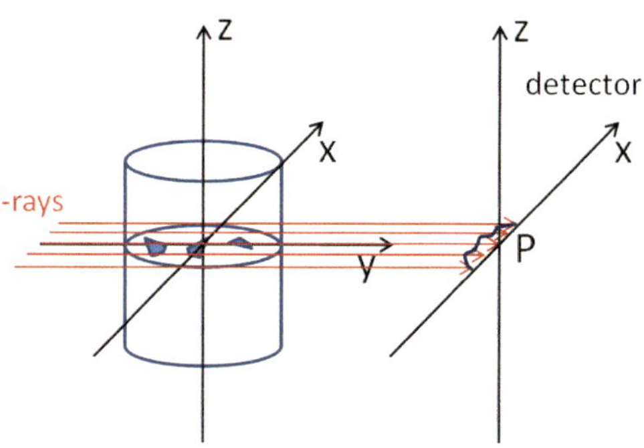

Figure 11.2 Sketch of a CT slice scan.

Due to the partial coherence of synchrotron radiation, attenuation-contrast projections have to be taken at a minimal sample-detector distance. Otherwise, interference effects at inner edges in the object lead to a contribution of edge enhancement in the signal (see Fig. 11.3 [9]). The X-ray beam energy has to be matched to the sample characteristics in order to obtain a sufficient absorption contrast in the radiographic images. For samples consisting mainly of the light elements or for samples consisting of elements with almost the same absorption coefficients, the more sensitive phase-contrast signal has to be used [10].

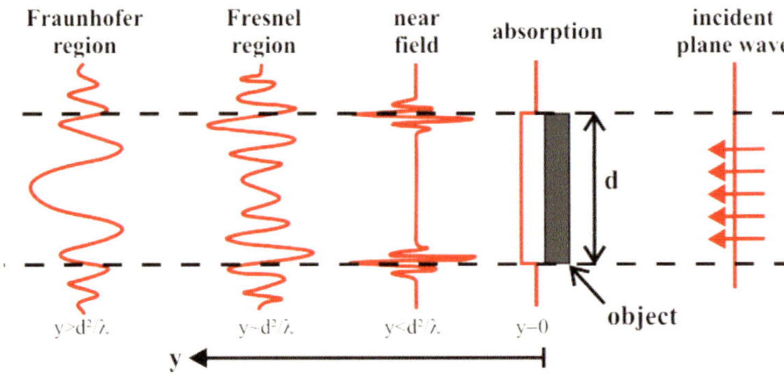

Figure 11.3 The partially coherent X-ray beam forms coherent images depending on the object-detector distance y, sample size d, and X-ray wavelength λ [9].

11.3 Phase-Contrast Tomography

Phase-contrast tomography methods are used for samples with less or no absorption contrast or for samples composed of matter with almost the same absorption contrast. The precondition for phase contrast is (partially) coherent X-ray radiation.

Modern insertion devices (such as undulators) in the synchrotron storage rings produce partially coherent X-rays. Because for phase-contrast imaging moderate coherence requirements are sufficient, this measurement method can be carried out at third-generation synchrotrons without additional X-ray optics for producing a coherent beam. Therefore, several phase-contrast X-ray imaging methods have become standard methods for tomography at third-generation synchrotrons.

Coherent X-rays, which interact with a sample, are attenuated and also shifted in phase. These can vary independently of each other. This behavior can be described by the complex index of refraction

$$n = 1 - \delta + i\beta$$

where the real part δ corresponds to the phase shift due to refraction and the imaginary part β belongs to absorption. Then, the key for phase-contrast imaging is to convert phase shifts into variations in the intensity.

Phase-contrast tomography measurement methods can be divided in qualitative phase-contrast imaging (i.e., phase-enhanced contrast, also called phase outline contrast) and quantitative phase-contrast methods (i.e., phase retrieval). In phase-enhanced imaging, the intensity is emphasized due to diffraction effects that occur at sharp changes in refractive index of the sample, that is, at inner interfaces of the sample. In this case, the phase contrast operated as an enhancement of the absorption contrast [11–13]. Such images can be used in a qualitative way to reveal shape information about inner structures in the sample. In contrast, in phase-retrieval-based tomography, the raw phase-contrast data are used as the basis for quantitative calculation of the complex refractive index of the sample [14,15]. The conversion of phase information into intensity contrast falls into two major classes:

1. Direct phase-contrast methods (interferometry).
2. Indirect phase-contrast methods (holotomography, differential phase-contrast tomography, diffraction-enhanced imaging).

11.3.1 Direct Phase-Contrast Methods

Interferometry has a long tradition for measuring phase shifts [16–18]. For the calculation of the phase shift the waves which are passing through a sample are superimposed with known reference waves. The phase difference between both leads to interference patterns. From these patterns, the phase shift generated by the sample can be recovered. The principle setup is shown in Fig. 11.4. Using the electron density σ, the phase shift ϕ is given by the following expression:

$$\phi \propto \sigma / E.$$

An X-ray interferometer allows only for measuring phase projections modulo 2π. To resolve the true phase shift, an analysis method known as phase unwrapping is required in which the ambiguities in the phase shift are resolved by using additional information, or additional assumptions about the measured data. A review about various procedures that were developed for interferometric phase-contract detection is given in Ref. [19].

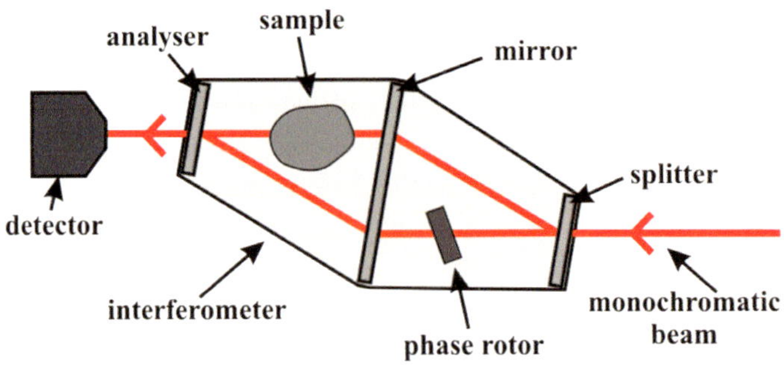

Figure 11.4 Sketch of an experimental setup for interferometric phase contrast. The interferometer is used to divide the monochromatic beam into two coherent beams. The phase rotor is used to create a defined phase shift. Therefore, the direct phase projection can be detected [9].

11.3.2 Indirect Phase-Contrast Methods

(a) Holo tomography

An indirect imaging method to measure phase projection is to record images for different sample-detector distances (see Fig. 11.5). Depending on the selected distance, different spatial frequencies in the projected phase images will vanish based on the Talbot effect. A holographic reconstruction algorithm then determines the phase projection [20]. For (partially) coherent X-ray radiation, this propagation method yields phase contrast without any additional optical elements.

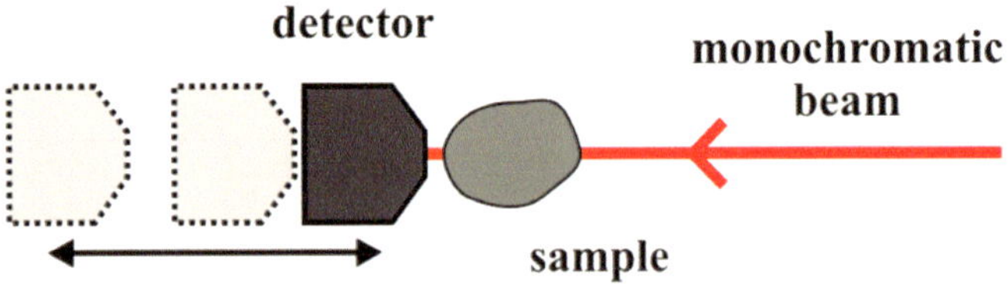

Figure 11.5 Schematic setup used for phase contrast imaging. By varying the sample-detector distance, different spatial frequencies can be detected [9].

(b) Grating interferometer phase-contrast imaging (DPCI)

The grating-based imaging is an indirect phase-contrast method using a beam-splitter grating and an analyzer grating to detect the

X-ray refraction in the sample (see Fig. 11.6). The refraction angle is related to the first derivative of the X-ray phase shift produced by the sample [21,22]. Remarkable about this method is, that in addition to the phase-shift gradient simultaneously the conventional attenuation contrast and the information about the small-angle scattering signal (called "dark-field signal" similar to the dark-field contrast in the visible light microscopy) can be detected quantitatively.

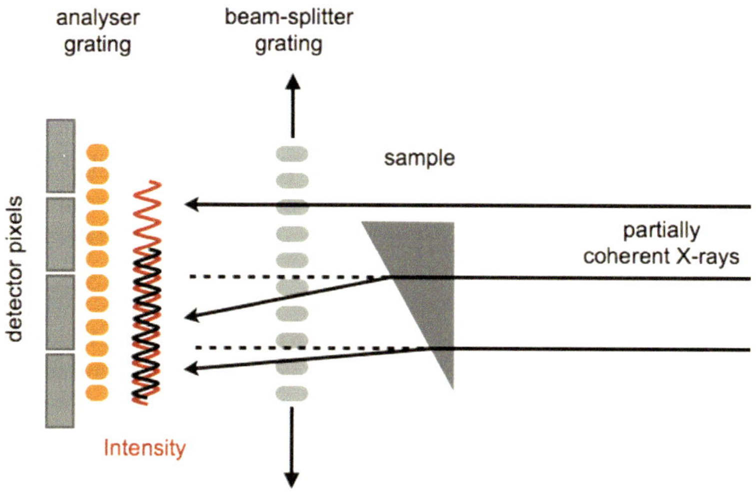

Figure 11.6 The beam-splitter grating produces an interference pattern in a certain distance due to the Talbot effect. This intensity pattern is shifted perpendicularly to the beam direction when a sample is refracting the incoming X-ray beam. The absorbing analyzer grating helps resolving the fine pattern with a conventional X-ray detector using a phase-stepping scan [23].

In the case of sufficiently coherent illumination, the beam-splitter grating splits the incoming X-ray beam into spatially not separated, overlapping diffraction orders. This overlap leads to interference and causes a periodic intensity pattern in certain distances downstream the beam-splitter grating according to the Talbot effect. The discrete Talbot distances can be calculated as follows:

$$D_n = n\frac{p_{eff}^2}{2\lambda} \quad \text{with} \quad p_{eff} = \frac{p}{\eta}.$$

where λ is the X-ray wavelength, D_{eff} is the effective aperture and *NA* is the numerical aperture of the X-ray optics. To illuminate the object with incoherent radiation, a diffuser has to be positioned into the beam in front of the sample.

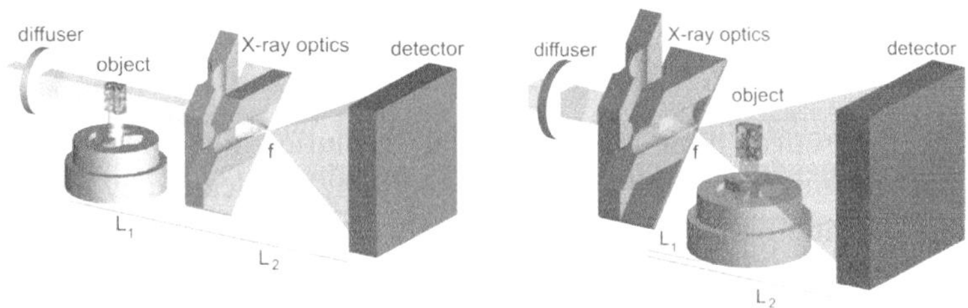

Figure 11.8 Schematic sketch of two concepts for nanotomography with magnifying optics.

In the right setup of Fig. 11.8, the magnification m of the transmitted images depends on the source-to-object distance L_1 and on the source to detector distance L_2, that is, $m = L_2/L_1$. Beside the magnification m and the detector pixel size, the spatial resolution for this setup depends strongly on the focus size of the small secondary source. Basic prerequisite for tomography with nanometer resolution is the mechanical stability of the experimental setup, especially of the sample rotation stage, which has to have an eccentricity and wobble error smaller than the requested spatial resolution.

11.5 Tomographic Reconstruction

For tomography, a detector records 2D projections of an object at many different angle positions θ. From this data, set a 3D image of the object structure can be calculated. How does it work? As shown in Fig. 11.9 (left), the projections are a set of the integrated values of absorption coefficient $\mu(x, y)$ of the object [39–41].

According Lambert–Beer law, these set of projections can be mathematically described as a set of line integrals of the function $\mu(x, y)$ along the X-ray path L.

$$I = I_0 \cdot \exp\left(-\int_L \mu(x,y)ds \right). \tag{11.1}$$

In the case of X-rays, L is not arbitrary but always straight. It can be modified only by translation or rotation. Therefore, the total attenuation

$$p(\theta,t)=\ln\left(\frac{I}{I_0}\right)=-\int_{L(t,\theta)}\mu(x,y)ds \tag{11.2}$$

can be written as a function of the sample rotation θ and the position t. Mathematically, it is equivalent to rotate the X-ray beam. A straight path is then given by (see also Fig. 11.9, right)

$$t = x\cos\theta + y\sin\theta. \tag{11.3}$$

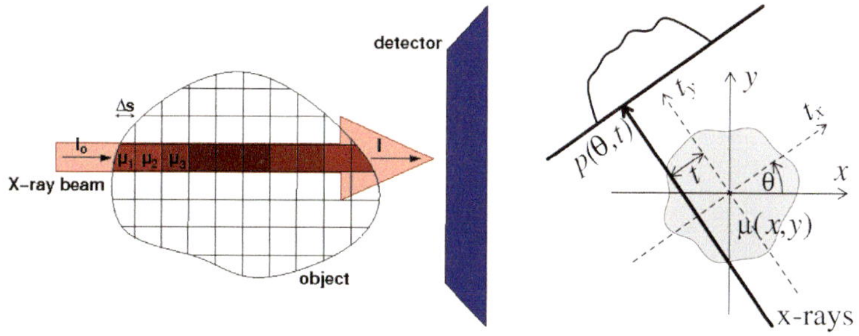

Figure 11.9 Left: Each 2D projection can be described as a set of line integrals through the object (parallel X-ray beam geometry as usual at synchrotron sources). Right: Parallel beam geometry. Each projection is made up of the set of line integrals through the object.

Placing this equation in Eq. (11.2), it can be rewritten as follows:

$$p(\theta,t)=-\int_{-\infty}^{\infty}\int_{-\infty}^{\infty}\mu(x,y)\delta(x\cos\theta+y\sin\theta-t)dxdy\,. \tag{11.4}$$

This function is called Radon transform (often also sinogram) of the function $\mu(x,y)$, that is, of the 2D object [42].

11.5.1 Fourier Slice Theorem

Equation (11.4) describes the Radon transform of the function $\mu(x,y)$. But, for the image reconstruction the inverse of the Radon transform is sought. This can be calculated by means of the so-called Fourier slice theorem. To obtain $\mu(x,y)$ from the above Eq. (11.4) means first finding the inverse Radon transform. The simplest example of the Fourier slice theorem is given for a projection at the angle $\mu = 0°$. Then, for $p(\theta, t)$ we find (see Fig. 11.9, right):

$$p(0°,t)=p(0°,x)=-\int_{-\infty}^{\infty}\mu(x,y)dy. \tag{11.5}$$

The 1D Fourier transform $p(0°, u)$ of $p(0°, x)$ is

$$P(0°,u)=\int_{-\infty}^{\infty}p(0°,x)\exp(-2\pi iux)dx. \tag{11.6}$$

This can be transformed into

$$\begin{aligned}P(0°,u)&=\int_{-\infty}^{\infty}\left(\int_{-\infty}^{\infty}\mu(x,y)dy\right)\exp(-2\pi iux)dx\\&=\int_{-\infty}^{\infty}\int_{-\infty}^{\infty}\mu(x,y)\exp(-2\pi i[ux+0\cdot y])dxdy.\end{aligned} \tag{11.7}$$

The right-hand side of this equation represents the one-dimensional Fourier transform of the projection $p(0°, t)$. Therefore, the following relationship between the vertical projection and the 2D transform of the object function can be found:

$$p(0°, u) \equiv F\{p(0°, x)\}. \tag{11.8}$$

This is the simplest form of the Fourier slice theorem. Of course, this result has to be independent of the orientation θ between the object and the coordinate systems. If the coordinate system (t_x, t_y) is rotated by θ, the Fourier transform of the projection is equal to the two-dimensional Fourier transform of the object along a line rotated by θ. Fig. 11.10 explains this method graphically. So the Fourier slice theorem can be defined as follows:

The Fourier transform of a parallel projection $p(\theta, t)$ of an image $\mu(x, y)$ taken at angle θ gives a slice of the two-dimensional transform, $F(u, v)$, subtending an angle θ with the u-axis. In other words, the Fourier transform of $p(\theta, t)$ gives the values of $F(u, v)$, (see Fig. 11.10) [40].

The 1D Fourier transform of the projections can by plotted as radial lines in $F(u, v)$. For projections at enough angles θ, the projections can be assembled into a widely complete estimate of the 2D Fourier transform of the measured object. This figure has to be inverted to achieve the estimate of the real object.

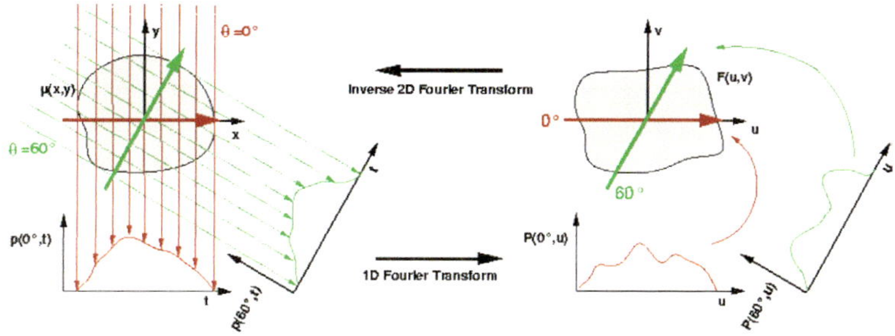

Figure 11.10 Fourier slice theorem for $\theta = 0°$ and for $\theta > 0°$. The 1D Fourier transformed projections $p(\theta, t)$ are plotted as radial lines in in the $F(u, v)$ diagram.

Nevertheless, usually in practice different approaches are used to reconstruct tomographic images. The algorithm that is currently being used at most in almost all applications of straight X-ray tomography is a stabilized and *discretized* version of the inverse Radon transform, known as filtered back projection algorithm. Several further algebraic reconstruction techniques are used. Here, the image slices are assumed as arrays of unknowns. Then algebraic equations are set up for the unknowns in terms of the measured projection data.

11.6 Image Artifacts

Although monochromatic synchrotron radiation helps minimize image artifacts, they cannot be neglected completely. In the following section an overview of some of the most occurring artifacts are presented.

Ring artifacts

Tomographic images are often superimposed by the so-called ring artifacts. They appear as concentric rings in the images around the center of rotation. Fig. 11.11 shows a cut through a tomographic image of a sea snail superimposed by such artifacts.

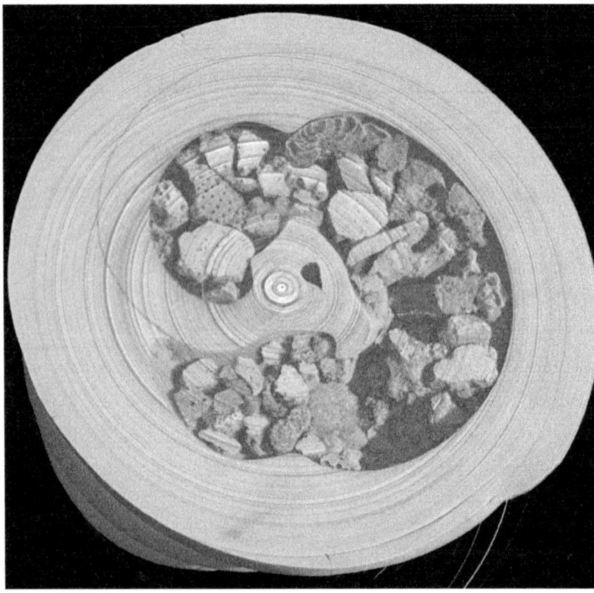

Figure 11.11 Tomographic image of a cut through a sea snail filled with sand and other particles. The image is clearly superimposed with ring artifacts.

The cause of these artifacts is usually imperfect detector pixels or impurities and scratches on the scintillator crystals. The common approach to reduce these artifacts is the flat-field correction. Thereby, the measured data are corrected by dividing the normalized radiographic images of the sample by background images (i.e., images without sample). But due to the intensity-dependent nonlinear response function of the detector and the scintillator and the time-dependent nonuniformities of the X-ray beam, this correction does not completely eliminate these artifacts.

A further, simple but effective method to avoid ring artifacts already during the data acquisition is to shift the sample or the detector for each radiographic image in defined horizontal and vertical steps [43,44]. Thereby, the characteristic of each detector element is averaged with neighboring elements, which leads to significantly reduced ring artifacts. Alternatively, it is possible to remove these artifacts by sinogram processing during the reconstruction or by postprocessing the images with means of size or shape filters (see Refs. [45–49]).

Image noise

The quality of CT images can be degraded by image noise appearing as graining in the images. This is caused by a low signal-to-noise ratio caused by an unfavorable conversion rate of the X-ray photons to visible light photons by the scintillator especially for higher X-ray energies [50,51]. To reduce this high throughput of visible light in the optical system, the lens aperture has to be reduced.

Nevertheless, if tomographic images are noisy, mean or median filters can be used to reduce the noise. Whereas mean filters reduce the spatial resolution of the images, median filters remove noise in smooth regions of the image, but preserve edges. Moreover, using conditional median filters preserve significantly more details in the images than the conventional median filter.

Motion artifacts

Motion artifacts are seen as blurring and/or streaking that are caused by movements of the sample being imaged or by the imprecise mechanic axes movement during the measurement. For example, usually the sample moves out of the beam in defined intervals during the measurement for taking flatfield images. Artifacts caused by the inexact movement of the translation axis can be detected in the sinogram as deviation from the ideal sinusoidal profile and can be corrected by a translational shifting of each sinogram line to the ideal sine curve. Motion artifacts caused by the sample tilting or shrinking can occur, for example, for native biological samples. There are several motion artifact compensation algorithms that help to reduce such artifacts but for larger movements the correction becomes difficult or even impossible (see Refs. [52,53]).

Beam hardening

Beam hardening means the phenomenon that polychromatic X-rays becomes more penetrating, that is, "harder," as it traverses through an object. The reason is that the attenuation process in matter is X-ray energy dependent [50]. Therefore, thicker, that is, more attenuating parts of the object attenuate softer X-rays of the original X-ray spectrum more effectively then thinner parts of the object. Because softer X-rays are more attenuated while harder X-ray are more penetrating, not only the intensity but also the spectral properties are changing during the beam propagation through the object. Due to this artifact, thicker sample regions in tomographic images appear denser than thinner sample regions although they differ only in thickness. The usual method to correct beam hardening effects is to measure reference objects and to find a correction function (linearization technique) [54–56].

Metal artifacts

Metal streak artifacts are an important problem in medicine CT. But also in synchrotron tomography this artifact occurs for small highly absorbing parts (metal) in a lower absorbing investigation object. These parts prevent or strongly reduce the transmission of the X-ray beam along the beam trajectories in which they are located so that information about other parts of the object along these trajectories are lost.

These nonlinear variations in the X-ray spectrum yield inconsistent radiographic images resulting in star-shaped artifacts around the small high-absorbing objects in the reconstructed 3D images. Metal artifacts are easily to detect as very bright sinusoidal lines in the corresponding sinogram. For correction of the metal-affected data, either a sinogram instauration or alternative reconstruction algorithms is used. Possible approaches are the attenuation of these bright lines by linear interpolation algorithms [57–59], by gradient information algorithms [60,61], or by only partially consideration of the measured values for reconstruction [62].

11.7 Applications and Quantitative 3D Image Analysis

Synchrotron tomography provides 3D images of a wide range of samples in many application fields including medicine, biology, geology, and engineering materials science with spatial resolutions in the micrometer or even submicrometer down to the nanometer range. Both, absorption and phase-contrast imaging methods were developed during the past decades. Further, the high flux of the modern synchrotron sources enables fast measurements, coupled with increasing computer power fast 3D in situ or even real-time investigations are now possible.

As example for an application of absorption synchrotron tomography, the quantitative image analysis of sintered copper powder will be presented [63,64]. Fig. 11.12 (left) shows a tomographic image of a small amount of this sintered copper powder.

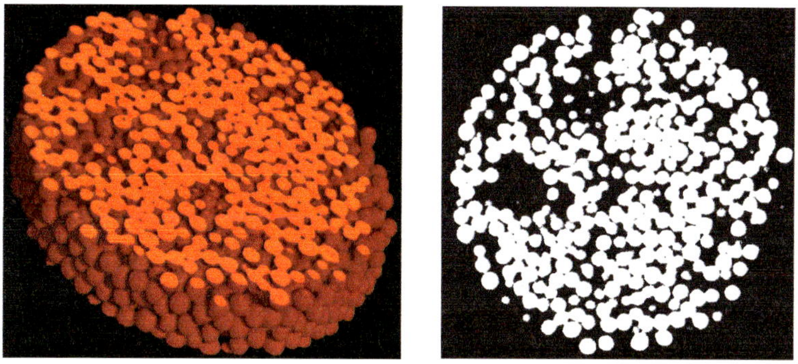

Figure 11.12 Left: Tomographic image of a small amount of sintered copper powder. Right: Boolean 2D slice of the 3D image.

The first step toward a quantitative analysis is to extract the parts of interest. This can be done, for example, by a separation according to the gray values of these parts and plotting them as a Boolean, that is, a black and white 3D image. As shown in Fig. 11.12 (right), all voxels (= 3D pixels) assigned to copper powder have the value one (white), all other voxels (background) have the value zero (black). As is evident from Fig. 11.12 (right), the powder particles are connected in the Boolean image. In order to isolate them numerically, a specific mathematical separation algorithm can be used. It is based on the Euclidean distance transformation followed by the so-called watershed transformation of Boolean image.

The Euclidean distance transformation is a morphological transformation that assign each white pixel in a Boolean image (foreground pixel) a new gray value as a function of the distance of this pixel from the nearest black pixel (background pixel). The watershed transformation interprets the resulting gradient image as a 3D topographic surface and floods this surface from its minima by "virtual water." During this flooding process at the positions, where adjacent catchment basins are flowing together due to rising water-level, a dividing line is drawn. So the image is portioned in catchment basins and watershed lines. Hence, connected areas can be separated (see Fig. 11.13).

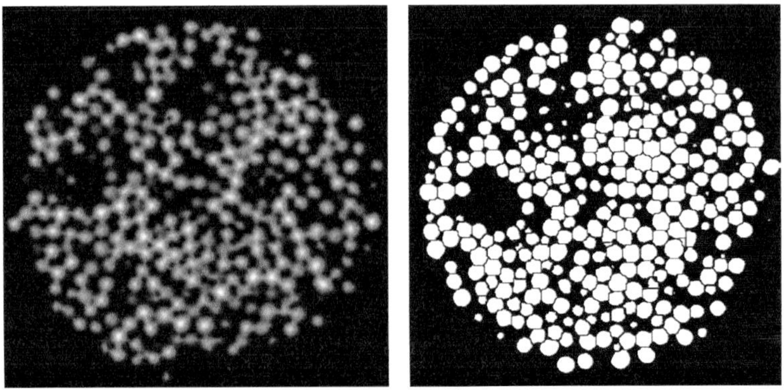

Figure 11.13 Left: Tomographic image of a small amount of sintered copper powder. Right: Boolean 2D slice of the 3D image.

After the separation, the now isolated copper particles can be labeled with numbers or by different colors. Now, it is possible to count all particles; determine volume and surface of each single particle; calculate the shape factor, the mean volume, and the size distribution for the particles; and to allocate the center of mass for each particle. Further, by subtraction the watershed transformed 3D image from the original 3D image, the sinter neck diameters of the sintered copper particles can be calculated (see Fig. 11.14).

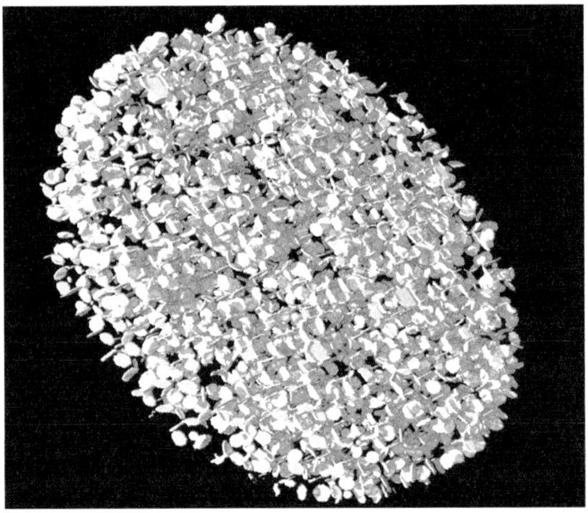

Figure 11.14 Sinter necks of the sintered copper powder visualized by subtraction of the watershed transformed 3D image from the original 3D image.

References

1. Moncton DE, Crosbie E, Shenoy GK (1989) Overview of the Advanced Photon *Source, Rev Sci Instrum*, **60**, 1403–1405.
2. Haensel R (1992) European Synchrotron Radiation Facility (ESRF), *Rev Sci Instrum*, **63**, 1571–1574.
3. Kamitsubo H (1992) The Jaeri-Riken SPring-8 Project, *Rev Sci Instrum*, **63**, 1586–1587.
4. Franz H, Leupold O, Röhlsberger R, Roth SV, Seeck OH, Spengler J, Strempfer J, Tischer M, Viefhaus J, Weckert E, Wroblewski T (2006) PETRA III: DESY's new high brilliance third generation synchrotron radiation source, *Synch Rad News*, **19**, 25–29.
5. Banhart J (2008) T*omographic Methods in Materials Research and Engineering*, Oxford University Press, Oxford, New York.
6. Reimers W, Pyzalla AR, Schreyer AK, Clemens H (2008) *Neutrons and Synchrotron Radiation in Engineering Material Science, From Fundamentals to Material and Component Characterization*, Wiley-VCH, Weinheim.
7. Graafsma H, Martin T (2008) Detectors for synchrotron tomography, in *Advanced Tomographic Methods in Materials Research and Engineering* (Banhart J, ed), Oxford University Press, Oxford, New York, pp. 227–300.
8. Scheuerlein C, Di Michiel M, Buta F (2009) Synchrotron Radiation Techniques for the Characterization of Nb3Sn Superconductors, I*EEE Trans Appl Supercond*, **19**, 2653–2656.
9. Beckmann F (2008) Neutron and synchrotron-radiation-based imaging for applications in materials science – from macro- to nanotomography, in *Neutrons and Synchrotron Radiation in Engineering Material Science, From Fundamentals to Material and Component Characterization* (Reimers W, Pyzalla AR, Schreyer AK, Clemens H, ed), Wiley-VCH, Weinheim, pp. 287–307.
10. Bonse U, Busch F (1996) X-ray computed microtomography (CT) using synchrotron radiation (SR), *Prog Biophys Molec Biol*, **65**, 133–169.
11. Raven C, Snigirev A, Snigireva I, Spanne P, Sourvorov A, Kohn V (1996) Phase-contrast microtomography with coherent high-energy synchrotron X-rays, *Appl Phys Lett*, **69**, 1826–1828.
12. Spanne P, Raven C, Snigireva I, Snigirev A (1999) In-line holography and phase-contrast microtomography with high energy X-rays, *Phys Med Biol*, **44**, 741–749.

13. Bravin A (2003) Exploiting the X-ray refraction contrast with analysers: the state of the art, *J Phys D: Appl Phys*, **36**, A24–A29.

14. Snigirev A, Snigireva I, Kohn V, Kuznetsov S, Schelokov I (1995) On the possibilities of X-ray phase contrast microimaging by coherent high-energy synchrotron radiation, *Rev Sci Instrum*, **66**, 5486–5492.

15. Cloetens P, Barrett R, Baruchel J, Guigay J-P, Schlenker M (1996) Phase objects in synchrotron radiation hard X-ray imaging, *J Phys D*, **29**, 133–146.

16. Momose A (1995) Demonstrating of phase-contrast X-ray computed tomography using an X-ray interferometer, *Nucl Instrum Methods A*, **352**, 622–628.

17. Beckmann F, Bonse U, Busch F, Günnewig O (1997) X-ray microtomography (μCT) using phase contrast for the investigation of organic matter, *J Comp Assist Tomogr*, **21**, 539–553.

18. Momose A (2002) Phase-contrast X-ray imaging based on interferometry, *J Synchrotron Rad*, **9**, 136–142.

19. Momose A (2003) Phase-sensitive imaging and phase tomography using X-ray interferometers, *Opt Express*, **11**, 2303.

20. Cloetens P, Ludwig W, Baruchel J, et al. (1999) Holotomography: quantitative phase tomography with micrometer resolution using hard synchrotron radiation X-rays, *Appl Phys Lett*, **75**, 2912–2914.

21. Weitkamp T, Diaz A, David C, et al. (2005) X-ray phase imaging with a grating interferometer, *Opt Express*, **13**, 6296–6304.

22. Pfeiffer F, Weitkamp T, Bunk O, David C (2006) Phase retrieval and differential phase-contrast imaging with low-brilliance X-ray sources, *Nat Phys*, **2**, 258–261.

23. Herzen J (2010) *A Grating Interferometer for Materials Science Imaging at a Second-Generation Synchrotron Radiation Source*, Dissertation, Department Physik der Universität Hamburg.

24. Lengeler B, Schroer CG, Richwin M, et al. (1999) A microscope for hard X-rays based on parabolic compound refractive lenses, *Appl Phys Lett*, **74**, 3924.

25. Takeuchi A, Uesugi K, Suzuki Y, Aoki S (2001) Hard X-ray microtomography using X-ray imaging optics, *Jpn J Appl Phys*, **40**, 1499.

26. Takeuchi A, Uesugi K, Takano H, Suzuki Y (2002) Submicrometer-resolution three-dimensional imaging with hard X-ray imaging microtomography, *Rev Sci Instrum*, **73**, 4246.

27. Schroer CG, Lengeler B (2005) Focusing hard x rays to nanometer dimensions by adiabatically focusing lenses, *Phys Rev Lett*, **94**, 054802.

28. Schroer CG, Benner B, Günzler TF, et al. (2004) Nanotomography using parabolic refractive X-ray lenses, in *Developments in X-Ray Tomography IV: Vol. 5535 of the Proceedings of the SPIE* (Bonse U, ed), Bellingham, pp. 701–708.

29. Lengeler B, Schroer CG, Kuhlmann M, et al. (2004) Beryllium parabolic refractive X-ray lenses, in *Design and Microfabrication of Novel X-Ray Optics II: Vol. 5539 of the Proceedings of the SPIE* (Snigirev AS, Mancini DC, ed), Bellingham, pp. 1–9.

30. Schroer CG, Cloetens P, Rivers M, Snigirev A, Takeuchi A, Yun W (2004) High resolution 3D imaging microscopy using hard X-rays, *MRS Bull*, **29**, 157–165.

31. Snigirev A, Kohn V, Snigireva I, Lengeler B (1996) A compound refractive lens for focusing high-energy X-rays, *Nature*, **384**, 49–51.

32. Snigirev A, Kohn V, Snigireva I, Souvorov A, Lengeler B (1998) Focusing high-energy X-rays by compound refractive lenses, *Appl Opt*, **37**, 653–662.

33. Grigoriev M, Shabelnikov L, Yunkin V, et al. (2001) Planar parabolic lenses for focusing high energy X-rays, *Proc. SPIE*, **4501**, 185–192.

34. Mimura H, Matsuyama S, Yumoto H, et al. (2005) Hard X-ray diffraction-limited nanofocusing with Kirkpatrick-Baez mirrors, *Jpn J Appl Phys*, **44**, L539.

35. Hignette O, Cloetens P, Rostaing G, Bernard P, Morawe C (2005) Efficient sub 100 nm focusing of hard X-rays, *Rev Sci Instrum*, **76**, 063709.

36. Hignette O, Cloetens P, Morawe C, Borel C, Ludwig W (2007) Nanofocusing at ESRF using graded multilayer mirrors, *AIP Conf Proc*, **879,** 792–795.

37. Takano H, Suzuki Y, Takeuchi A (2003) Sub-100 nm hard X-ray microbeam generation with Fresnel zone plate optics, *Jpn J Appl Phys*, **43**, L132.

38. Suzuki Y, Takeuchi A, Takano H, Takenaka H (2005) Performance test of Fresnel zone plate with 50 nm outermost zone width in hard X-ray region, *Jpn J Appl Phys*, **44**, 1994.

39. Hermann GT (1980) *Image Reconstruction from Projections. The Fundamentals of Computerized Tomography*, Academic Press, New York.

40. Kak AC, Slaney M (1988) *Principles of Computerized Tomographic Images*, IEEE Press, New York.
41. Natterer F (1986) *The Mathematics of Computerized Tomography*, B. G. Teubner, Stuttgart and John Wiley & Sons, Chichester.
42. Radon J (1927) Über die Bestimmung von Funktionen durch ihre Integralwerte längs gewisser Mannigfaltigkeiten, *Ber Ver Sächs Akad Wiss Leipzig Math-Phys Kl*, **69**, 262–277.
43. Jenneson PM, Gilboy WB, Morton EJ, Gregory PJ (2003) An X-ray micro-tomography system optimised for the low-dose study of living organisms, *Appl Radiat Isot*, **58**, 177–181.
44. Davis GR, Elliott JC (1997) X-ray microtromography scanner using time-delay integration for elimination of ring artefacts in the reconstructed image, *Nuc Instrum Meth Phys Res A*, **394**, 157–162.
45. Sijbers J, Postnov A (2004) Reduction of ring artifacts in high resolution micro-CT reconstructions, *Phys Med Biol*, **49,** 247–253.
46. Raven C (1998) Numerical removal of ring artifacts in microtomography, *Rev Sci Instrum*, **69**, 2978–2980.
47. Antoine C, Nygard P, Gregersen Ø, Holmstad R, Weitkamp T, Rau C (2002) 3D images of paper obtained by phase-contrast X-ray microtomography: image quality and binarisation, *Nucl Instrum Meth Phys Res A*, **490**, 392–402.
48. Boin M, Haibel A (2006) Compensation of ring artefacts in synchrotron tomographic images, *Opt Express*, **14**, 12071–12075.
49. Axelson M, Svensson S, Borgefors G (2006) Reduction of ring artifacts, in *High Resolution X-Ray Microtomography Images, in Pattern Recognition* (Franke K, Müller K-R, Nickolay B, Schäfer R, ed), Springer-Verlag, Berlin, Heidelberg, pp. 61–70.
50. Di Michiel M, Merino JM, Fernandez-Carreiras D, et al. (2005) Fast microtomography using high energy synchrotron radiation, *Rev Sci Instrum*, **76**, 043702.
51. Marion A (1991) *An Introduction to Image Processing*, Chapmann and Hall, London.
52. Lu W, Mackie TR (2002) Tomographic motion detection and correction directly in sinogram space, *Phys Med Biol*, **47**, 1267.
53. Ritchie CJ, Crawford CR, Godwin JD, King KF, Kim Y (1996) Correction of computed tomography motion artifacts using pixel-specific back-projection, *IEEE Trans Med Imaging*, **15**, 333–342.

54. Hammersberg P, Mångård M (1998) Correction for beam hardening artefacts in computerised tomography, *J X-Ray Sci Tech*, **8**, 75–93.

55. Chen C-Y, Chuang K-S, Wu J, Lin H-R, Li M-J (2001) Beam hardening correction for computed tomography images using a post reconstruction method and equivalent tissue concept, *J Digit Imaging*, **14**, 54–61.

56. Van de Casteele E, Van Dyck D, Sijbers J, Raman E (2002) An energy-based beam hardening model in tomography, *Phys Med Biol*, **47**, 4181–4190.

57. Kalender WA, Hebel R, Ebersberger J (1987) Reduction of CT artifacts caused by metallic implants, *Radiology*, **164**, 576–577.

58. Watzke O, Kalender WA (2004) A pragmatic approach to metal artifact reduction in CT: merging of metal artifact reduced images, *Eur Radiol Phys*, **14**, 849–856.

59. Roeske JC, Lund C, Pelizzari CA, Pan X, Mundt AJ (2003) Reduction of computed tomography metal artifacts due to the Fletcher-Suit applicator in gynecology patients receiving intracavitary brachytherapy, *Brachytherapy*, **2**, 207–214.

60. Bertram M, Rose F, Schäfer D, Wiegert J, Aach T (2004) Directional interpolation of sparsely sampled cone-beam CT sinogram data, *Proc IEEE ISBI*, 928–931.

61. Oehler M, Kratz B, Knopp T, Müller J, Buzug TM (2008) Evaluation of surrogate data quality in sinogram-based CT metal-artefact reduction, *SPIE Symposium on Optical Engineering - Image Reconstruction from Incomplete Data Conference*, **707607**, 1–10.

62. Lemmens C, Faul D, Nuyts J (2009) Suppression of metal artifacts in CT using a reconstruction procedure that combines MAP and projection completion, *IEEE Trans Med Imaging*, **28**, 250–260.

63. Nöthe M, Schulze M, Grupp R, Kieback B, Haibel A (2007) Investigation of sintering of spherical copper powder by micro focus computed tomography (μ-CT) and synchrotron tomography, *Mater Sci Forum*, **539–543**, 2657–2662.

64. Ohser J, Schladitz K (2009) *3D Images of Materials Structures Processing and Analysis*, Wiley-VCH, Weinheim, Berlin.

Chapter 12

Coherent X-Ray Diffraction Imaging of Nanostructures

Ivan A. Vartanyants[a,b] **and Oleksandr M. Yefanov**[a]

[a]*Deutsches Elektronen Synchrotron (DESY), Notkestraße 85, 22607 Hamburg, Germany*

[b]*National Research Nuclear University "MEPhI", 115409 Moscow, Russia*

ivan.vartaniants@desy.de

12.1 Introduction

Coherent X-ray diffractive imaging (CXDI) is a relatively new imaging method that can produce an image of a sample without using optics between the sample and detector (see Fig. 12.1). This differs from conventional microscopy schemes that use objective lenses to produce an image of an object. Taking into account the difficulties of producing lenses at hard X-ray energies that are both highly resolving and efficient, we see clearly the advantages of the so-called "lensless" microscopy techniques. After its first demonstration [1–4], CXDI was successfully applied at third-generation synchrotron sources for imaging micron- and nanometer-size samples (see Refs. [5–10] for recent reviews).

The conventional CXDI experiment is performed with an isolated sample illuminated by a coherent, plane wave (Fig. 12.1). The

X-Ray Diffraction: Modern Experimental Techniques
Edited by Oliver H. Seeck and Bridget M. Murphy

ISBN 978-981-4303-59-0 (Hardcover), 978-981-4303-60-6 (eBook)
www.panstanford.com

incident wave may be described by a complex field (with a real and an imaginary part) of uniform magnitude and phase. The radiation interacts with the sample, which affects both the amplitude and phase of this field. The scattered radiation from the sample propagates to a two-dimensional (2D) detector in the far-field, and the diffracted intensities are measured. The detector can be positioned either in the forward direction, or in the case of a crystalline sample at the Bragg angle positions (Fig. 12.1). It will be shown in the following sections that in the limit of kinematical scattering, which is a good approximation for scattering on nanostructures, the amplitude of the scattered field can be expressed as the Fourier transform (FT) of the electron density of a sample (see also Chapter1). However, the measurement of diffracted intensities exclusively is insufficient to unambiguously determine the electron density of a sample, as the phase information is lost during the measurement process (the measured quantity is the intensity and not the complex amplitude). Fortunately, with some additional knowledge of constraints on the sample in object space, the structure of the sample can be reconstructed using phase retrieval algorithms based on an iterative approach [11–13] (see also Ref. [14] for a review of iterative methods).

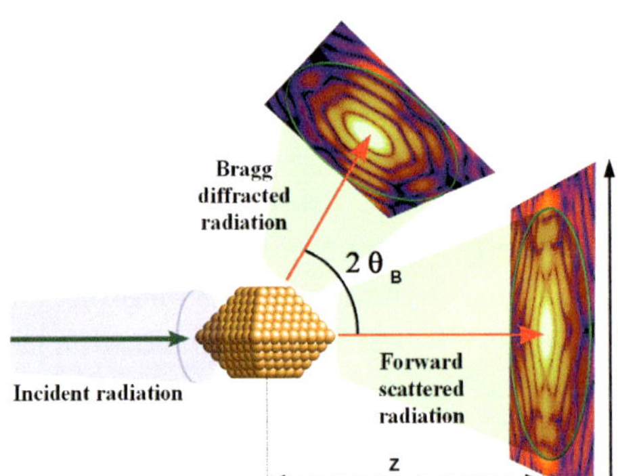

Figure 12.1 Schematic of a coherent X-ray diffractive imaging (CXDI) experiment. Incident radiation illuminates a crystalline sample of a finite size (from the left). The forward scattered and diffracted radiation are measured in the far-field at a distance *z* by an area detector of size *d*. The detector can be positioned either downstream from the sample position or at some Bragg angle position.

More formally, in CXDI experiments, the modulus squared of the scattered amplitude, $A(\mathbf{q})$, where $\mathbf{q}$ is the momentum transfer, is measured. Note that the amplitude $A(\mathbf{q})$ is a complex function, while we can only measure its modulus squared. The image reconstruction process begins with assigning random phases to $\phi(\mathbf{q})$ the known magnitudes $|A^{\exp}(\mathbf{q})| = \sqrt{I^{\exp}(\mathbf{q})}$ in reciprocal space. This complex wave in the far-field $A'(\mathbf{q}) = |A^{\exp}(\mathbf{q})|e^{i\phi(\mathbf{q})}$ is then inverse Fourier transformed to real space giving the first guess of the object in real space, $s'(\mathbf{r})$. This first guess is unlikely to be the correct solution and constraints in the object space, most importantly the finite extent of the object, have to be taken into account. This typically involves setting the values of $s'(\mathbf{r})$ outside some boundary to zero, known as the error reduction (ER) method, or forcing them toward zero, most commonly the hybrid input–output (HIO) method [12]. After the constraints have been applied, the updated function $s(\mathbf{r})$ is then Fourier transformed to the far-field. The magnitude $|A(\mathbf{q})|$ of this new far-field guess is replaced by the measured intensities $|A^{\exp}(\mathbf{q})| = \sqrt{I^{\exp}(\mathbf{q})}$ while the phases $\phi(\mathbf{q})$ are kept. This process is then repeated typically for thousands of iterations until it converges (see Fig. 12.2). The resulting function $s(\mathbf{r})$ in kinematical approximation is the electron density of the sample.

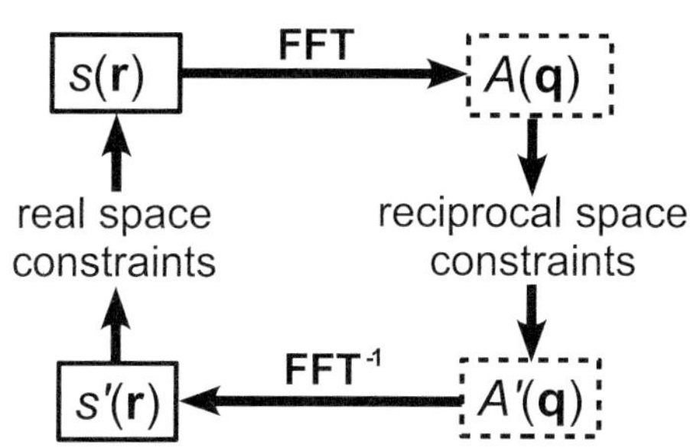

Figure 12.2 Block diagram of the iterative reconstruction method.

A necessary condition for the successful reconstruction of the electron density of a sample from a diffraction pattern is the appropriate sampling of the pattern [15]. A useful experimental rule of thumb is that at least two measurement points per fringe in the diffraction pattern are required for adequate sampling. This means that the autocorrelation function of the data is correctly sampled according to Shannon's sampling theorem [16], which is twice what is required to sample the fully complex wavefield. This essential sampling consideration leads this method to be sometimes referred to as the "oversampling" method. We discuss these sampling conditions with application to reconstruction of crystalline structures in more detail in the following sections.

There are many variations of the standard iterative methods, one of which is known as the guided HIO (GHIO) method [17]. In this method, N different HIO reconstructions (called families) are performed in parallel, each with different random starting phases and typically continuing for a few thousand iterations. Subsequently, convergence criteria are applied to the results and the best reconstruction is used to seed another set of reconstructions through the expression $s^{g+1,n}(\mathbf{r}) = (s^{g,template}(\mathbf{r}) \times s^{g,n}(\mathbf{r}))^{1/2}$, where n denotes a given family in the reconstruction and g is termed the generation. The resulting N terms $s^{g+1,n}(\mathbf{r})$ are used as inputs for the next "generation" of HIO reconstructions. This process is repeated a number of times until convergence.

One of the strong limitations of conventional CXDI is the restriction of imaging only very small objects. A modification of the CXDI method known as "ptychography" [18] removes this limitation. Ptychography is similar to scanning microscopy, where the sample is scanned through the X-ray beam and a far-field diffraction pattern is collected from each point of the sample scanned. Importantly, the illuminated regions should overlap in the sample plane. The relationship between each scan point is used to reconstruct the exit surface wave. This algorithm requires an overlap of the single illumination areas to yield a more stable reconstruction, and, most importantly, to avoid the finite object constraint of classical CXDI. A successful ptychographic phase and amplitude reconstruction of both the probe and an artificial gold zone plate with a resolution of tens of nanometers has been published [19] (see also Ref. [20]). The ability to image a field of view larger than the probe makes this an attractive technique for future application in hard condensed matter [21,22] and biological [23,24] systems.

As stated earlier, CXDI requires no optics between the sample and the detector, which is a great experimental simplification. However, there are also intrinsic limitations of this method. Typically, a CXDI iterative procedure requires some thousands of iterations for an image to be reconstructed, meaning that the imaging is certainly not performed instantaneously. Moreover, CXDI is a photon hungry method requiring increasingly higher flux to achieve higher resolutions. The reason is that the scattered intensity is a function of momentum transfer and scales as $I(q) \propto q^{-k}$, where k is between three and four depending upon the sample [25–27]. This means that for a generic object, a factor of three or four orders of magnitude in flux is required to improve the resolution by a single order; therefore, very bright photon beams are needed. Further, a beam stop or a hole in the detector is required to prevent damage through the intense direct beam or the strong beam scattered at Bragg angles. This means a lack of data at these positions. Consequently, if this missing data region is too large, iterative reconstructions may fail [28]. Finally, the beam needs to be sufficiently coherent over the sample area, both transversely and longitudinally, for a successful reconstruction [29–31].

This chapter is organized as follows: In the following section, the basics of coherent scattering on a finite size crystal will be discussed. The difference between conventional crystallography applied to large samples and coherent scattering on a finite size samples will be outlined. Then the formalism of coherent scattering from a finite size crystal with a strain field will be developed. In the subsequent section coherent scattering conditions will be relaxed and partially coherent illumination of a crystalline sample will be considered. In the final section, some recent experimental examples demonstrating applications of CXDI to the study of crystalline structures on the nanoscale will be given.

12.2 Coherent and Partially Coherent Scattering on Crystals

The scattering from an isotropic sample is quite different from that of a periodic crystal, where the intensity peaks at the Bragg positions. R. Millane [32] was the first to discuss in detail similarities and differences of the phase retrieval problem in crystallography and optics. Here, we will give a short overview of this problem extending it to the case of finite size and strained crystals (see also Ref. [29]) where the link with the phase retrieval applied to noncrystallographic samples will be most evident. We will also discuss how this problem is connected with Shannon's sampling theorem and the possibilities of using the results of this theorem for phase retrieval in crystallography.

12.2.1 Coherent Scattering from a Finite Size Crystal

It is well known (see Refs. [33,34]) that the scattering amplitude $A(\mathbf{q})$ of coherent monochromatic radiation from an infinite crystal in kinematical approximation[1] is equal to

$$A(\mathbf{q}) = \int \rho(\mathbf{r}) e^{-i\mathbf{q}\cdot\mathbf{r}} d^3r, \tag{12.1}$$

where $\rho(\mathbf{r})$ is the electron density at the point $\mathbf{r}$, $\mathbf{q} = \mathbf{k}_f - \mathbf{k}_i$ is the momentum transfer and $\mathbf{k}_i$ and $\mathbf{k}_f$ are the incident and scattered wave vectors ($|\mathbf{k}_i| = |\mathbf{k}_f| = 2\pi/\lambda$, where λ is the wavelength; see also Chapter 1). The electron density of a finite size crystal can be written as follows:

$$\rho(\mathbf{r}) = \rho_{uc}(\mathbf{r}) \otimes [\rho_\infty(\mathbf{r}) \cdot s(\mathbf{r})], \tag{12.2}$$

where the sign $\otimes$ denotes the convolution. Here, $\rho_{uc}(\mathbf{r})$ is the electron density of a unit cell:

$$\rho_{uc}(\mathbf{r}) = \sum_j \rho_j(\mathbf{r} - \mathbf{r}_j),$$

where $\mathbf{r}_j$ is a coordinate and $\rho_j(\mathbf{r})$ is the electron density of individual atoms in a unit cell. To define an infinite ideal lattice, we introduce the periodic function

$$\rho_\infty(\mathbf{r}) = \sum_{n=1}^{\infty} \delta(\mathbf{r} - \mathbf{R}_n),$$

where $\mathbf{R}_n = n_1\mathbf{a}_1 + n_2\mathbf{a}_2 + n_3\mathbf{a}_3$ is the position of the unit cell and $\mathbf{a}_1$, $\mathbf{a}_2$, $\mathbf{a}_3$ are the lattice vectors. In Eq. (12.2), we have also introduced a shape function $s(\mathbf{r})$ equal to the unity inside the volume V of the crystal and zero outside (the so-called Ewald function [39]) where $s(\mathbf{r})$ stands for the finite size of the sample:

[1] In this chapter, we assume that the kinematical approximation for the description of X-ray scattering on crystalline samples is valid. This is a good approximation for scattering of X-rays in the range of 10 keV and submicron crystal sizes. However, if crystalline particles reach few micron size, multiple scattering, or dynamical effects [35,36] could become important [37]. For these crystal sizes, refraction effects should also be considered [38].

$$s(\mathbf{r})=\begin{cases}1 & \text{for } \mathbf{r}\in V \\ 0 & \text{for } \mathbf{r}\notin V\end{cases}. \tag{12.3}$$

The shape function $s(\mathbf{r})$ leads to a "spreading" of the δ-type intensity distributions around the Bragg peaks that characterize an infinite crystal. The scattering amplitude $A(\mathbf{q})$ in Eq. (12.1) can now be conveniently calculated by the means of the approach originally proposed by von Laue [40], which reduces the sum over the points of the ideal lattice within the volume of the finite crystal to an integral over all space.

Now substituting Eq. (12.2) for the electron density into Eq. (12.1) and using convolution theorem ,we get for the scattered amplitude:

$$A(\mathbf{q}) = F(\mathbf{q}) \cdot \rho_\infty(\mathbf{q}) \otimes s(\mathbf{q}). \tag{12.4}$$

Here

$$F(\mathbf{q})=\int \rho_{uc}(\mathbf{r})e^{-i\mathbf{q}\cdot\mathbf{r}}d^3r=\sum_j f_j(\mathbf{q})e^{-i\mathbf{q}\cdot\mathbf{r}_j} \tag{12.5}$$

is the structure factor of the unit cell and

$$f_j(\mathbf{q})=\int \rho_j(\mathbf{r})e^{-i\mathbf{q}\cdot\mathbf{r}}d^3r \tag{12.6}$$

is the atomic scattering factor of the atom j in the unit cell and integration is performed over the volume of the unit cell. It is also assumed that the structure factors of the different cells are identical, as it is in general for perfect crystals. Usually, the structure factor $F(\mathbf{q})$ is a complex function. In Eq. (12.4),

$$s(\mathbf{q})=\int s(\mathbf{r})e^{-i\mathbf{q}\cdot\mathbf{r}}d^3r \tag{12.7}$$

is the Fourier transform of the shape function $s(\mathbf{r})$ and $\rho_\infty(\mathbf{q})$ is the Fourier transform of the lattice function that reduces to the sum of δ-functions

$$\rho_\infty(\mathbf{q})=\int \rho_\infty(\mathbf{r})e^{-i\mathbf{q}\cdot\mathbf{r}}d^3r=\frac{(2\pi)^3}{v}\sum_n \delta(\mathbf{q}-\mathbf{h}_n), \tag{12.8}$$

where v is the volume of the unit cell, $\mathbf{h}_n = h\mathbf{h}_1 + k\mathbf{h}_2 + l\mathbf{h}_3$ being the reciprocal lattice vectors and the summation is carried out over all reciprocal lattice points. In summary, we obtain for the scattered amplitude (12.4)

$$A(\mathbf{q})=\frac{F(\mathbf{q})}{v}\sum_n A_n(\mathbf{q}-\mathbf{h}_n), \tag{12.9}$$

with $A_n(\mathbf{q} - \mathbf{h}_n) = s(\mathbf{q} - \mathbf{h}_n)$ being the amplitude scattered in the vicinity of the reciprocal lattice vector $\mathbf{h}_n$. From this expression, we can see that the scattering amplitude is directly connected with the FT of the "shape" function $s(\mathbf{r})$. Here, it is also important to note that the structure factor $F(\mathbf{q})$ is extended in reciprocal space over many reciprocal lattice points. In the limit of the infinite crystal FT of the shape function $s(\mathbf{q})$ (12.7) reduces to the δ-function, and we get for the amplitude (12.9)

$$A(\mathbf{q}) = \frac{F(\mathbf{q})}{v} \sum_n \delta(\mathbf{q} - \mathbf{h}_n). \tag{12.10}$$

It is a well-known result in crystallography that the scattering amplitude of the infinite periodic object is sampled at its reciprocal lattice or Bragg points and, in principle, no information is available between these sampling points. Taking the inverse FT of (12.10), we obtain a well-known crystallographic formula for the electron density of the unit cell expressed through the Fourier components of structure factors

$$\rho(\mathbf{r}) = \frac{1}{v} \sum_{\mathbf{h}_n} F(\mathbf{h}_n) e^{i\mathbf{h}_n \cdot \mathbf{r}}. \tag{12.11}$$

Unfortunately, only the amplitudes $|F(\mathbf{h}_n)|$ of the structure factors can be measured in experiment and there is no direct phase information available (the so-called phase problem in crystallography).

For a crystal of macroscopic dimensions $D >> a$, where a is the size of a unit cell, the function $s(\mathbf{q})$ has appreciable values only for $\Delta q \sim 2\pi/D$ much smaller than the reciprocal lattice parameters $h_n = (2\pi/a)n$. Thus, according to Eq. (12.9) and neglecting the small cross terms, the intensity scattered by the crystal of finite dimensions will be determined by the sum over reciprocal lattice points:

$$I(\mathbf{q}) = |A(\mathbf{q})|^2 = \frac{|F(\mathbf{q})|^2}{v^2} \sum_n |A_n(\mathbf{q} - \mathbf{h}_n)|^2. \tag{12.12}$$

In the vicinity of the reciprocal point $\mathbf{h}_n = \mathbf{h}$, $\mathbf{q} \approx \mathbf{h}$, the intensity distribution can be written as follows:

$$I_h(\mathbf{Q}) = \frac{|F(\mathbf{h})|^2}{v^2} |A_h(\mathbf{Q})|^2, \tag{12.13}$$

where $\mathbf{Q} = \mathbf{q} - \mathbf{h}$ and $A_h(\mathbf{Q}) = s_h(\mathbf{Q})$.

In the case of the infinite crystal, we have from Eq. (12.10)

$$I(\mathbf{q}) = |A(\mathbf{q})|^2 = \frac{|F(\mathbf{q})|^2}{v^2}\sum_n \delta(\mathbf{q}-\mathbf{h}_n). \tag{12.14}$$

According to this mathematics, several important points have to be outlined. First of all, according to Eqs. (12.10) and (12.14) in the case of the infinite crystal, the scattering amplitude (or structure factor of the unit cell), and thus the intensity, is sampled at fixed points in reciprocal space. These points correspond to the nodes of the reciprocal lattice $\mathbf{h}_n$. Generally, no experimental information is available in between these sampling points, and there is no way to measure continuous diffraction patterns from infinite crystal. Hence, there is no possibility to oversample diffraction patterns from infinite crystals. This was noted first by D. Sayre [41] in early 1950s. It was also noted by the same author that sampling of the reciprocal space at Bragg points corresponds exactly to Nyquist sampling of the electron density of the unit cell of size a (for simplicity, we will limit our discussion with 1D case). In other words, if we have an infinite periodic density $\rho(x)$ with the unit cell size a (see Fig. 12.3), the corresponding reciprocal space amplitude $F(q)$ will be sampled at Bragg points $h_n = (2\pi/a)n$. At the same time according to the Shannon theorem [16], the minimum sampling distance for the electron density of a unit cell with size a will be $\Delta q = 2\pi/a$, which is identical to the sampling at the Bragg positions. According to the same theorem, there is no additional information between these sampling points. If all values of the amplitude $F(q)$ at the sampling points are known, then a continuous amplitude could be constructed *uniquely* as follows:

$$F(q) = \frac{2\pi}{a}\sum_{h_n} F(h_n)\frac{\sin[a(q-h_n)/2]}{\pi(q-h_n)}. \tag{12.15}$$

If the complex structure factors $F(h_n)$ (including the phases) could be measured, it would be possible to construct a continuous function $F(q)$ according to Eq. (12.15). Then, applying the inverse FT, the electron density in the unit cell could be reconstructed. However, as mentioned earlier, the phases of $F(h_n)$ are missing in the measurements. Therefore, in the case of the infinite periodic sample, it appears impossible to obtain a continuous diffraction pattern and consequently to apply an oversampling method for phase retrieval methods. The situation looks even worse when the intensity distribution $I(\mathbf{q})$ (12.14) is analyzed. The intensity distribution can be presented as a FT of the autocorrelation function of the electron density. For the electron density of the unit cell of size a, the corresponding autocorrelation function would have an extension of $2a$ with corresponding minimum Nyquist sampling frequency $\Delta q = \pi/a$. That means that we cannot make use of the sampling theorem (12.15) to reconstruct continuous intensity distribution function $I(\mathbf{q})$, because half of the necessary data points are missing. As a consequence, phase problem in crystallography is *twice* underdetermined. This makes phase retrieval for crystallography even more difficult problem then in optics.

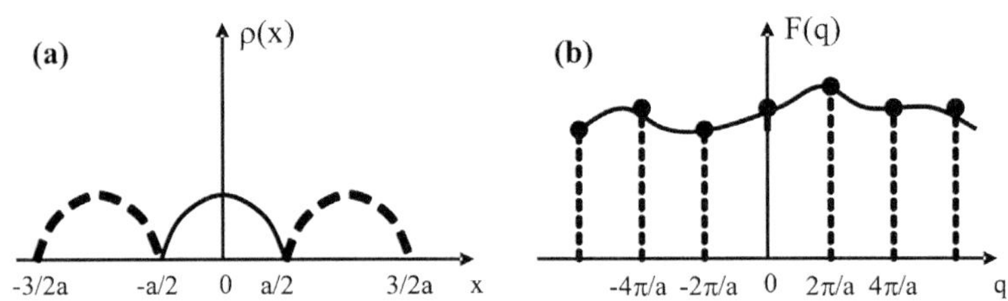

Figure 12.3 (a) Periodic electron density $\rho(x)$ with the unit cell size a. (b) Due to a periodicity of the electron density structure factor $F(q)$ is sampled at Bragg points $h_n = (2\pi/a)n$. No additional data points can be measured in between these sampling points, so in traditional crystallography oversampling is not possible.

The situation is quite different if, instead of an infinite crystal, a crystal of finite size is illuminated with coherent beams. Then the intensity distribution is given by Eq. (12.12) and is therefore the FT of a shape function $s(\mathbf{r})$ around *each* of the Bragg points. It is clear that in this case, the intensity distribution around each Bragg point will be a continuous function and can be, in principle, oversampled to the necessary level. Hence, a unique reconstruction of the crystal shape is possible.

Some general properties of this intensity distribution have to be outlined. For an arbitrary shape of an unstrained crystal, the intensity distribution (12.12) is a periodic function of $\mathbf{q}$ (see Fig. 12.4b). The intensity distribution is locally centrosymmetric around *each* $\mathbf{h}_n$ and has the same shape for *each* reciprocal lattice point $\mathbf{h}_n$. It takes its maximum value of $|F(\mathbf{h}_n)|^2\, V^2/\mathrm{v}^2$, where V is the volume of the crystal, if the scattering vector is exactly equal to $\mathbf{q} = \mathbf{h}_n$, and this point is the center of symmetry of the intensity distribution $I(\mathbf{q})$ (since according to (12.7) $s(-\mathbf{q}) = s^*(\mathbf{q})$). As follows from Eq. (12.12), the simplest picture of identical repeated distributions arises in unstrained crystals of any arbitrary shape. The detailed 3D shape of this distribution is determined by the FT of the crystal shape function $s(\mathbf{r})$ (see Eq. (12.7)). The intensity distribution measured by the 2D detector depends also on the Bragg angle and on the deviation from the exact Bragg conditions (the detector plane is always perpendicular to $\mathbf{k}_f$ vector). If the **z**-axis in reciprocal space is directed along the $\mathbf{k}_f$ vector and the detector is positioned at a Bragg angle, then we have from Eqs. (12.7)–(12.13) the following distribution of the amplitude in reciprocal space:

$$A(Q_x, Q_y) = \frac{F(\mathbf{h})}{v} \int s_z(x, y) e^{-iQ_x x - iQ_y y} dxdy\,, \tag{12.16}$$

where $s_z(x,y) = \int s(x,y,z)dz$ is the *projection* of the crystal shape on the (x, y)-plane that is defined as perpendicular to the $\mathbf{k}_f$ vector. Obviously, the inverse FT of the amplitude distribution $A(Q_x, Q_y)$ will recover this projection of the crystal shape. It is also clear that for a general crystalline sample, this projection is equivalent to the projection of the *electron density* of a sample to the same plane.

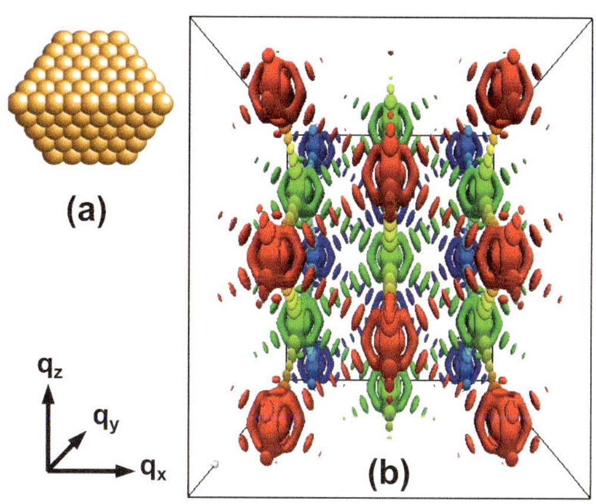

Figure 12.4 A finite size crystal (a) and its representation in reciprocal space (b). Different colors correspond to different locations of Bragg peaks in q_y direction.

In Fig. 12.5, diffraction patterns calculated for different crystal shapes are presented. The cut of the reciprocal space for the values of Q_z other than $Q_z = 0$ can also be calculated using Eq. (12.16). Simply, it requires the replacement of the real function $s_z(x, y)$ by a complex valued function $s_z(x,y)=\int s(x,y,z)\exp(iQ_z z)dz$. Examples of such calculations are presented in Fig. 12.6 for different values of Q_z and the crystal shape is shown in the left panel of Fig. 12.5. We have to note here that due to the presence of the phase factor $\exp(iQ_z z)$, each cut of the reciprocal space by the Ewald sphere at $Q_z \neq 0$ will produce noncentrosymmetric intensity distribution even for an unstrained crystal (see Fig. 12.6). However, the 3D intensity distribution will be centrosymmetric around each Bragg point (see Fig. 12.4).

In principle, the whole 3D intensity distribution $I(\mathbf{Q})$ can be measured by a 2D detector and by scanning the sample's angle near Bragg position or changing the incident energy. This 3D distribution can also be directly inverted by giving the shape of the crystalline part of the sample in 3D (see for the first demonstration [42]). Remarkably, in the case of the 3D phase retrieval, the necessary conditions on oversampling in the third dimension are quite relaxed as was noted by Millane [43]. In practice, few tens of angular positions in reciprocal space are sufficient to invert the 3D diffraction pattern of a micron-size crystalline particle (see for details Refs. [42,44]).

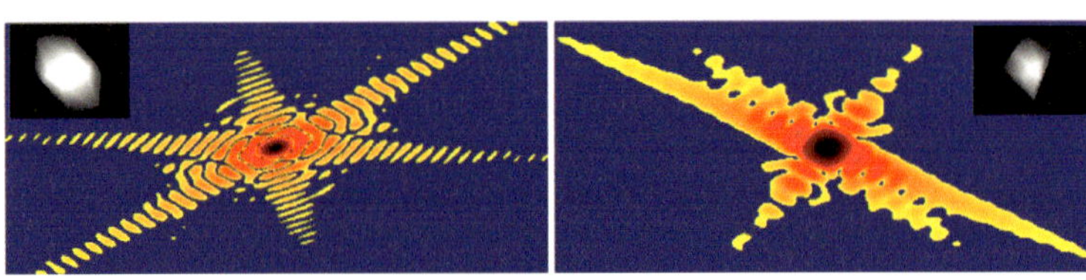

Figure 12.5 Projection of different crystal shapes on the plane perpendicular to $\mathbf{k}_f$ and the corresponding diffraction pattern calculated at the exact Bragg position. Adapted from Ref. [29].

As was proposed by von Laue [40], the Green's theorem can be applied to Eq. (12.7) and the volume integral can be transformed to an integral over the external surface area (S) of the crystal:

$$s(\mathbf{q}) = \frac{i}{q^2}\int_S (\mathbf{q}\cdot\mathbf{n})e^{-i\mathbf{q}\cdot\mathbf{r}}d\sigma, \tag{12.17}$$

where the unit vector **n** is an outward normal to the crystal. The maximum of this distribution for the flat surface is along directions normal to the surface. The existence of these flares was predicted by von Laue [40] and given the name "Stacheln" (spike). They have been experimentally observed later in the studies of the surface diffraction in the form of crystal truncation rods (CTRs) [45], or asymptotic Bragg diffraction [46]. In the case of a crystal with a center of symmetry and with a pair of identical opposite facets, we obtain from Eq. (12.17)

$$s(\mathbf{q}) = 2\frac{(\mathbf{q}\cdot\mathbf{n})}{q^2}\int_S \sin(\mathbf{q}\cdot\mathbf{r})d\sigma\,. \tag{12.18}$$

If the distance between the facets is equal to D, then for the direction of **q** perpendicular to the facets we finally get

$$s(q) = \frac{2}{q}S\sin(qD/2)\,. \tag{12.19}$$

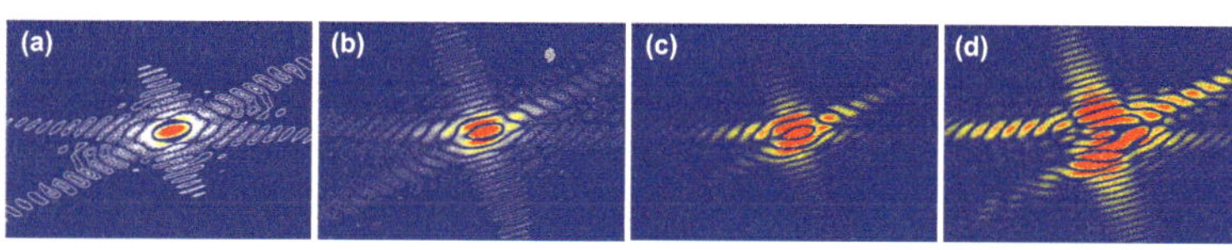

Figure 12.6 The cross-section of the reciprocal space of the diffraction pattern calculated from the crystal shape shown in the inset of Fig. 12.5 (left panel) for different q_z values: (a) $q_z = 0$, (b) $q_z = 0.375q_D$, (c) $q_z = 0.476q_D$, and (d) $q_z = 1.19q_D$. Here, q_D correspond to the fringe spacing $q_D \sim 2\pi/D$. Intensity in figure is rescaled for clarity. Adapted from Ref. [29].

It follows immediately that for two opposite facets and coherent illumination *interference*, patterns appear in the intensity distribution rather than a smooth q^{-2} decrease of intensity for a single surface (see Fig. 12.5). This interference pattern is similar to the fringes from slit scattering in optics, when the slit is illuminated by coherent light. The width of this intensity distribution in reciprocal space is equal to $\delta q = 2\pi/D$. This leads to a rod-like shape intensity distribution for crystals shaped like a compressed disc. Such behavior was, for example, observed in the study of thin films of Cu_3 Au [47]. In the case of a flat surface, the same Green's theorem can be applied once more to Eq. (12.17) transforming the surface integral to an integral around the boundary of the facet S. Now, diffraction from the edges will produce, instead of truncation rods, crystal truncation planes (CTPs) (see Ref. [48] for the first observation of CTP).

In conclusion, we can summarize that any pair of opposite facets and corresponding edges of unstrained crystals in a coherent beam produces an interference pattern with the maximum intensity distribution along the normal to the facets (CTR), or perpendicular to the opposite surface edges (CTP).

Equation (12.12) corresponds to the situation when *one* particle is illuminated by a coherent beam. In the case when two or more crystallites are located at some distance from each other and are illuminated by the same coherent beam, additional interference terms will appear in the expression for the intensity (12.12). Especially interesting (but not discussed here) is the case when a small particle is well separated from a big one and illuminated by the same coherent beam. This is similar to the principles of Fourier holography, when the object can be found as one term in the autocorrelation [49–51].

Up to now, we discussed the scattering geometry when diffraction patterns are measured locally around a fixed Bragg point for a crystal of finite size. However, another kind of measurement can be suggested, when diffraction patterns are measured simultaneously around several reciprocal lattice points. Taking into account that this distribution is measured for a finite number of reciprocal lattice points up to $q \sim q_{max}$, the complex amplitude distribution can be written as follows:

$$A(\mathbf{q}) = B(\mathbf{q}) \cdot F(\mathbf{q}) \cdot [\rho_{\infty}(\mathbf{q}) \otimes s(\mathbf{q})], \tag{12.20}$$

where $B(\mathbf{q})$ is an envelope function with the effective size $q \sim q_{max}$. Inverting this relationship with the inverse FT yields in real space an electron density:

$$\rho(\mathbf{r}) = [b(\mathbf{r}) \otimes \rho_{uc}(\mathbf{r})] \otimes [\rho_{\infty}(\mathbf{r}) \cdot s(\mathbf{r})], \tag{12.21}$$

where $b(\mathbf{r})$ is the inverse Fourier transform of $B(\mathbf{q})$. This electron density peaks at the regular positions of the unit cell due to the function $\rho_{\infty}(\mathbf{r})$ and has an overall shape of the sample $s(\mathbf{r})$. Most importantly, it contains the position of the atoms in the unit cell due to the reconstruction of the electron density function of a unit cell $\rho_{uc}(\mathbf{r})$. This means the following: if the continuous intensity distribution around several Bragg peaks is measured simultaneously and phase retrieval methods are applied to get the phase, then, in principle, the electron density with *atomic* resolution may be obtained. The resolution in real space around each atomic position in such experiments will be determined by the area accessed in the measurements in reciprocal space and can be estimated to be about $\Delta r \sim 2\pi/q_{max}$.

In order to map several Bragg peaks simultaneously with one detector, different approaches can be applied. For conventional crystalline samples with a unit cell of the size of a few angstroms hard X-rays in the range of 100 keV could be used [52]. Another approach is to use long-period crystalline samples such as colloidal crystals with a typical unit cell sizes on the order of a few hundred nanometers. The latter was successfully realized in CXDI experiments on colloidal 2D and 3D samples [53,54] (see Section 12.3.1). High energy electron beams in nanodiffraction experiments with the conventional transmission electron microscope (TEM) can be also used for realization of these ideas [55–57].

12.2.2 Coherent Scattering from a Finite-Size Crystal with a Strain

In the case of a deformed crystal, we can write the electron density as the sum of terms corresponding to individual atoms:

$$\rho(\mathbf{r}) = \sum_{n=1}^{N}\sum_{j=1}^{S}\rho_{nj}\left(\mathbf{r}-\mathbf{R}_{nj}-\mathbf{u}(\mathbf{R}_{nj})\right), \tag{12.22}$$

where $\mathbf{R}_{nj} = \mathbf{R}_n - \mathbf{r}_j$ and $\mathbf{u}(\mathbf{R}_{nj})$ is the displacement from the ideal lattice point. Summation in Eq. (12.22) is performed over N unit cells that contain S atoms. We assume here that the whole crystal is coherently illuminated. Substituting this expression for the electron density into Eq. (12.1) and changing variables in each term, the scattering amplitude can be written as a sum over the unit cells:

$$A(\mathbf{q}) = \sum_{n=1}^{N} F_n(\mathbf{q}) e^{-i\mathbf{q}\cdot\mathbf{u}(\mathbf{R}_n)} e^{-i\mathbf{q}\cdot\mathbf{R}_n}, \tag{12.23}$$

where $F_n(\mathbf{q})$ is the complex valued structure amplitude (see Eq. (12.5)) of the n-th cell. Here, it is assumed that all atoms in the unit cell are displaced uniformly $\mathbf{u}(\mathbf{R}_{nj}) \equiv \mathbf{u}(\mathbf{R}_n + \mathbf{r}_j) = \mathbf{u}(\mathbf{R}_n)$. It is important to note that Eq. (12.23) is also valid for the more general case allowing different displacements of atoms in different unit cells but with another definition of the structure amplitude $F_n(\mathbf{q})$ [58].

Now we will consider, as before, the scattering of X-rays on a crystal with finite size. The scattering amplitude $A(\mathbf{q})$ (12.23) for the crystal of finite dimensions will be calculated using the same approach as described in the previous section. According to this approach Eq. (12.23) can be identically rewritten in the form

$$A(\mathbf{q}) = F(\mathbf{q})\int \rho_\infty(\mathbf{r})S(\mathbf{r})e^{-i\mathbf{q}\cdot\mathbf{r}}d^3r, \tag{12.24}$$

where it is assumed that the structure factors of the different unit cells are identical with $F_n(\mathbf{q}) = F(\mathbf{q})$ and integration is carried out over the whole space. In this equation, we have introduced a *complex* function:

$$S(\mathbf{r}) = s(\mathbf{r})\exp(-i\mathbf{q}\cdot\mathbf{u}(\mathbf{r})), \tag{12.25}$$

with the shape function $s(\mathbf{r})$ (12.3) as an amplitude and the phase $\phi(\mathbf{r}) = \mathbf{q}\cdot\mathbf{u}(\mathbf{r})$, where the deformation field $\mathbf{u}(\mathbf{r})$ is included. It is important to note that no restrictions on the shape of the crystal and the deformation field apply.

Now performing the same calculations as in the previous section, we will obtain for the amplitude (see Eq. (12.4)):

$$A(\mathbf{q}) = F(\mathbf{q})\cdot S(\mathbf{q}) \otimes \rho_\infty(\mathbf{q}), \tag{12.26}$$

where $S(\mathbf{q})$ is the Fourier integral of $S(\mathbf{r})$

$$S(\mathbf{q}) = \int S(\mathbf{r})e^{-i\mathbf{q}\cdot\mathbf{r}}d^3r, \tag{12.27}$$

and the integration over d^3r is carried out over the whole space.

Now using the expression for the FT of $\rho_\infty(\mathbf{r})$ (12.8), we can derive the amplitude (12.26):

$$A(\mathbf{q}) = \frac{F(\mathbf{q})}{V}\sum_n A_n(\mathbf{q}-\mathbf{h}_n), \tag{12.28}$$

with $A_n(\mathbf{q}-\mathbf{h}_n) = S(\mathbf{q}-\mathbf{h}_n)$, now containing the deformation values. From this expression, we can see that the scattering amplitude is directly connected with the FT of the complex "shape" function $S(\mathbf{r})$ and its phase for the fixed reciprocal lattice point $\mathbf{h}$ is a sum of phases of the structure factor $F(\mathbf{h})$ and function $S(\mathbf{q})$.

For a crystal of macroscopic dimensions, we will again get a periodic intensity distribution:

$$I(\mathbf{q}) = |A(\mathbf{q})|^2 = \frac{|F(\mathbf{q})|^2}{v^2} \sum_n |A_n(\mathbf{q} - \mathbf{h}_n)|^2 , \tag{12.29}$$

where each term gives intensity values close to reciprocal point,

$$I_h(\mathbf{Q}) = \frac{|F(\mathbf{h})|^2}{v^2} |A_{\mathbf{h}}(\mathbf{Q})|^2 . \tag{12.30}$$

Here, the amplitudes $A_{\mathbf{h}}(\mathbf{Q})$ are defined as follows:

$$A_{\mathbf{h}}(\mathbf{Q}) = \int s(\mathbf{r}) e^{-i\mathbf{h}\cdot\mathbf{u}(\mathbf{r})} e^{-i\mathbf{Q}\cdot\mathbf{r}} d^3 r, \tag{12.31}$$

and as before $\mathbf{Q} = \mathbf{q} - \mathbf{h}$.

In the following, we will outline the differences between the coherent diffraction pattern of an unstrained crystal and a crystal with a deformation field. In this case for any arbitrary form of the crystal and the strain field, the intensity distribution (12.29) is still localized around reciprocal lattice vectors $\mathbf{h}_n$. However, in contrast to the unstrained case, for samples with arbitrary strain, the intensity distribution *locally* is not centrosymmetric around $\mathbf{h}_n$ and the shape differs at every reciprocal lattice point $\mathbf{h}_n$. Effects associated with the strain $\mathbf{u}(\mathbf{r})$ lead to different distributions near different reciprocal-lattice points.

This is illustrated in Fig. 12.7, where a two-dimensional example of the diffraction from a strained object is shown. In Fig. 12.7a, a real envelope function $s(\mathbf{r})$, defining a crystal shape is shown. It is obtained from a scanning-electron microscope image of a partially annealed array of gold nanocrystals on a glass substrate. In Fig. 12.7b, its Fourier transform, which shows the expected local symmetry is depicted. Figure 12.7c is a representation of a complex envelope function $S(\mathbf{r})$, where the amplitude is the same as before and the phase varies quadratically as a function of the radius from the object's center. The shaped (colored) rings represent phase reversals. The FT (Fig. 12.7d) shows the diffraction pattern expected from such a strained particle. Clearly, the local symmetry of the unstrained case on the left is distorted but still the main features, the flares, and the fringes resemble. This example shows the main point of our discussion, that the *shape* of a small crystal is related to the symmetric part of its coherent diffraction pattern, while its internal *strain* appears as an asymmetry in the diffraction.

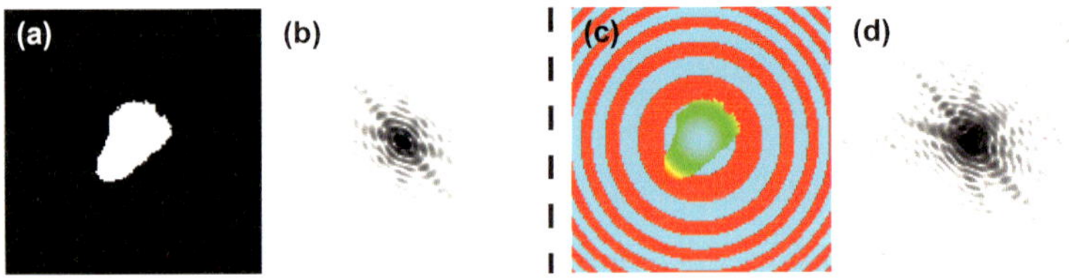

Figure 12.7 Illustration of the effects of strain on the diffraction from a 2D crystal, (a) is an unstrained object and (b) its calculated coherent diffraction pattern, (c) is the object with the addition of a real space strain increasing quadratically with radius from the center of the object and (d) its diffraction pattern. The alternate circular shading denotes positions for which the phase lies between 0 and π or between π and 2π. Adapted from Ref. [59].

Examples of successful experimental realization of these ideas in mapping the strain field in thin crystalline layers, or in isolated crystalline particles can be found in Refs. [44,60–63] (see also Ref. [8] for review).

12.2.3 Partially Coherent Scattering from a Finite-Size Crystal

In the previous sections the case of fully coherent incident radiation was considered. Now, we will assume that the incoming beam is partially coherent. The general properties of partially coherent radiation are discussed in detail in a number of textbooks [64–66]. We will reformulate the general results for the special case of scattering of partially coherent X-ray radiation on small crystalline particles (see also Refs. [29,30,67], for the case of partially coherent X-ray scattering on surfaces, see Ref. [68,69] and for CXDI with the partially coherent X-rays in the forward direction [31]). The whole problem is divided into two parts. For the first part, we consider scattering of radiation with an arbitrary state of coherence from a small crystalline particle. For the second part, a special realization of the incoherent source with a Gaussian intensity distribution will be discussed.

Let us assume the incident radiation to be a quasimonochromatic wave with one polarization state of the electric field,

$$E_{in}(\mathbf{r}, t) = A_{in}(\mathbf{r}, t)e^{i\mathbf{k}_i \cdot \mathbf{r} - i\omega t}, \tag{12.32}$$

where $|\mathbf{k}| = 2\pi/\lambda$. Here, λ and ω are the average wavelength and frequency of the beam. The amplitude $A_{in}(\mathbf{r}, t)$ is a slowly varying function with spatial variations much bigger than the wavelength λ and time scales much larger than $1/\omega$. Consequently, according to the Huygens–Fresnel principle [64] in the limits of kinematical scattering, the amplitude of the wavefield $E_{out}(\mathbf{v}, t)$, after being scattered from the sample to position $\mathbf{v}$ at the detector (see Fig. 12.8a), can be written as follows:[2]

$$E_{out}(\mathbf{v},t) = \int \rho(\mathbf{r}) \frac{A_{in}(\mathbf{r}, t - \tau_r)}{l_r} e^{i\mathbf{k}_i \cdot \mathbf{r} - \omega(t - \tau_r)} d^3 r, \tag{12.33}$$

where l_r is the distance between points $\mathbf{r}$ and $\mathbf{v}$ with the origins in the sample center and detector plane respectively, $\tau_r = l_r/c$ is the time delay for the radiation propagation between these points and c is the speed of light. In this expression for the scattered wavefield we have neglected the absorption assuming a small sample size and set the obliquity factor to unity.

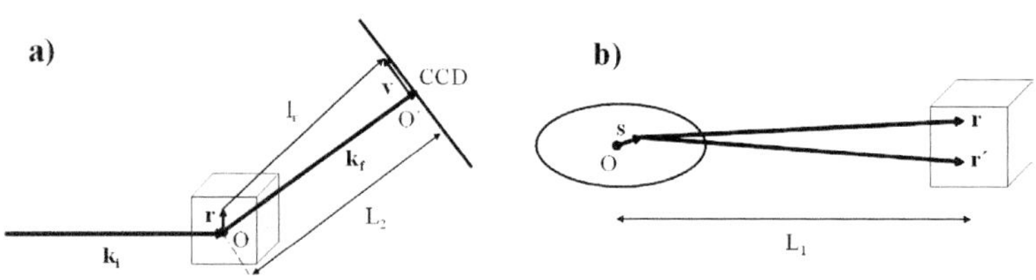

Figure 12.8 Definition of notations used for the scattering geometry in calculation of scattering of partial coherent radiation. (a) The X-ray beam is scattered on a small crystal particle and intensity is measured at the distance L_2 from the sample by 2D detector (e.g., CCD detector). (b) A synchrotron source produces an incoherent beam at the distance L_1 from the sample. Adapted from Ref. [29].

We will define the amplitude of the scattered wave $A(\mathbf{v}, t)$ in the usual way by $E_{out}(\mathbf{v}, t) = A(\mathbf{v}, t)(e^{ikL_2}/L_2)e^{i\omega t}$. We will also assume that the distance from the object to detector is $L_2 >> D$, where D is a typical size of an object. Then, in the limits of paraxial approximation for the distance l_r between points $\mathbf{r}$ and $\mathbf{v}$, we can use an expansion $l_r \approx L_2 -$

[2]In this expression and below, we omit all not essential integral prefactors.

$\mathbf{n}_f \cdot \mathbf{r} + (\mathbf{v} - \mathbf{r})^2/(2L_2)$, where $\mathbf{n}_f = \mathbf{k}_f/|\mathbf{k}_f|$. Here, we also assume that the detector plane is perpendicular to $\mathbf{k}_f$. Substituting this expansion into Eq. (12.33) we obtain for the scattering amplitude:

$$A(\mathbf{v},t) = \int \rho(\mathbf{r}) A_{in}(\mathbf{r}, t-\tau_r) P_{L_2}(\mathbf{v}-\mathbf{r}) e^{-i\mathbf{q}\cdot\mathbf{r}} d^3 r, \tag{12.34}$$

where $P_{L_2}(\mathbf{v} - \mathbf{r})$ is the Green's function (or propagator) that describes the propagation of radiation in free space. In the frame of the same theory, this function is equal to

$$P_{L_2}(\mathbf{v}-\mathbf{r}) = \frac{1}{i\lambda L_2} e^{i(k/2L_2)(\mathbf{v}-\mathbf{r})^2}. \tag{12.35}$$

We are interested in the far-field (or Fraunhofer) limit of Eq. (12.34), when the condition $kD^2/(2L_2) << 1$ is satisfied.[3] For the typical X-ray coherent experiment with radiation energy $E_\gamma \approx 8$ keV and detector at a distance $L_2 \approx 3$ m, this condition limits the size of the particles to $D << 10$ μm. In this limit, we can neglect the $(k/2L_2)$ $\mathbf{r}^2$ term in the exponent (12.35) and have

$$\lim_{(kD^2)/L_2 \to 0} P_{L_2}(\mathbf{v}-\mathbf{r}) \to \frac{1}{i\lambda L_2} \exp[i(k/2L_2)\mathbf{v}^2] \exp(-\mathbf{q}_v \cdot \mathbf{r}),$$

for the propagator, where $\mathbf{q}_v = (k/L_2)\mathbf{v}$. So, in the far-field we obtain for the amplitude (12.34):

$$A(\mathbf{q}',t) = \int \rho(\mathbf{r}) A_{in}(\mathbf{r}, t-\tau_r) e^{-i\mathbf{q}'\mathbf{r}} d^3 r, \tag{12.36}$$

where $\mathbf{q}' = \mathbf{q} + \mathbf{q}_v$. Here, we have omitted the phase term $\exp[i(k/2L_2)$ $\mathbf{v}^2]$ before the integral as it will cancel while calculating intensities at the same point $\mathbf{v}$ at the detector plane. We note here that this expression coincides with the coherent amplitude of Eq. (12.1) in the limit $A_{in}(\mathbf{r}, t - \tau_r) \to 1$. Now, we will consider the case when the scattering particle is a crystalline sample with a periodic electron density function and the amplitude $A_{in}(\mathbf{r}, t - \tau_r)$ is a slow varying function on the size of the unit cell. Under these considerations performing the same transformations as in the previous section, we finally obtain

$$A(\mathbf{q}',t) = \frac{F(\mathbf{q})}{v} \sum_n A_n(\mathbf{q}' - \mathbf{h}_n, t), \tag{12.37}$$

[3]Near-field Fresnel conditions can be also considered in the frame of the same theory. This will bring just to appearance of the additional phase factor $\exp[i(k/2L_2)\mathbf{r}^2]$ in the integrand of Eq. (12.36) (see for details [30]).

where

$$A_n(\mathbf{q},t) = \int s(\mathbf{r}) A_{in}(\mathbf{r}, t-\tau_r) e^{-i\mathbf{q}\cdot\mathbf{r}} d^3r \tag{12.38}$$

is a scattering amplitude in the vicinity of reciprocal vector $\mathbf{h}_n$. Here, as before, $s(\mathbf{r})$ is a shape function of the crystal and for simplicity, we assume that crystal is unstrained. However, this result can also be generalized for the case of a strained crystal by adding the exponential factor $\exp[-i\mathbf{q}\cdot\mathbf{u}(\mathbf{r})]$ to the integral (12.38).

According to Eqs. (12.37) and (12.38), the intensity of the scattered radiation measured at the position $\mathbf{v}$ of the detector near one of the Bragg points $\mathbf{h}_n = \mathbf{h}$ (we denote the corresponding amplitude as $A_h(\mathbf{q}, t)$) is equal to

$$\begin{aligned} I(\mathbf{Q}) &= < A(\mathbf{Q},t) A^*(\mathbf{Q},t) >_T = \frac{|F(\mathbf{h})|^2}{v^2} <|A_h(\mathbf{Q},t)|^2>_T \\ &= \frac{|F(\mathbf{h})|^2}{v^2} \iint s(\mathbf{r}) s(\mathbf{r}') \Gamma_{in}(\mathbf{r},\mathbf{r}',\Delta\tau) e^{-i\mathbf{Q}\cdot(\mathbf{r}-\mathbf{r}')} d^3r d^3r', \end{aligned} \tag{12.39}$$

where $\mathbf{Q} = \mathbf{q}' - \mathbf{h} = \mathbf{q}_v + \mathbf{q} - \mathbf{h}$, $\Delta\tau = (l_r - l_{r'})/c$ is a time delay and

$$\Gamma_{in}(\mathbf{r}, \mathbf{r}', \tau) = < A_{in}(\mathbf{r}, t) A_{in}^*(\mathbf{r}', t+\tau) >_T \tag{12.40}$$

is the mutual coherence function. The averaging $<>_T$ in Eqs. (12.39) and (12.40) is conducted for a time T, which is much longer than the time of fluctuation of the X-ray field, and it is assumed that the incoming radiation is ergodic and stationary.

For the case of cross-spectral pure light, we can write the mutual coherence function as a product [65,66]:

$$\Gamma_{in}(\mathbf{r},\mathbf{r}',\tau) = \sqrt{I(\mathbf{r})}\sqrt{I(\mathbf{r}')}\gamma_{in}(\mathbf{r},\mathbf{r}')F(\tau). \tag{12.41}$$

Here, $I(\mathbf{r}) =< |A_{in}(\mathbf{r}, t)|^2 >_T$ and $I(\mathbf{r}') =< |A_{in}(\mathbf{r}', t)|^2 >_T$ are the averaged intensities of the incoming radiation at points $\mathbf{r}$ and $\mathbf{r}'$, $\gamma_{in}(\mathbf{r}, \mathbf{r}')$ is the complex degree of coherence and $F(\tau)$ is the time autocorrelation function.

We will make some more simplifying considerations to have an explicit form for the mutual coherence function (12.41). We will assume that the incident radiation comes from a planar incoherent source with Gaussian distribution of intensity, which is located at a distance L_1 from the sample (Fig. 12.8b). This will be an approximation for the actual 3D X-ray source from a synchrotron storage ring. We will also consider that the distance L_1 is much larger than the size of the particle D and, than the average size of the source S. According to the van Cittert–Zernike theorem [65,66], the complex degree of coherence at the sample position can be obtained in the same limit of paraxial approximation (see Ref. [70] for the generalization of this approach):

$$\gamma_{in}(\mathbf{r},\mathbf{r}')=\frac{e^{i\psi}\int I(\mathbf{s})e^{i\mathbf{s}\cdot(\mathbf{r}-\mathbf{r}')k/L_1}d^3s}{\int I(\mathbf{s})d^3s}, \tag{12.42}$$

where the phase factor is $\psi = (k/2L_1)(r^2 - r'^2)$, $I(\mathbf{s})$ is the intensity distribution of the incoherent source and integration is performed over the whole area of the incoherent source. It is interesting to note here that Eq. (12.42) is exact up to second-order terms in $\mathbf{s}$ for an incoherent source. For the typical CXD experiment at a synchrotron source with a distance from source to sample of $L_1 \approx 40$ m and energy of $E_\gamma \approx 8$ keV, the far-field condition $kD^2/(2L_1) << 1$ can easily be satisfied, giving the upper limit for the size of the particle as $D << 40$ μm. In this far-field limit, we can neglect the phase prefactor $e^{i\psi}$ in Eq. (12.42) and consider the complex degree of coherence $\gamma_{in}(\mathbf{r}, \mathbf{r}')$ (12.42) as a real valued function. Following this model of an incoherent source in the far-field limit, the intensity of the incoming radiation at points $\mathbf{r}$ and $\mathbf{r}'$ at the sample can be calculated as $I(\mathbf{r})$ $I(\mathbf{r}) \approx I(\mathbf{r}') = I_0 = (\lambda / L_1)^2 \int I(\mathbf{s})d^2s$.

It is common to describe the intensity distribution of the synchrotron source by the Gaussian function:

$$I(s_x, s_y) = \frac{I_0}{2\pi\sigma_x\sigma_y} e^{-(s_x^2/\sigma_x^2 + s_y^2/\sigma_y^2)/2}, \tag{12.43}$$

where σ_x and σ_y are the rms values of the intensity distribution in x and y directions. Due to the fact that Eq. (12.42) is an FT, we also get a Gaussian form for the complex degree of coherence:

$$\gamma_{in}(\mathbf{r}_\perp - \mathbf{r}'_\perp) = \exp\left(-\frac{(\mathbf{r}_\perp - \mathbf{r}'_\perp)^2}{2\xi_\perp^2}\right) = \exp\left(-\frac{(x-x')^2}{2\xi_x^2} - \frac{(y-y')^2}{2\xi_y^2}\right). \tag{12.44}$$

$\mathbf{r}_\perp$ and $\mathbf{r}'_\perp$ are projections of $\mathbf{r}$ and $\mathbf{r}'$ across the beam propagation direction and $\xi_{x,\,y} = L_1/(k\sigma_{x,\,y})$ are usually defined as the two transverse coherence lengths. For a high brilliance third-generation synchrotron source as PETRA III [71] with the rms source size $\sigma_x \approx$

36 μm, $\sigma_y \approx 6$ μm corresponding to a low-β operation of the storage ring at a distance $L_1 = 60$ m from the source and X-ray energy $E_\gamma \approx 12$ keV, we obtain $\xi_x \approx 30$ μm and $\xi_x \approx 200$ μm for the coherence length in both directions [70].

We will further assume that the time autocorrelation function $F(\tau)$ in Eq. (12.41) has a pure exponential form:

$$F(\tau) = F_0 \exp(-\tau/\tau_{||}), \tag{12.45}$$

which would be the exact result for a Lorentzian power spectral density of the source [66]. The characteristic time $\tau_{||}$ of the decay of the time autocorrelation function defines the longitudinal coherence length $\xi_{||} = c\tau_{||}$. It can easily be shown [64–66] that the coherence length $\xi_{||}$ is determined by the bandwidth $(\Delta\lambda/\lambda)$ of the incoming radiation and for exponential autocorrelation function $F(\tau)$ it is equal to $\xi_{||} = (2/\pi)(\lambda^2/\Delta\lambda)$. For a Si111 double-crystal monochromator with a bandwidth $\Delta\lambda/\lambda \approx 3 \times 10^{-4}$ and the wavelength $\lambda \approx 1.5$Å (which corresponds to $E_\gamma \approx 8$ keV) we get $\xi_{||} \approx 0.32$ μm for the longitudinal coherence length.

In the far-field limit the time autocorrelation function $F(\Delta\tau)$ is given by the following expression:

$$F(\Delta\tau) = F(|\mathbf{r}_{||} - \mathbf{r}'_{||}|) = F_0 \exp(-|l_r - l_{r'}|/\xi_{||}) = F_0 \exp(-|\mathbf{r}_{||} - \mathbf{r}'_{||}|/\xi_{||}), \tag{12.46}$$

where $\mathbf{r}_{||}$ and $\mathbf{r}'_{||}$ are the components of $\mathbf{r}$ and $\mathbf{r}'$ along the beam, and we have neglected small perpendicular contribution.

Substituting Eqs. (12.40)–(12.46) into (12.39), we obtain the intensity

$$I(\mathbf{Q}) = \frac{|F(\mathbf{h})|^2}{v^2} \iint s(\mathbf{r}) s(\mathbf{r}') \gamma_{in}(\mathbf{r}_\perp - \mathbf{r}_\perp') F(\mathbf{r}_{||} - \mathbf{r}_{||}') e^{-i\mathbf{Q}\cdot(\mathbf{r}-\mathbf{r}')} d^3r d^3r', \tag{12.47}$$

where the complex degree of coherence $\gamma_{in}(\mathbf{r}_\perp - \mathbf{r}'_\perp)$ is defined by (12.44) and the autocorrelation function $F(|\mathbf{r}_{||} - \mathbf{r}'_{||}|)$ by Eq. (12.46). This expression can be simplified further by changing the variables:

$$I(\mathbf{Q}) = \frac{|F(\mathbf{h})|^2}{v^2} \int \phi_{11}(\mathbf{r}) \gamma_{in}(\mathbf{r}_\perp) F(\mathbf{r}_{||}) e^{-i\mathbf{Q}\cdot\mathbf{r}} d^3r, \tag{12.48}$$

where $\phi_{11}(\mathbf{r}) = \int s(\mathbf{r}') s(\mathbf{r}' + \mathbf{r}) d^3r'$ is the autocorrelation function of the shape function $s(\mathbf{r})$.

In the coherent limit, the transverse and longitudinal coherence lengths $\xi_\perp$, $\xi_{||}$ become infinite and that simultaneously leads to the following limits: $\gamma_{in}(\mathbf{r}_\perp)\to 1$ for the complex degree of coherence and $F(\mathbf{r}_{||}) \to 1$ for the autocorrelation function. In this case, we get the intensity of the coherently scattered radiation:

$$I_{coh}(\mathbf{Q}) = \frac{|F(\mathbf{h})|^2}{v^2}\int \phi_{11}(\mathbf{r})e^{-i\mathbf{Q}\cdot\mathbf{r}}d^3r = |A(\mathbf{Q})|^2, \tag{12.49}$$

where $A(\mathbf{Q}) = (F(\mathbf{h})/v)\int s(\mathbf{r})\exp(-i\mathbf{Q}\cdot\mathbf{r})d^3r$ is the kinematically scattered amplitude from the crystal with shape function $s(\mathbf{r})$. This result completely coincides (assuming zero strain field $\mathbf{u}(\mathbf{r}) = 0$) with the coherent limit of Eqs. (12.30) and (12.31) discussed earlier.

Applying the convolution theorem, the intensity $I(\mathbf{Q})$ in Eq. (12.48) can be written in the form of convolution of two functions:

$$I(\mathbf{Q}) = \frac{1}{(2\pi)^3}\int I_{coh}(\mathbf{Q}')\tilde{\Gamma}(\mathbf{Q}-\mathbf{Q}')d^3Q' = I_{coh}(\mathbf{Q})\otimes\tilde{\Gamma}(\mathbf{Q}), \tag{12.50}$$

where $I_{coh}(\mathbf{Q})$ is intensity of coherently scattered radiation (12.49) and $\tilde{\Gamma}(\mathbf{Q})$ is the Fourier transform:

$$\tilde{\Gamma}(\mathbf{Q}) = \int \gamma_{in}(\mathbf{r}_\perp)F(|\mathbf{r}_{||}|)e^{-i\mathbf{Q}\cdot\mathbf{r}}d^3r\,. \tag{12.51}$$

Now we will consider orthogonal coordinates with the z axis along the diffracted beam propagation direction and the x, y axes perpendicular to this direction as already used in the previous sections. In this coordinate system for the exact Bragg position $\mathbf{q} = \mathbf{h}$ and $\mathbf{Q} = \mathbf{q}_v$, we can write intensity distribution (12.48) in the detector plane as follows

$$I(\mathbf{q}_v) = \frac{|F(\mathbf{h})|^2}{v^2}\int \phi_{11}^z(\mathbf{x})\gamma_{in}(\mathbf{x})e^{-i\mathbf{q}_v\cdot\mathbf{x}}d^2x, \tag{12.52}$$

where $\mathbf{x}$ is a 2D vector $\mathbf{x} = (x, y)$ and $\phi_{11}^z(\mathbf{x}) = \int\phi_{11}(\mathbf{r})\exp(|-z|/\xi_{||})dz$. It is interesting to note that the intensity distribution (12.52) can also be calculated as a convolution of the functions $\phi_{11}^z(\mathbf{q})$ and $\gamma_{in}(\mathbf{q})$. For the case of large longitudinal length $\xi_{||} >> D$, the function $\phi_{11}^z(\mathbf{q})$ gives just the projection of the 3D autocorrelation function to the plane $\mathbf{x}$. Smaller values of $\xi_{||} \le D$ reduce the real space volume of the scattering object along the propagating beam that contribute coherently in the diffraction pattern.

In Fig. 12.9, we present calculations of 2D diffraction patterns obtained from Eq. (12.52) for the crystal shape shown in the left inset of Fig. 12.5 with different values of transverse coherence lengths ξ_x and ξ_y. For simplicity, it was assumed here that the longitudinal coherence length is long enough $\xi_{||} >> D$ and therefore it was not considered in the simulations. From this figure, it is obvious that decreasing the values of parameters ξ_x and ξ_y leads to a decrease in the contrast of the diffraction pattern. In the right-hand side column of Fig. 12.9, a reconstruction of the crystal projected electron density from the corresponding diffraction patterns is presented (see Refs. [29, 30] for details). Not surprisingly, the quality of the reconstructions becomes poor with small degree of coherence.

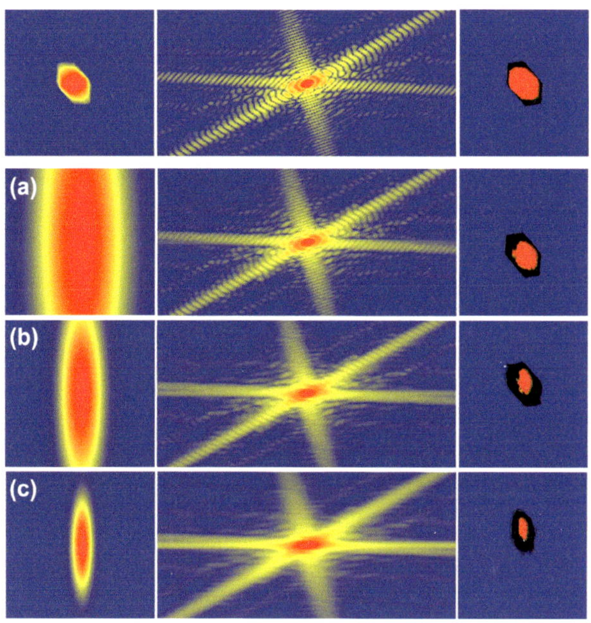

Figure 12.9 Complex degree of coherence $\gamma_{in}(x, y)$ (left column) used for calculations of diffraction intensity patterns (central column) Reconstructed real space images are shown in right column. For comparison the top row shows the case of coherent illumination with the coherence length $\xi_x = \xi_y = \infty$. The values of the coherence lengths from top down: (a) $\xi_x = 91$ pixels, $\xi_y = 367$ pixels, (b) $\xi_x = 45$ pixels, $\xi_y = 183$ pixels, and (c) $\xi_x = 22$ pixels, $\xi_y = 91$ pixels. Adapted from Ref. [29].

12.3 Experimental Examples

In this section, we give several experimental examples of applications of coherent X-ray diffractive imaging for different systems.

12.3.1 Coherent X-Ray Imaging of Defects in Colloidal Crystals

First, we present results of CXDI applied to reveal the structure of a regular part and also a part containing a defect of a colloidal 2D crystal (see Ref. [53] for details).

Self-organized colloidal crystals are an attractive material for modern technological devices. They can be used as the basis for novel functional materials such as photonic crystals, which may find applications in future solar cells, LEDs, lasers, or even as the basis for circuits in optical computing and communication. For these applications, the crystal quality is crucial and monitoring the defect structure of real colloidal crystals is essential [72].

The experiment [53] was performed at the microoptics test bench at the ID06 beamline of the European Synchrotron Radiation Facility (ESRF) using an incident X-ray energy of 14 keV. The geometry of the experiment allows for rotation of the sample around the vertical axis, which is perpendicular to the incident X-ray direction (see Fig. 12.10). A 6.9 μm pinhole was positioned at a close distance in front of the colloidal crystal. The pinhole selects a highly coherent part of the beam and produces a finite illumination area. The initial orientation of the sample (with azimuthal angle $\varphi = 0°$) corresponds to the direction of the incident X-rays along the [111] direction of the face-centered cubic (fcc) colloidal crystal and was perpendicular to the surface normal of the sample. Rotating the sample around the x-axis allows the measurement of different sets of diffraction planes. Particularly important was the direction of the incident X-rays along the [110] direction of the colloidal sample lattice at $\varphi = 35°$, when the set of (111) planes was aligned along the incident beam. The diffraction data were recorded using a CCD with 4,005 × 2,671 pixels with a resolution of $\Delta q = 0.16\ \mu m^{-1}$ per pixel. In the experiment, a thin film of a colloidal crystalline sample on a glass substrate was used. It was grown by the convective assembly technique using polystyrene microspheres (diameter 425 nm, relative standard deviation 5 percent). The grown crystalline films have an fcc structure and were typically 20–30 layers thick.

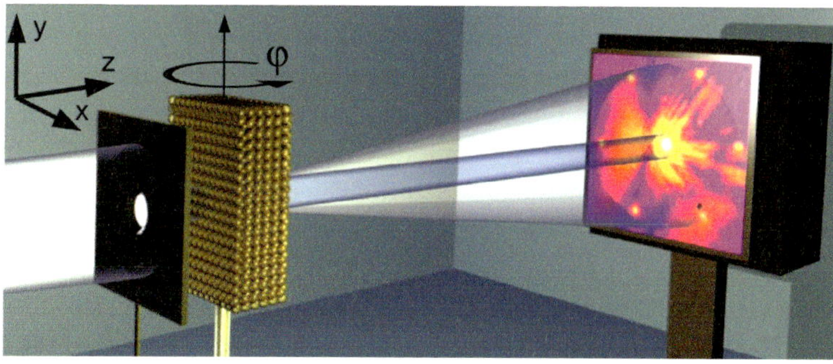

Figure 12.10 Schematic of the experiment with colloidal crystals. Adapted from Ref. [53].

Positioning the sample just after the aperture yields strong fringes typical of an Airy pattern [64] from a circular aperture, centered at $q = 0$ (Fig. 12.11a,d). In addition, due to the long-range order in the colloidal crystal, several orders of Bragg peaks are easily visible in the diffraction patterns. The strongest are the hexagonal set of 220 Bragg peaks typical for scattering from an fcc structure. Each of these Bragg peaks contains a few orders of diffraction fringes similar to those at $q = 0$, due to the finite aperture in front of the sample. In addition to the allowed 220 Bragg peaks, we also observed much weaker forbidden peaks (2/3, 2/3, 4/3 in our case). Their appearance is an indication of defects in the crystal.

The measured diffraction data were inverted by applying the guided hybrid input–output GHIO algorithm [17]. To improve the quality of the reconstruction, a scaled diffraction pattern of the pinhole was subtracted from the diffraction pattern of the sample Fig. 12.11a. After this procedure, the 220 Bragg peaks, and especially the fringes around them, are clearly visible against the background Fig. 12.11b. Negative values, shown in black in the difference diffraction pattern in Fig. 12.11b, were left to evolve freely in the reconstruction procedure. To stabilize the reconstruction process, the central region (with $q < 5.4\ \mu m^{-1}$) of the reconstructed diffraction pattern was kept fixed after 20 initial iterations.

A real space image of a colloidal sample obtained as a result of reconstruction of the diffraction pattern shown in Fig. 12.11b is presented in Fig. 12.11c. This image represents a *projection* of the "atomic" structure of the colloidal crystal along the [111] direction. The hexagonal structure is clear across the whole illuminated region, with only slightly lower intensity values of the image around the edges of the pinhole aperture. As a consequence of the image being a projection of 3D arrangement of colloidal particles, the periodicity does not correspond to the colloidal interparticle distance d in a *single*-crystalline layer. Due to ABC ordering in fcc crystals, a reduced periodicity of $d/\sqrt{3}$ is measured in this geometry.

The major differences from the results of previous work with CXDI on crystalline samples [44] are clearly demonstrated in Fig. 12.11c. Instead of a continuous shape and strain field reconstructed from the measurements of diffraction patterns around a single Bragg peak, the hexagonal structure shown in Fig. 12.11c gives the projected *positions* of the colloidal particles as it was discussed in details in Section 12.2.1

An estimate of the resolution by performing line scans through the reconstructed image in Fig. 12.11c and by measuring the widths of the peaks by Gaussian fit gives a full-width half maximum (FWHM) of 95 nm at the particle positions in Fig. 12.11c.

The diffraction patterns measured at an angle of $\varphi = 35°$ (see Fig. 12.11d) were especially intriguing. They show strong streaks of varying intensity originating at the Bragg peaks with an angle of 55° to the horizontal direction (Fig. 12.11d,e). It is well known from previous studies [73] of similar colloidal systems that such streaks in reciprocal space are induced by stacking faults in the fcc structure in the (111) planes.

These diffraction patterns were reconstructed using the same procedure as described earlier and the result of this reconstruction is presented in Fig. 12.11f. The "atomicity" of the colloidal crystal sample is again present in the reconstruction. In addition, a stacking fault appears (indicated by arrows in Fig. 12.11f) as a break in the "correct" ABC ordering [33]. One can see a stacking fault, which consists of two hcp planes, and two fcc domains with the same stacking direction. The effect of the stacking fault is a translation of the two fcc crystals relative to each other. This "sliding" can be seen in Fig. 12.11f as a "break" of the lines of bright spots at the defect. It was recently suggested that, over a large (submillimeter) sample area, these double-stacking defects consisting of two hcp planes is a common imperfection in convectively assembled colloidal crystals [74].

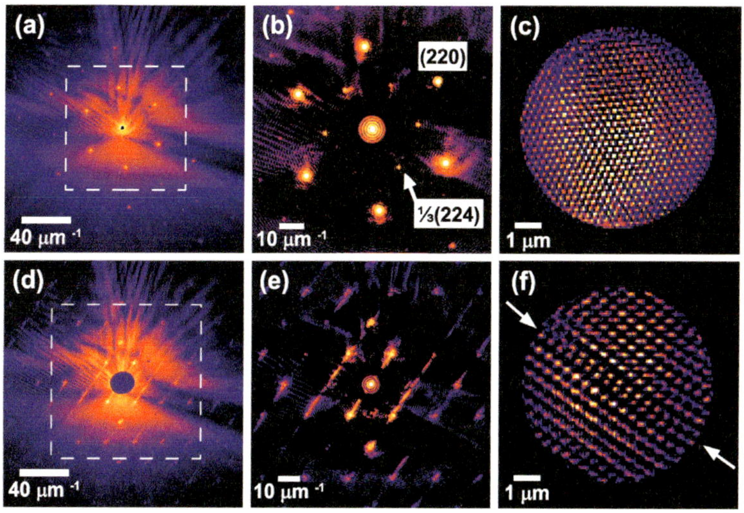

Figure 12.11 (a,d) Measured diffraction patterns from the pinhole and the sample. Marked regions in (a,d) correspond to the area used for the reconstruction. (b,e) Difference diffraction patterns obtained as a result of the subtraction of the scaled diffraction pattern of the pinhole from the measured diffraction patterns of the sample. Diffraction patterns are shown on a logarithmic scale. (c,f) Reconstruction of the colloid sample from the diffraction patterns. The arrows in (f) point to the defect in the crystal. First row (a,b,c) measurement at the azimuthal angle $\varphi = 0$ and the second row (d,e,f) at the angle $\varphi = 35°$. Adapted from Ref. [53].

Next we demonstrated that the simple and nondestructive mechanism of coherent X-ray diffractive imaging opens a unique route to determine the structure of mesoscopic materials such as colloidal crystals. CXDI has the potential to provide detailed information about the local defect structure in colloidal crystals. This is especially important for imaging photonic materials when refraction index matching is not possible or the sizes of colloidal particles are too small for conventional optical microscopy. To extend this method to larger fields of view scanning methods such as ptychography [18–20] can be used, while tomographic methods such as coherent X-ray tomography [75] have the potential to visualize the atomic structure of the defect core in 3D (see recent publication [54]).

12.3.2 Coherent Diffraction Tomography of Nanoislands from Grazing Incidence Small-Angle X-Ray Scattering

In our second example, we show how tomographic methods can be combined with CXDI to provide 3D images of nanocrystaline materials (see Ref. [75] for details).

Tomography and especially X-ray tomography has become one of the most important tools for investigating 3D structures in condensed matter [76]. When projected absorption contrast or phase contrast measurements are carried out in conventional X-ray transmission tomography, the achievable resolution is limited by the spatial resolution of the area detector that can be about one micrometer. CXDI represents a possible solution to this dilemma. As no lenses are required in this imaging technique and the resolution is given by the scattered signal, in principle, the resolution limits of conventional X-ray transmission tomography can be overcome.

To obtain a 3D image of a noncrystallographic object in the forward scattering geometry by the CXDI technique the sample is usually mounted on a $Si_3 N_4$ membrane (see Ref. [77]) and then rotated with fixed azimuthal angular steps (see inset (a) in Fig. 12.12). Unfortunately, in this approach not all angles for a full 3D scan are accessible due to the positioning of the object on a membrane. Obviously, the inaccessible part of the reciprocal space is not available for the tomographic reconstruction, resulting in a certain loss of features that are actually present in the original object. Instead, as it was first proposed in [48,75], a sample can be positioned on a flat thick substrate and tomographic scans can be performed by collecting successive coherent scattering diffraction patterns at different azimuthal positions of a sample in a grazing-incidence small-angle X-ray scattering (GISAXS) geometry [48] (see Fig. 12.12). With this approach, there are no limitations on the angle of rotation. Consequently, large areas in reciprocal space can be measured with sufficient resolution and without missing wedge. The feasibility of this approach was tested and proven by a number of simulations [78]. Below experimental realization of this coherent diffraction tomographic technique is reported [75].

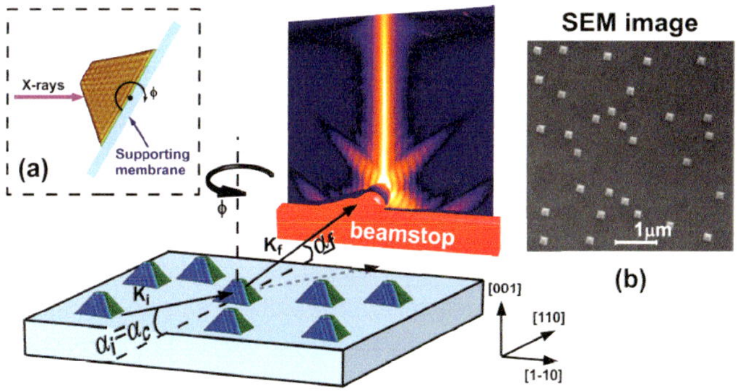

Figure 12.12 Schematic diagram of the GISAXS scattering geometry on a group of nanoislands in the form of a truncated pyramid with a square base. The incident wave vector $\mathbf{k}_i$ at grazing incidence angle $\alpha_i = \alpha_c$ is shown and the scattered wave vector $\mathbf{k}_f$ at angles $\alpha_f \geq \alpha_c$. The sample is rotated around the surface normal (azimuth angles). Inset (a): Schematic diagram of a conventional CXDI tomography when the sample is positioned on a supporting membrane. Inset (b): A scanning electron microscopy image of the nanoislands. Adapted from Ref. [75].

As a model samples SiGe islands of 200 nm average base size grown by liquid phase epitaxy were used. All islands were coherently grown on a (001) Si surface and exhibit a truncated pyramidal shape with a square base (see inset (b) in Fig. 12.12). In addition, they exhibit a narrow-size distribution (~10 percent FWHM) and the same crystallographic orientation on the Si surface (Fig. 12.12).

Experiments [48,75,79] were performed at the ID01 beamline of the ESRF in Grenoble. The incidence angle was taken equal to the critical angle for total external reflection of the Si substrate that corresponds to $\alpha_i = \alpha_c = 0.224°$ for the chosen X-ray energy of 8 keV. This particular angle was used because at these conditions the scattering may be considered as predominantly kinematical [80]. The coherently scattered signal was measured up to $q_{||} = \pm 0.56\ \text{nm}^{-1}$ in reciprocal space in the transverse direction. However, due to a certain noise level, only a limited part of reciprocal space up to $q_{||} =$

±0.36 nm^{-1} was considered for the reconstruction, which provides 17.4 nm resolution in real space. An azimuth scan was performed from −5° to 50° with an angular increment of 1°. Due to the four-fold and mirror symmetry of {111} facetted islands, such scans cover the whole reciprocal space. During the azimuthal scan, the incidence angle was kept constant at the critical angle α_c.

The diffraction patterns at each azimuthal angle position represent slices through the 3D reciprocal space on a pseudopolar grid and can be combined to produce a 3D intensity distribution in reciprocal space (see Fig. 12.13). An iso-intensity surface of this 3D representation of the scattered intensity is shown in Fig. 12.13c. Strong crystal truncation rods (CTR) along the <111> directions and much weaker crystal truncation planes (CTP) connecting the CTRs can be observed [48]. For comparison we performed calculations of GISAXS diffraction patterns in the framework of the distorted-wave Born approximation (DWBA) theory [81,82] for similar pyramids. The corresponding 3D representation of the scattered intensity is shown in Fig. 12.13a. Strong interference fringes due to the coherent scattering of the X-ray beam from a small pyramid-shaped object can be observed in this figure. These interference fringes are smeared out in our experimental data (Fig. 12.13c) partly due to a finite-size distribution of the islands and partly due to a lack of sufficient counting statistics. In order to model the experimental results, we added a Gaussian mask to the calculated set of data. The corresponding 3D representation of the scattered intensity is shown in Fig. 12.13b.

Results of the island shape reconstruction from the experimental GISAXS diffraction patterns are presented in Fig. 12.13f. For a comparison, results of the reconstruction of the island shape from simulated data are also shown in Fig. 12.13d,e. The electron densities of the islands obtained as a vertical section through the center of the islands are presented in Fig. 12.13g–i. From these results, it can be seen that the shape of the islands is reconstructed correctly for the experimental data set (Fig. 12.13f). However, for the electron density inside the island we observe artifacts in the form of low density regions in the bottom of the islands (Fig. 12.13i). Reconstructions performed with the simulated data sets show that the scattering data obtained in the DWBA conditions correctly reproduce the shape (Fig. 12.13d) and electron density (Fig. 12.13g) of an island.

However, when the modified theoretical data set with the Gaussian mask is used for reconstruction, artifacts similar to those from the experimental data set appear.

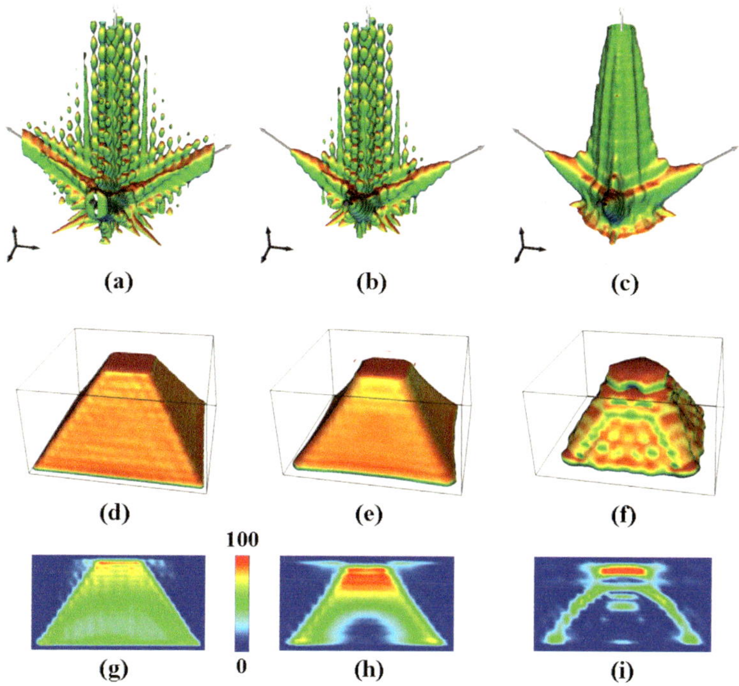

Figure 12.13 Left column [(a), (d), (g)]: simulations in the framework of the DWBA theory. Middle column [(b), (e), (h)]: simulations in the framework of the DWBA theory with an additional Gaussian mask (see text for details). Right column [(c), (f), (i)]: experiment. [(a), (b),(c)]: 3D plot of an isointensity surface in the reciprocal space. RGB colors correspond to the z-projection of the iso-surface normal. Gray arrows indicate directions along the crystallographic planes (001) top and {111} on the side. Black arrows indicate q_x, q_y, q_z directions in reciprocal space. The length of each black arrow corresponds to 0.1 nm^{-1}. [(d), (e), (f)]: Reconstructed shape of the islands. The transparent box indicates the size of the support. [(g), (h), (i)]: Electron density of the islands obtained as a vertical section through the center of each island. Adapted from Ref. [75].

Here, we have demonstrated how this approach of coherent diffraction GISAXS can be used to obtain the 3D electron density of nanometer-sized islands. This was achieved by performing tomographic azimuth scans in a GISAXS geometry on many identical islands and subsequent phase retrieval which yields the tomographic information, such as the shape and the electron density. It is important to note that this approach does not depend on the crystalline structure of such an island and may be applied to any material system.

12.3.3 Coherent-Pulse 2D Crystallography at Free-Electron Lasers

As our third example, we show how ultrabright coherent pulses of a new free-electron laser (FEL) sources can be applied to determine the structure of 2D crystallographic objects (see for details Ref. [83], and also for reviews of CXDI experiments at FEL source [5,6,9]).

Crystallization and radiation damage is presently a bottleneck in protein structure determination. As it was first proposed in Ref. [83], 2D finite crystals and ultrashort FEL pulses can be effectively used to reveal the structure of single molecules. This can be especially important for membrane proteins that in general do not form 3D crystals, but easily form 2D crystalline structures. In this chapter, single pulse train coherent diffractive imaging was demonstrated for a finite 2D crystalline sample, and it was concluded that this alternative approach to single-molecule imaging is a significant step towards revealing the structure of proteins with subnanometer resolution at the newly built XFEL sources.

Revealing the structure of protein molecules is mandatory for understanding the structure of larger biological complexes. The major progress in uncovering the structure of proteins in past decades was due to the development of phasing methods [84] allowing the determination of the structure of complex molecules that crystallize. One new approach to overcome these difficulties is based on the use of ultrashort pulses of X-ray FELs (XFEL) [85–87]. This elegant idea is based on measuring a sufficiently sampled diffraction pattern from a single molecule illuminated by an FEL pulse [88,89]. However, despite the extreme intensity of the FEL pulses, a diffraction pattern from only one molecule will not be sufficient to obtain a high-resolution diffraction pattern. Many reproducible copies will need to be measured to get a sufficient signal to noise ratio for each projection necessary for 3D imaging at subnanometer spatial resolution.

FELs are especially well suited for such coherent 2D crystallography. They provide femtosecond coherent pulses [70,90,91] with extremely high power. Only the combination of all of these unique properties will allow the realization of 2D crystallographic X-ray imaging on biological systems. Brilliant, ultrashort pulses could overcome the radiation damage problem [88,92], which is a severe limitation of conventional crystallography at third generation synchrotron sources [93]. Higher luminosity and hence improved statistics for such experiments can be obtained by the use of pulse trains that can be provided by FLASH (Free-electron LASer in Hamburg) [94].

A finite 2D crystallography was demonstrated by using a microstructured crystal array that was prepared on a 100 nm thick silicon nitride membrane substrate coated with 600 nm of gold, and 200 nm of palladium. The finite crystal sample was manufactured by milling holes in the film in a regular array pattern using a focused ion beam (FIB). The "unit cell" of our crystal consists of a large hole of 500 nm diameter (representing a "heavy atom" in conventional crystallography) and a smaller hole of 200 nm diameter (representing a "light atom"). The whole structure was composed of five unit cells in each direction, making the total structure size about 10 μm × 10 μm.

The diffraction data were measured at FLASH on the PG2 monochromator beamline [95] with a fundamental wavelength of 7.97 nm. An exposure time of 0.2 s was used to collect a series of single pulse train data from our sample. FLASH was operated in a regime producing 21 bunches of electrons per pulse train, with a pulse train repetition rate of 5 Hz. The bunches within each pulse train were spaced at 1 MHz. The average pulse energy was 15 μJ which is equivalent to 6×10^{11} photons per pulse or 1.3×10^{13} photons per train at the source. The coherent flux on the sample area was 1.5×10^{10} photons per pulse train.

A typical data set is shown in Fig. 12.14a. The diffraction pattern as measured fills the whole detector, which corresponds to a minimum feature size of 220 nm (Fig. 12.14a). We note that all expected features of a finite, crystalline structure as they were discussed in the previous sections are observed in this diffraction pattern. The Bragg peaks due to the regular array are clearly seen, as are the oscillations between the Bragg peaks that are the result of the finite extent and coherent illumination of our sample. Also seen is the form factor from the individual elements—the large holes—that can be observed as a radial intensity modulation across the measured diffraction pattern.

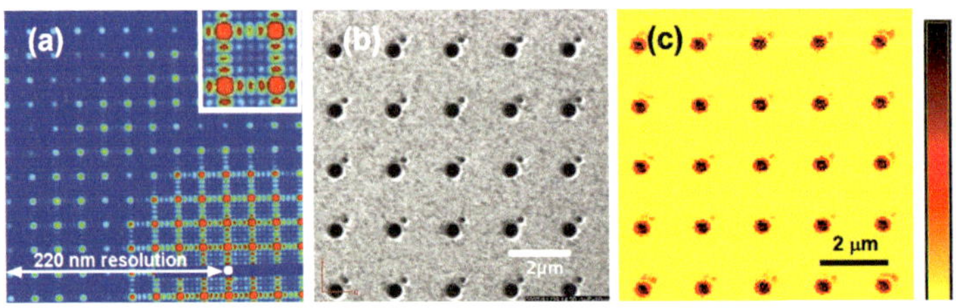

Figure 12.14 (a) Far-field diffraction data measured from a single train of 21 femtosecond pulses from the FEL. (Inset: Enlarged region of diffraction pattern). (b) SIM image of the finite, periodic structure. (c) The reconstructed image using the original data binned 5 × 5. Adapted from Ref. [83].

Due to the limited signal to noise ratio of the initial data set and the symmetry of the unit cell the initial reconstructions stagnated with two equivalent solutions superimposed. One is with the small dots appearing at the top right of the larger dots, and the other is with the dots appearing at the bottom left. To solve this problem, the data were binned 5 × 5 and a more constrained support of 25 rectangular boxes each centered on the positions of the unit cell was used. This data set was used for reconstruction by applying the HIO [12] iterative phase retrieval algorithm. By increasing the signal-to-noise ratio and reducing the symmetry in real space, it was possible to improve the reconstruction to resolve the smallest features in the sample (Fig. 12.14c). A scanning ion micrograph (SIM) image of the object under investigation is shown in Fig. 12.14b for comparison. The resolution in real space in the reconstructed image was better than 240 nm. This compares favorably with the measured maximum momentum transfer corresponding to a 220 nm resolution.

Summarizing the XFEL experiment, it was demonstrated that single pulse train coherent diffractive imaging is possible for a finite

2D crystalline sample with the reconstructed image exhibiting resolution commensurate with the measured data. In this experiment, the *crystalline* structure was essential in providing the necessary information to determine the structure of the unit cells. If only a *single* unit cell had been used simulations suggest that a successful reconstruction would be impossible with the resolution presented in the example. This alternative approach to single molecule imaging is a significant step toward revealing the structure of proteins with subnanometer resolution at the newly built XFEL sources.

12.4 Summary

Coherent X-ray diffractive imaging gives us a high-resolution imaging tool to reveal the electron density and strain in nanocrystalline samples. Progress is still ongoing. We foresee that in future, it will reach a resolution of approximately a few nanometers at synchrotron sources, and a few angstroms for the protein nanocrystals imaged with the ultrashort FEL pulses [96]. Nonreproducible objects may be imaged with a few nanometer resolution at XFELs [9,26,97]. There are several technological developments that are incremental for the future progress of the CXDI technique. The ultimate hard X-ray storage rings [98] with a few tens picometer emittance are under discussion. The advances in hard X-ray optics will allow reaching sub-ten nanometer focus sizes [99]. Also, hard X-ray FEL sources now become online. Finally, the mathematical tools for interpretation of data from coherent X-ray scattering experiments are continuously improving. Thus, we can expect CXDI to become eminently important for the investigation of the structure of materials on the nanoscale and to grow rapidly in the near future.

References

1. Vartanyants IA, Pitney JA, Libbert JL, Robinson IK (1997) Reconstruction of surface morphology from coherent X-ray reflectivity, *Phys Rev B*, **55**, 13193–13202.

2. Miao J, Charalambous P, Kirz J, Sayre D (1999) Extending the methodology of X-ray crystallography to allow imaging of micrometre-sized non-crystalline specimens, *Nature*, **400**, 342–344.

3. Robinson IK, Libbert JL, Vartanyants IA, et al. (1999) Coherent X-ray diffraction imaging of silicon oxide growth, *Phys Rev B*, **60**, 9965–9972.
4. Robinson IK, Vartanyants IA, Williams GJ, Pfeifer MA, Pitney JA (2001) Reconstruction of the shapes of gold nanocrystals using coherent X-ray diffraction, *Phys Rev Lett*, **87**, 195505.
5. Vartanyants IA, Robinson IK, McNulty I, David C, Wochner P, Tschentscher Th (2007) Coherent X-ray scattering and lensless imaging at the European XFEL facility, *J Synchrotron Rad*, **14**, 453–470.
6. Vartanyants IA, Mancuso AP, Singer A, Yefanov OM, Gulden J (2010) Coherence measurements and coherent diffractive imaging at FLASH, *J Phys B: At Mol Opt Phys*, **43**, 194016/1–10.
7. Nugent KA (2010) Coherent methods in the X-ray sciences, *Adv Phys*, **59**, 1–99.
8. Robinson IK, Harder R (2009) Coherent X-ray diffraction imaging of strain at the nanoscale, *Nat Mater*, **8**, 291–298.
9. Mancuso AP, Yefanov OM, Vartanyants IA (2010) Coherent diffractive imaging of biological samples at synchrotron and free electron laser facilities, *J Biotechnol*, **149**, 229–237.
10. Quiney HM (2011) Coherent diffractive imaging using short wavelength light sources, *J Mod Opt*, **57**, 1109–1149.
11. Gerchberg RW, Saxton WO (1972) A practical algorithm for the determination of phase from image and diffraction plane pictures, *Optik*, **35**, 237–246.
12. Fienup JR (1982) Phase retrieval algorithms: a comparison, *Appl Opt*, **21**, 2758–2769.
13. Elser V (2003) Phase retrieval by iterated projections, *J Opt Soc Am A*, **20**, 40–55.
14. Marchesini S (2007) A unified evaluation of iterative projection algorithms for phase retrieval, *Rev Sci Instr*, **78**, 049901.
15. Bates RHT (1982) Fourier phase problems are uniquely solvable in more than one dimension. I: underlying theory, *Optik*, **61**, 247–262.
16. Shannon CE (1949) Communication in the presence of noise, *Proc IRE*, **37**, 10–21.
17. Chen C-C, Miao J, Wang CW, Lee TK (2007) Application of optimization technique to noncrystalline X-ray diffraction microscopy: guided hybrid input-output method, *Phys Rev B*, **76**, 064113.
18. Rodenburg JM, Hurst AC, Cullis AG, et al. (2007) Hard-X-ray lensless imaging of extended objects, *Phys Rev Lett*, **98**, 034801.

19. Thibault P, Dierolf M, Menzel A, Bunk O, David Ch, Pfeiffer F (2008) High-resolution scanning X-ray diffraction microscopy, *Science*, **321**, 379–382.

20. Schropp A, Boye P, Feldkamp JM, et al. (2010) Hard X-ray nanobeam characterization by coherent diffraction microscopy, *Appl Phys Lett*, **96**, 091102/1–3.

21. Godard P, Carbone G, Allain M, et al. (2011) Three-dimensional high-resolution quantitative microscopy of extended crystals, *Nat Comm*, **2**, 568.

22. Takahashi Y, Zettsu N, Nishino Y, et al. (2010) Three-dimensional electron density mapping of shape-controlled nanoparticle by focused hard X-ray diffraction microscopy, *Nano Lett*, **10**, 1922–1926.

23. Giewekemeyer K, Pierre Thibault P, Kalbfleisch S, et al. (2010) Quantitative biological imaging by ptychographic X-ray diffraction microscopy, *Proc Natl Acad Sci USA*, **107**, 529–534.

24. Dierolf M, Menzel A, Thibault P, et al. (2010) Ptychographic X-ray computed tomography at the nanoscale, *Nature*, **467**, 436–439.

25. Shen Q, Bazarov I, Thibault P (2004) Diffractive imaging of nonperiodic materials with future coherent X-ray sources, *J Synchrotron Rad*, **11**, 432–438.

26. Bergh M, Huldt G, Timneanu N, Maia FRNC, Hajdu J (2008) Feasibility of imaging living cells at subnanometer resolutions by ultrafast X-ray diffraction, *Q Rev Biophys*, **41**, 181–204.

27. Huang X, Miao H, Steinbrener J, et al. (2009) Signal-to-noise and radiation exposure considerations in conventional and diffraction microscopy, *Opt Express*, **17**, 13541–13553.

28. Thibault P, Elser V, Jacobsen C, Shapiro D, Sayre D (2006) Reconstruction of a yeast cell from X-ray diffraction data, *Acta Cryst A*, **62**, 248–261.

29. Vartanyants IA, Robinson IK (2001) Partial coherence effects on the imaging of small crystals using coherent X-ray diffraction, *J Phys: Condens Matter*, **13**, 10593–10611.

30. Vartanyants IA, Robinson IK (2003) Origins of decoherence in coherent X-ray diffraction experiments, *Opt Commun*, **222**, 29–50.

31. Williams GJ, Quiney HM, Peele AG, Nugent KA (2007) Coherent diffractive imaging and partial coherence, *Phys Rev B*, **75**, 104102.

32. Millane RP (1990) Phase retrieval in crystallography and optics, *J Opt Soc Am A*, **7**, 394–411.

33. Warren BE (1990) *X-Ray Diffraction*, Dover Publ. Inc., New York.

34. Als-Nielsen J, McMorrow D (2011) *Elements of Modern X-Ray Physics*, John Wiley and Sons, New York.
35. Batterman BW, Cole H (1964) Dynamical diffraction of X rays by perfect crystals, *Rev Mod Phys*, **36**, 681–717.
36. Authier A (2003) *Dynamical Theory of X-Ray Diffraction*, 2nd ed, Oxford University Press, Oxford.
37. Shabalin A, Yefanov OM, Nosik VL, Vartanyants IA (XXXX) Dynamical effects in coherent X-ray scattering on a finite crystal. (In preparation).
38. Harder R, Pfeifer MA, Williams GJ, Vartanyants IA, Robinson IK (2007) Orientation variation of surface strain, *Phys Rev B*, **76**, 115425/1–4.
39. Ewald PP (1940) X-ray diffraction by finite and imperfect crystal lattices, *Proc Phys Soc*, **52**, 167.
40. von Laue M (1936) Die äußere Form der Kristalle in ihrem Einfluß auf die Interferenzerscheinungen an Raumgittern, *Ann d Physik*, **26**, 55.
41. Sayre D (1952) Some implications of a theorem due to Shannon, *Acta Cryst*, **5**, 843.
42. Williams GJ, Pfeifer MA, Vartanyants IA, Robinson IK (2003) Three-dimensional imaging of microstructure in Au nanocrystals, *Phys Rev Lett*, **90**, 175501.
43. Millane RP (1996) Multidimensional phase problems, *J Opt Soc Am A*, **13**, 725.
44. Pfeifer MA, Williams GJ, Vartanyants IA, Harder R, Robinson IK (2006) Three-dimensional mapping of a deformation field inside a nanocrystal, *Nature*, **442**, 63–66.
45. Robinson IK (1986) Crystal truncation rods and surface roughness, *Phys Rev B*, **33**, 3830–3836.
46. Afanas'ev AM, Aleksandrov PA, Imamov RM, Lomov AA, Zavyalova AA (1985) Three-crystal diffractometry in grazing Bragg-Laue geometry, *Acta Cryst A*, **41**, 227.
47. Pitney JA, Robinson IK, Vartanyants IA, Appelton R, Flynn CP (2000) Streaked speckle in Cu_3Au coherent diffraction, *Phys Rev B* **62**, 13084–13088.
48. Vartanyants IA, Zozulya AV, Mundboth K, et al. (2008) Crystal truncation planes revealed by three-dimensional reconstruction of reciprocal space, *Phys Rev B* **77**, 115317.
49. Eisebitt S, Luning J, Schlotter WF, et al. (2004) Lensless imaging of magnetic nanostructures by X-ray spectro-holography, *Nature*, **432**, 885–888.

50. Stadler L-M, Gutt Ch, Autenrieth T, et al. (2008) Hard X-ray holographic diffraction imaging, *Phys Rev Lett*, **100**, 245503.

51. Chamard V, Stangl J, Carbone G, et al. (2010) Three-dimensional X-ray Fourier transform holography: the Bragg case, *Phys Rev Lett*, **104**, 165501.

52. Gulden J, Yefanov OM, Weckert E, Vartanyants IA (2011) Imaging of nanocrystals with atomic resolution using high-energy coherent X-rays, *The 10th International Conference on X-Ray Microscopy, AIP Conf Proc*, **1365**, 42–45.

53. Gulden J, Yefanov OM, Mancuso AP, et al. (2010) Coherent X-ray imaging of defects in colloidal crystals, *Phys Rev B*, **81**, 224105.

54. Gulden J, Yefanov OM, Mancuso AP, et al. (2012) Three-dimensional structure of a single colloidal crystal grain studied by coherent X-ray diffraction, *Opt Express*, **20**, 4039–4049.

55. Zuo J-M, Vartanyants IA, Gao M, Zhang R, Nagahara LA (2003) Atomic resolution imaging of a carbon nanotube from diffraction intensities, *Science*, **300**, 1419–1421.

56. Huang WJ, Zuo JM, Jiang B, Kwon KW, Shim M (2009) Sub-angstrom-resolution diffractive imaging of single nanocrystals, *Nat Phys*, **5**, 129–133.

57. Dronyak R, Liang KS, Stetsko YP, et al. (2009) Electron diffractive imaging of nano-objects using a guided method with a dynamic support, *Appl Phys Lett*, **95**, 111908.

58. Krivoglaz MA (1969) *Theory of X-Ray and Thermal-Neutron Scattering by Real Crystals*, Plenum Press, New York.

59. Robinson IK, Vartanyants IA (2001) Use of coherent X-ray diffraction to map strain fields in nanocrystals, *Appl Surf Sci*, **182**, 186–191.

60. Vartanyants IA, Ern C, Donner W, Dosch H, Caliebe W (2000) Strain profiles in epitaxial films from X-ray Bragg diffraction phases, *Appl Phys Lett*, **77**, 3929–3931.

61. Minkevich AA, Gailhanou M, Micha J-S, Charlet B, Chamard V, Thomas O (2007) Inversion of the diffraction pattern from an inhomogeneously strained crystal using an iterative algorithm, *Phys Rev B*, **76**, 104106/1–5.

62. Newton MC, Leake SJ, Harder R, Robinson IK (2010) Three-dimensional imaging of strain in a single ZnO nanorod, *Nat Mater*, **9**, 120–124.

63. Watari M, McKendry RA, Vögtli M, et al. (2011) Differential stress induced by thiol adsorption on facetted nanocrystals, *Nat Mater*, **10**, 862–866.

64. Born M, Wolf E (2000) *Principles of Optics*, Cambridge University Press, Cambridge.
65. Mandel L, Wolf E (1995) *Optical Coherence and Quantum Optics*, Cambridge University Press, Cambridge.
66. Goodman JW (1985) *Statistical Optics*, Wiley, New York.
67. Szöke A (2001) Diffraction of partially coherent X-rays and the crystallographic phase problem, *Acta Cryst*, A**57**, 586–603.
68. Sinha SK, Tolan M, Gibaud A (1998) The effects of partial coherence on the scattering of X-rays by matter, *Phys Rev B*, **57**, 2740–2758.
69. Gutt C, Ghaderi T, Tolan M, Sinha SK, Grübel G (2008) Effects of partial coherence on correlation functions measured by X-ray photon correlation spectroscopy, *Phys Rev B*, **77**, 094133/1–10.
70. Vartanyants IA, Singer A (2010) Coherence properties of hard X-ray synchrotron sources and X-ray free-electron lasers, *New J Phys*, **12**, 035004.
71. Balewski K, Brefeld W, Decking W, Franz H, Röhlsberger R, Weckert E (2004) *PETRA III: A Low Emittance Synchrotron Radiation Source. Technical Design Report*, DESY, Hamburg.
72. Lodahl P, van Driel AF, Nikolaev IS, et al. (2004) Controlling the dynamics of spontaneous emission from quantum dots by photonic crystals, *Nature*, **430**, 654.
73. Hilhorst J, Abramova VV, Sinitskii A, et al. (2009) Double stacking faults in convectively assembled crystals of colloidal spheres, *Langmuir*, **25**, 10408.
74. Napolskii KS, Sapoletova KS, Gorozhankin DF, et al. (2010) Fabrication of artificial opals by electric-field-assisted vertical deposition, *Langmuir*, **26**, 2346.
75. Yefanov OM, Zozulya AV, Vartanyants IA, et al. (2009) Coherent diffraction tomography of nanoislands from grazing-incidence small-angle X-ray scattering, *Appl Phys Lett*, **94**, 123104.
76. Banhart J (2008) *Advanced Tomographic Methods in Materials Research and Engineering*, Oxford University Press, New York.
77. Miao J, Chen C-C, Song Ch, et al. (2006) Three-dimensional GaN-Ga_2O_3 core shell structure revealed by X-ray diffraction microscopy, *Phys Rev Lett*, **97**, 215503.
78. Yefanov OM, Vartanyants IA (2009) Three dimensional reconstruction of nanoislands from grazing-incidence small-angle X-ray scattering, *Eur Phys J Special Topics*, **167**, 81–86.

79. Zozulya AV, Yefanov OM, Vartanyants IA, et al. (2008) Imaging of nano-islands in coherent grazing-incidence small-angle X-ray scattering experiments, *Phys Rev B*, **78**, 121304.

80. Vartanyants I, Grigoriev D, Zozulya A (2007) Coherent X-ray imaging of individual islands in GISAXS geometry, *Thin Solid Films*, **515**, 5546.

81. Sinha SK, Sirota EB, Garoff S, Stanley HB (1988) X-ray and neutron scattering from rough surfaces, *Phys Rev B*, **38**, 2297.

82. Rauscher M, Paniago R, Metzger H, et al. (1999) Grazing incidence small angle X-ray scattering from free- standing nanostructures, *J Appl Phys*, **86**, 6763.

83. Mancuso AP, Schropp A, Reime B, et al. (2009) Coherent-pulse 2D crystallography using a free-electron laser X-ray source, *Phys Rev Lett*, **102**, 035502.

84. Woolfson M, Hai-Fu F (1995) *Physical and Non-Physical Methods of Solving Crystal Structures*, Cambridge University Press, Cambridge.

85. Emma P, Akre R, Arthur J, et al. (2010) First lasing and operation of an ångstrom-wavelength free-electron laser, *Nat Photon*, **4**, 641.

86. Tanaka H, Yabashi M, et al. (2012) A compact X-ray free-electron laser emitting in the sub-angstrom region, *Nature Photonics*, **6**, 540.

87. Altarelli M, et al. (2006) *XFEL: The European X-Ray Free-Electron Laser. Technical Design Report DESY 2006-097*, http://xfel.desy.de/tdr/tdr/.

88. Neutze R, Wouts R, van der Spoel D, Weckert E, Hajdu J (2000) Potential for biomolecular imaging with femtosecond X-ray pulses, *Nature*, **406**, 752.

89. Gaffney KJ, Chapman HN (2007) Imaging atomic structure and dynamics with ultrafast X-ray scattering, *Science*, **316**, 1444–1448.

90. Singer A, Vartanyants IA, Kuhlmann M, Düsterer S, Treusch R, Feldhaus J (2008) Transverse-coherence properties of the free-electron-laser FLASH at DESY, *Phys Rev Lett*, **101**, 254801.

91. Vartanyants IA, Singer A, Mancuso AP, et al. (2011) Coherence properties of individual femtosecond pulses of an X-ray free-electron laser, *Phys Rev Lett*, **107**, 144801/1–5.

92. Howells MR, Beetz T, Chapman HN, et al. (2009) An assessment of the resolution limitation due to radiation-damage in X-ray diffraction microscopy, *J Electron Spectrosc Relat Phenom*, **170,** 4–12.

93. Nave C, Garman EF (2005) Towards an understanding of radiation damage in cryocooled macromolecular crystals, *J Synchrotron Rad*, **12**, 257.

94. Ackermann W, et al. (2007) Operation of a free electron laser in the wavelength range from the extreme ultraviolet to the water window, *Nat Photon*, **1**, 336–342.
95. Wellhoefer M, Martins M, Wurth W, Sorokin AA, Richter M (2007) Performance of the monochromator beamline at FLASH, *J Opt A: Pure Appl Opt*, **9**, 749.
96. Chapman HN, et al. (2011) Femtosecond X-ray protein nanocrystallography, *Nature*, **470**, 73–77.
97. Seibert M, et al. (2011) Single mimivirus particles intercepted and imaged with an X-ray laser, *Nature*, **470**, 78–81.
98. Bei M, Borland M, Cai Y, et al. (2010) The potential of an ultimate storage ring for future light sources, *Nucl Instrum Meth Phys Res A*, **622**, 518-535.
99. Mimura H, Handa S, Kimura T, et al. (2010) Breaking the 10 nm barrier in hard-X-ray focusing, *Nat Phys*, **6**, 122–125.

Chapter 13

X-Ray Photon Correlation Spectroscopy

Christian Gutt[a,b] and Michael Sprung[a]
[a]*Deutsches Elektronen Synchrotron (DESY), Notkestraße 85, 22607 Hamburg, Germany*
[b]*Department of Physics, University of Siegen, Emmy-Noether-Campus, Walter-Flex-Str. 3, 57072 Siegen, Germany*
christian.gutt@uni-siegen.de, michael.sprung@desy.de

13.1 Introduction

X-ray diffraction is an important tool as a structural probe for condensed matter on atomic length scales. Since its early beginning, it allowed for structural determination of a wide variety of materials such as hard- and soft-condensed matter. Inherent in conventional X-ray diffraction experiments is an ensemble averaging process over the illuminated volume of the sample. The measured structure factor $S(Q)$ is such an ensemble averaged quantity according to $S(Q) = <\tilde{\rho}(Q)\tilde{\rho}^*(Q)>_E$ with $\tilde{\rho}(Q)$ denoting the Fourier transform of the sample's electron density $\rho(r)$ and Q is the wave vector transfer. Quite often, such an averaging process is desirable as it allows atomic-scale quantities to be measured over a region of micro- to millimeters with great statistical precision. However, the incoherent averaging process leads to a loss of information, which is not desirable whenever samples are, for example, nonperiodic or evolve in time.

X-Ray Diffraction: Modern Experimental Techniques
Edited by Oliver H. Seeck and Bridget M. Murphy

ISBN 978-981-4303-59-0 (Hardcover), 978-981-4303-60-6 (eBook)
www.panstanford.com

If the X-ray waves exhibit a high degree of coherence over the illuminated sample volume, the measured quantity $S(Q) = \tilde{\rho}(Q)\tilde{\rho}^*(Q)$ is without any statistical averages. Thus, if $\tilde{\rho}(Q)$ has built-in a static randomness or disorder, the structure factor $S(Q)$ exhibits sharp and random fluctuations about some average value giving rise to the phenomenon of speckle [1,2]. As the sample is illuminated with coherent waves, the resulting diffraction pattern reflects the exact spatial arrangement of all the particles in the beam. Thus, if particles are fluctuating, as they do, for example, in a liquid suspension of colloids, the speckle pattern fluctuates accordingly. Those intensity fluctuations can be measured using autocorrelation techniques revealing the dynamic structure factor of the investigated system.

With the advent of the first visible laser sources in the late 1960s, this technique—known as dynamic light scattering (DLS)—became popular as it allows in a rather simple way the determination of, for example, particle sizes and diffusion constants [3]. Visible light, however, is limited in spatial resolution by its long wavelength and often by practical aspects such as multiple scattering. Here, X-rays offer distinct advantages as they can—at least in principle—provide atomic resolution and can easily penetrate matter. Under grazing incidence conditions, X-rays are surface sensitive and by tuning photon energies to distinct absorption edges the scattering mechanism is highly element specific and sensitive to magnetic properties.

Modern third-generation synchrotron sources provide sufficient coherent photons to implement X-ray scattering techniques, which specialized in the use of coherence. X-ray photon correlation spectroscopy (XPCS), the X-ray analogue of DLS, was a natural candidate and was developed as soon as third-generation sources became available [4,5]. Over the past 15 years, significant improvements to the XPCS technique have been achieved, and today many successful XPCS experiments are reported and several review articles are available [6–12].

Even the most brilliant third-generation synchrotron sources are intrinsically incoherent sources and the (partially) coherent X-ray photons must be obtained by cutting out the small coherent portion of the beam at the cost of available total X-ray flux. The advent of the X-ray free-electron laser sources changes the X-ray landscape dramatically [13]. They (e.g., FLASH and the future EuXFEL in Germany, the LCLS in the United States or the SCSS in Japan) provide highly intense, short pulses with a high degree of transverse coherence and of unprecedented peak brightness [14,15]. They allow performing XPCS experiments with time resolutions of pico- to nanoseconds giving completely new insights into the equilibrium and nonequilibrium properties of condensed matter.

This chapter presents an overview of the current status of XPCS at the third-generation synchrotron sources and possible extensions to free-electron laser sources. First, a theoretical introduction into the technique of XPCS with the accessible quantities is given. Expressions of the correlation functions for equilibrium and nonequilibrium dynamics are derived, followed by an introduction into the new field of X-ray cross-correlation analysis (XCCA). The design of successful XPCS experiments is discussed next with special emphasis on coherence properties and detector issues. A typical XPCS beamline setup is given by the example of station 8ID-I/XOR of the APS. Next, several scientific highlights illustrate the power of correlation techniques and the chapter closes with a short summary.

13.2 Theory

13.2.1 Equilibrium Fluctuations

Scattering experiments measure density correlation functions. In the simplest case, the static structure factor $S(Q)$ represents the Fourier transform of the density–density correlation function $\int \rho(r)\rho(r+R)dr = <\rho(0)\rho(R)>$ or, in other words, the correlation of the density at point r with the density at a point $r + R$. XPCS experiments seek to find the time-dependent density correlation function $<\rho(0, 0)\ \rho(R, \tau)>$.

For this purpose, XPCS experiments measure the intensity–intensity time autocorrelation function (ACF). This function describes the correlation of the X-ray intensity $I(r_1, t_1)$ measured at time t_1 at position r_1 with the intensity $I(r_2, t_2)$ measured at a later time t_2 and eventually at a different position r_2 in space

$$G_2(r_1, r_2, t_1, t_2) = < I(r_1, t_1)\ I(r_2, t_2)>. \tag{13.1}$$

We can express the intensity correlation function with the aid of the scattered fields $U_S(r, t)$ as

$$G_2(r_1, r_2, t_1, t_2) = < U_S(r_1, t_1)\, U_S^*(r_1, t_1)\, U_S(r_2, t_2)\, U_S^*(r_2, t_2)> \tag{13.2}$$

in which $U_S(r, t)$ is a complex scalar representation of the scattered field. The rigorous evaluation of Eq. (13.2) is nontrivial. One way to approach a solution is based on the statistical properties of the scattered field. If the scattered fields are of zero mean, obey Gaussian statistics and depend on time differences only, the Gaussian momentum theorem can be applied, yielding a modified version of the well-known Siegert relation. To proceed, we assume the fields at the detector to be approximately Gaussian, so we can apply the momentum theorem and obtain

$$G_2(r_1, r_2, \tau) = < I(r_2, 0)>< I(\mathrm{r}2, \tau) > + |< U_S^*(r_1, 0)\, U_S(r_2, \tau) >|^2, \tag{13.3}$$

with $\tau = t_2 - t_1$. In the next step, we have to determine the mutual coherence function (MCF) $\Gamma_s(r_1, r_2, \tau) = <U_S^*(r_1, 0)\, U_S(r_2, \tau) >$ of the scattered fields. In the first-order Born approximation, the MCF depends on both the electron density correlations $C_{rr}(r_1, r_2, \tau) = < \rho(r_1, 0)\rho(r_2, \tau) >$ of the sample and the MCF of the incident fields. One can show explicitly that the MCF of the scattered fields can be written as a convolution integral of the Fourier transform of the electron density correlation function and a resolution function containing all the effects of partial coherence. This finally leads to an expression for the normalized intensity ACF of

$$g_2(Q,\tau) = 1 + \frac{\iint \tilde{C}(Q',\tau)\tilde{C}(Q'',\tau)F(Q,Q',Q'')\,dQ'dQ''}{(\int \tilde{C}(Q',0)H(Q,Q')dQ')^2}, \tag{13.4}$$

with F and H denoting generalized resolution functions [16]. The full expression is needed for, for example, evaluating oscillating or heterodyne correlation functions as has been observed for surface systems. Luckily enough, the coherence properties and the properties of the density correlation function are often such that we can separate both and approximate Eq. (13.4) as

$$g_2(Q,\tau) \approx 1 + \frac{\iint F(Q,Q',Q'')dQ'dQ''}{(\int H(Q,Q')dQ')^2} \times \left|\frac{\tilde{C}(Q,\tau)}{\tilde{C}(Q,0)}\right|^2 = 1 + \beta|f(Q,\tau)|^2, \tag{13.5}$$

with $f(Q, \tau)$ denoting the intermediate scattering function and β a parameter describing the speckle contrast. The absolute square in Eq. (13.5) is only important for complex scattering functions with $k_B T << \hbar\omega$.

Equation (13.5) shows that time-averaged intensity ACFs yield the intermediate scattering function. In the well-known example of Brownian diffusion of particles in a solution, the scattering function is a single exponential function with $f(Q, \tau) = \exp(-\Gamma\tau)$ and the damping constant $\Gamma = DQ^2$ depends on the diffusion constant D and the wave vector transfer Q.

The contrast factor is given by $\beta = \beta(Q)/(1 + (P/S)^2)$, where $\beta(Q)$ is a factor depending on the coherence lengths of the X-ray radiation [17–19]. P denotes the pixel size of the detector and S the speckle size given by $S = \lambda d/R$ with d denoting the beam size at the sample position and R the sample-detector distance. For a high contrast factor (signal-to-noise ratio, SNR_{g2}), the pixel size should match the speckle size of the experiment. The contrast or visibility can be determined from the intensity fluctuations of a speckle pattern according to

$$\beta(Q) = \frac{\sqrt{<I^2> - <I>^2}}{<I>}. \tag{13.6}$$

The SNR of an XPCS experiment is given by [19]

$$SNR_{g2}(Q) \propto \beta(Q) <I> (Q)(N_p T \tau_a)^{1/2}. \tag{13.7}$$

with the mean count rate per pixel $<I>$, the number of pixels N_P, the total experimental duration T and the accumulation time τ_a.

13.2.2 Two-Time Correlation Functions

Dynamic properties in equilibrium are usually assumed to be homogenous, that is, all spatial regions of the sample relax in the same way and on the same time scales. If, however, the system becomes undercooled, glassy, or jammed, then heterogeneous dynamics become increasingly important. Heterogeneity implies that different regions of the sample relax on different time scales. Many soft matter and hard matter materials display such as nonequilibrium dynamics and time-resolved experiments are able to monitor such heterogeneous behavior. However, in contrast to the previously described approach, the time correlation analysis must now be performed explicitly as a function of the absolute time or age of the sample. The usual time average in the Siegert relation is then replaced by an ensemble average taken over equivalent wave vectors in reciprocal space. Such an approach requires the use of area detectors, which allow measuring the intensity of many different speckles simultaneously. The measured quantity is

$$G(Q,t_1,t_2)=\frac{<I(Q,t_1)I(Q,t_2)>_\phi}{<I(Q,t_1)>_\phi<I(Q,t_2)>_\phi}, \qquad (13.8)$$

where the subscript ϕ denotes the averaging over an ensemble of pixels corresponding to an (almost) equivalent set of wave vector transfers [20,21]. The two-time correlation function is usually plotted as a 2d plot with axis t_1 and t_2 and the degree of correlation G indicated by a color scale. Nonequilibrium dynamics is immediately visible in a nonhomogenous behavior of G along the diagonal (see Fig. 13.1).

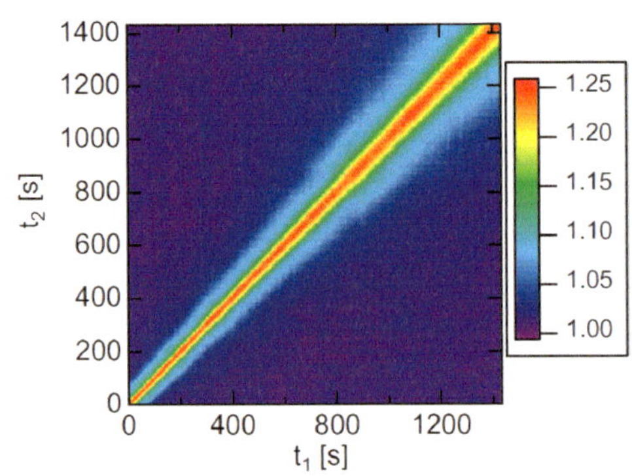

Figure 13.1 Two-time correlation function $G(Q = 0.16\ \text{nm}^{-1}, t_1, t_2)$ for a concentrated depletion gel after the application of strong shear [22]. The measurement was initiated approximately 700 s after the application of the shear, so that the waiting time t_d can be taken as $t_d = 700 + (t_1, t_2)/2$ s. Figure taken from Ref. [23].

The nonequilibrium nature of the dynamics can be quantified by measuring the fluctuations in G via the normalized variance:

$$\chi(Q,\tau)=\frac{<G^2(Q,t_1,\tau)>_{t1}-<G(Q,t_1,\tau)>^2_{t1}}{(g_2(Q,0)-1)^2}, \qquad (13.9)$$

which is similar to the dynamical four-point susceptibility χ_4, that is, widely studied in glassy materials to characterize spatial heterogeneity [23–26]. An increase of χ_4 has been identified as a key signature of a growing dynamical correlation length in supercooled liquids near the glass transition and granular materials near the jamming transition. Although χ characterizes temporal heterogeneity and χ_4 spatial heterogeneity, both of them can possibly be connected.

13.3 XPCS via Split and Delay Techniques at XFEL Sources

XPCS at third-generation synchrotron sources allows one to measure slow collective dynamics (typically 10^{-2} to 10^{6} Hz), whereas the access to ultrafast dynamics on nanometer length scales is prohibited by the limited coherent photon flux at storage ring-based sources. In contrast, X-ray free-electron lasers based on self-amplified spontaneous emission (SASE) provide uniquely intense, coherent, polarized, short-pulse radiation. At the European XFEL, for example, up to 3,000 pulses of 20–100 femtoseconds duration will be delivered with a temporal spacing of 200 ns between each pulse. The overall repetition rate will be 10 Hz [13]. This peculiar time structure of SASE-based sources excludes the classical sequential way of measuring intensity of ACFs with arbitrary lag times τ.

Therefore, a different scheme based on a split-pulse technique has been proposed [27]. The concept of this technique is to split each X-ray pulse into two equal-intensity pulses separated in time. The time separation is achieved via a delay path similar to delay line units used in optical laser technology (see Fig. 13.2). The scattering from the two pulses will be collected during the same exposure on an area detector making ultrafast time resolution of area detectors unnecessary. Instead, an analysis of the speckle contrast of the recorded images as a function of the different delay times yields direct information about the intermediate scattering function of the sample. The intensity of the two pulses separated by the delay time τ is $S(\tau) = I(t_i) + I(t_i + \tau)$.

The contrast of a speckle pattern is the normalized variance

$$c_2(\tau) = \frac{< S^2(\tau) > - < S(\tau) >^2}{< S(\tau) >^2}. \tag{13.10}$$

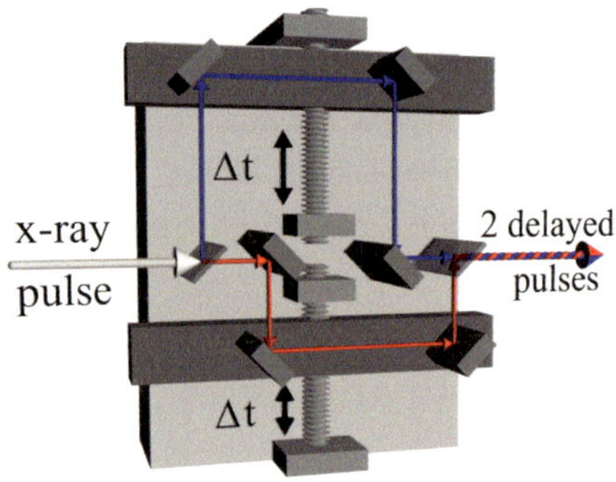

Figure 13.2 Schematics of the split- and delay line. Each XFEL pulse is split into two pulses by a very thin semi-transparent Bragg monochromator. The reflected (blue) and the transmitted (red) beams travel different paths resulting into a time delay. Finally, the both beams are merged by another thin Bragg monochromator [28,29].

If the correlation time of the sample is much larger than the delay time between the two pulses, the two speckle patterns are identical and the contrast of the summed image is close to 1. In contrast, if the correlation time of the sample is much shorter than the delay time the two speckle patterns are uncorrelated and the speckle contrast is close to 0.5. Thus, by varying the delay time, one can map out the dynamics by measuring the speckle contrast [30]. Using the Siegert relation for Gaussian signals, Eq. (13.10) can be expressed as

$$c_2(\tau) = \frac{1}{2}(1 + |f(Q,\tau)|^2). \tag{13.11}$$

This shows that the split and delay technique allows to measure the intermediate scattering function.

A hard X-ray split and delay line based on single crystal beam splitters and mixers has recently been commissioned and time delays in the range of nano- to picoseconds have been realized [28,29].

13.4 X-Ray Cross-Correlation Analysis: Local Bond Order in Liquids and Glasses

As mentioned in the Introduction, a speckle pattern is related to the exact spatial arrangement of the particles in the sample. Therefore, it carries information about local structural properties of the particles, which can be uncovered by correlating intensities in reciprocal space instead of time—a method called XCCA. An important materials science problem is that disordered matter—such as liquids and glasses—does not exhibit long-range translational order. In turn, it is able to accommodate different local symmetries in the same system [31–33]. Very prominent among those local symmetries in liquids is the icosahedral structure, which cannot fill space. Those symmetries are believed to play a major role in the existence of undercooling of liquids and the glassy state.

Conventional X-ray diffraction techniques fail to detect local structures—at least in the bulk—because their intrinsic spatial and temporal averaging mechanism yields only the pair distribution function $g(r) = n_0^{-2} < \rho(0)\rho(r)>$. The associated averaged structure factor $S(Q) = 1 + n_0 \int (g(r)-1)e^{iQr}dr$ depends on the momentum transfer Q and shows rather similar features for all disordered structures, thus carrying only limited information, that is, the probability to find another atom in a certain distance from a given average atom. In particular, $S(q)$ provides no direct answer on the local symmetries, which are intimately related to the local bonding and angles.

To go beyond the pair distribution function, correlations in the angular distribution of the X-ray speckles must be detected. At first sight, this is not evident, since the angular variation of the intensity exhibits an isotropic distribution as anticipated for amorphous systems. However, it has been shown that by applying the generic angular dependent intensity–intensity cross correlation

$$C_Q(\Delta) = \frac{< I(Q,\varphi)I(Q,\varphi+\Delta) >_\varphi}{< I(Q,\varphi) >_\varphi^2}, \tag{13.12}$$

information on local bond order and symmetries in disordered materials can be obtained [34–36].

13.5 Designing XPCS Experiments

There are many issues, which have to be considered in order to perform an XPCS experiment successfully: the sample needs to be illuminated coherently, the detector system needs to spatially resolve the speckles, the time resolution of the detector system needs to match the dynamics of the sample, the sample needs to scatter sufficiently in the investigated Q range, and the sample should not be damaged during the required exposure time of the measurement. In the following, these points are discussed in more detail.

A sample is coherently illuminated if the collimating pinhole or slit is smaller than the vertical and horizontal coherence lengths $\xi_{t,\,v}$, $\xi_{t,\,h}$ and the path length difference within the sample is smaller than the longitudinal coherence length ξ_l. The transverse coherence lengths are given by the distance R to the source, the wavelength λ, and the transverse source size $\Theta_{v,h}$ via $\xi_{t,v,h} = 0.4 \cdot \lambda R/\Theta_{v,h}$. The longitudinal coherence length depends on the wavelength and the energy spread according to $\xi_l = \lambda(\lambda/\Delta\lambda)$.

Using typical values for the distance $R \approx 50 \ldots 100$ m between the undulator source and the collimating slit position near the sample and typical source sizes of third-generation synchrotrons of $\Theta_v \approx 10 \ldots 40$ μm and $\Theta_h \approx 80 \ldots 1000$ μm, the transverse coherence lengths can be estimated to be several hundred micrometers in the vertical direction and to be significantly smaller in the horizontal direction (<50 μm) for 8 keV X-rays. This means that the beam becomes transversely coherent by limiting the illumination area of the sample to the transverse coherent portion of the beam using small pinholes or slits.

The longitudinal coherence length of a standard Si111 double crystal monochromator using an X-ray energy of 8 keV is $\xi_l \approx 1$ μm. The path length difference within the sample depends on the scattering angle θ, absorption coefficient μ, sample thickness W, and beam size d_{slit}. The following requirements for coherent illumination can be obtained [7]:

1. reflection geometry: $\xi_l > 2\mu^{-1}\sin^2\theta$
2. transmission geometry: $\xi_l > 2W\sin^2\theta + d_{\text{slit}}\sin 2\theta$.

This leads to maximum wave vector transfers of $Q_{\text{max,refl}} \approx 3\text{Å}^{-1}$ (for $1/\mu = 15$ μm) and $Q_{\text{max,trans}} \approx 0.3\text{Å}^{-1}$ ($W = 1$ mm and $d_{\text{slit}} = 10 \times 10$ μm^2). Those values indicate that modern third-generation synchrotrons are sufficiently coherent to perform XPCS experiments at a large variety of Q values.

Area detectors have proven to be advantageous for performing XPCS experiments as they increase the SNR. One of the key detector parameters is the pixel size, which needs to roughly match the speckle size for obtaining a good SNR_{g2}. To use a modern fast Pilatus detector with a pixel size of 172 μm at a reasonable sample to detector distance of 5 m and an X-ray energy of 8 keV, the collimating slit size must be smaller than $4 \times 4\ \mu m^2$. This is much smaller than the actual transverse coherence lengths. It means that focusing devices are necessary at XPCS beamlines in order to use the fully available coherent photon flux. For example, the coherence Beamline P10 at PETRA III would use only a fraction of 1/2,500 of its coherent flux in a low beta configuration if no focusing schemes were applied [13].

XPCS experiments use either a point detector (e.g., scintillation counter, avalanche photo diode) in combination with a detector slit and a hardware correlator (e.g., Correlator.com) or an area detector (e.g., PI-LCX or Pilatus 300 K) and software-based analysis tools. The first combination is used quiet successfully to study relatively fast dynamic phenomena on timescales of 10^{-7} ...10^{0} s. However, a big drawback is that only a single Q value can be measured at a time.

The use of 2D detectors and multi-speckle analysis tools to study slow dynamics has advanced the XPCS field considerable [19,37]. However, there are still issues when using 2D detectors. To measure a meaningful correlation function, the investigated time window needs to cover at least two to three orders of magnitude around the characteristic correlation time of the sample. Only in this way it is ensured that the contrast, the baseline, and the shape of the temporal decay of the correlation function can be precisely determined. Using 2D cameras, the time window is defined by the time between two successive frames and the number of measured frames. The fastest X-ray cameras can run up to ~1 kHz limiting the lower end of the accessible time window for 2D detectors to $\approx 10^{-3}$ s.

Time-resolved experiments need a high scattering intensity per time interval. As a rule of thumb, one can estimate that the required intensity in a single speckle/pixel is proportional to the frequency of the dynamic phenomena under investigation. This means that samples with fast dynamic phenomena need to scatter much stronger than samples with slow dynamics and explains the success of XPCS to study the glass or jamming transition.

One major feature of XPCS experiment is that the speckle pattern is directly related to the exact spatial distribution within the sample. However, this also means that the beam has to stay on the exact same spot of the sample during the duration of the measurement. During this time, the sample must be undisturbed and unchanged by the irradiation. This condition is very challenging for many soft-condensed matter systems. Nevertheless, it is possible to apply sophisticated measurement protocols to overcome some of this problem. As a guideline, most XPCS experiments have used an incident flux density of 10^7 ...10^8 photons/s/μm^2 and the sample must be stable under these irradiation conditions for a minimum of several seconds.

13.6 Experimental XPCS Setup

The Troïka beamline ID10 at the European Synchrotron Radiation Facility (ESRF) and the beamline 8ID/XOR pioneered the XPCS technique at modern synchrotron sources, and many features of these beamlines need to be considered to design a successful experimental station [18]. A synchrotron is an incoherent source, and only due to small source sizes and high brilliances of modern synchrotrons are the transverse coherence lengths and the coherent flux sufficient in the medium to hard X-ray regime for XPCS experiments. Nevertheless, the coherent portion of the X-ray beam at 8 keV is still only of the order of 0.5–2.0 percent at third-generation sources.

The performance capabilities of XPCS beamlines are mainly defined by their source properties, that is, the brilliance B defined by the combination of storage ring and undulator and the wavelength λ since the coherent flux is given by [38]

$$I_c \propto \left(\frac{\lambda}{2}\right)^2 B\,. \tag{13.13}$$

If one looks at newer beamlines such as the beamline P10 at PETRA III or the upgrades plans of ESRF or APS respectively, one finds that the tendency is to optimize this available brightness by using long undulators. Low beta or high beta sources work equally well for XPCS setups.

In the beginning, the statement "No optics are good optics" was certainly valid for XPCS or coherent beamlines in general. However, over the past decade, X-ray optics has improved significantly. Therefore, the use of optics is now advantageous, if great care is used to select and implement the optical components. In general, one needs to differentiate between reflecting and transmitting optical elements. Reflecting optical elements are mostly monochromators or mirrors, whereas transmitting optical elements are mainly vacuum windows, focusing lenses or absorbing filters. Today, it is possible to produce reflecting X-ray optics of such a high surface quality that the beam profile is (almost) undisturbed by speckles coming out of these surfaces [39]. However, these elements are often exposed to large heat loads. Thus, cooling schemes have to be applied, which can induce vibrations and beam motions and subsequently lead to enlarged effective source sizes and reduced available coherent flux. The angular stability of these elements needs to be of the order or better than 50 nrad, which is very challenging. A reduction of the heat load is possible by using frontend slits, since only a small fraction of the central cone of the undulator radiation is coherent. Transmitting optical elements should be selected such that (almost) no speckles are imposed to the beam profile, that is, these elements need good surface qualities and no internal microstructures. This excludes, for example, metal foils as absorbing filters.

In the following, a short overview of the end station 8ID-I/XOR of the APS is given [18]. This station serves mainly the small-angle scattering XPCS experiments. The beamline uses an APS undulator A (2.4 m length; 72 poles) as radiation source. The asymmetric source sizes and divergences are 660 μm and 11.8 μrad in the horizontal and 30 μm and 4.6 μrad in the vertical direction. The first optical component in the optical hutch is a "white beam" capable of collimating aperture of roughly 300 μm diameter placed at 27 m from the source. This aperture selects only the central cone of the undulator radiation and thus reduces the heat load on the following components significantly. The next optical component situated at ~29 m is a horizontally reflecting flat (R_{bend} > 100 km) mirror. The heat load is removed by a gravity fed closed-cycle water cooling system to minimize the introduction of vibrations. The mirror serves two purposes: It reduces the higher harmonic content of the undulator spectrum and it steers the "pink" beam out of the bremsstrahlung cone. A few meters further downstream a 250-μm thick Beryllium window (grade IF1; both sides polished) is used to separate the storage ring vacuum from the beam line vacuum. This is the only vacuum window along the beamline in normal operation. Finally, the pink beam is transported from the optics hutch to the experimental station.

In the experimental station 8ID-I, a vertical reflecting pseudo channel cut Ge111 monochromator can be used to select a smaller bandwidth of the radiation (pink: ~2 percent $\Delta E/E$ or mono: ~0.03 percent $\Delta E/E$). The heat load of the monochromator is removed by a gravity-fed water cooling system. Finally, a pair of vacuum slits is used to select a coherent portion of the beam (collimating slit) and to reduce the slit scattering effects (guard slit) before the beam hits the sample. Normally, the sample is mounted inside a vacuum chamber, which is connected via bellows to the slits and to the following 3–4 m long flight path. As a final object, a high precision translation stage is placed at ~70 m from the source to carry the detector systems.

13.7 Examples

13.7.1 Surface Dynamics of Thin Polymer Films

Liquid surfaces fluctuate with thermally induced height fluctuations (capillary waves), which arise from the collective hydrodynamic motion of the bulk liquid [40]. These waves play an important role at surfaces and interfaces in viscous polymeric liquids. They account for extra roughening of the immiscible polymer–polymer interfacial width, and they influence kinetic processes such as spinodal dewetting and droplet coalescence. Capillary modes on viscous polymer films are overdamped modes characterized by an exponential decay. The ability to measure the dispersion of these modes in a noninvasive manner using XPCS allows studying viscoelasticity and entanglement effects in thin polymer films and from interfaces [41–43]. In addition, dynamics of glassy bulk liquids and wetting films have been investigated [42,44–50].

In a pioneering experiment, Kim et al. reported on capillary waves on polymer films measured via XPCS [41]. The experimental setup is shown in Fig. 13.3a. The incidence angle of the X-ray beam is below the critical angle of total external reflection leading to a restricted penetration depth of the X-rays of approximately 9 nm, far less than the film thicknesses studied. The off-specular diffuse scattering from the polymer surface is detected with a CCD camera.

With α_i denoting the incident angle and α_f the exit angle, the wave vector transfer parallel to the surface is given by

$$Q_{||} = \frac{2\pi}{\lambda}(\cos\alpha_i - \cos\alpha_f). \quad (13.14)$$

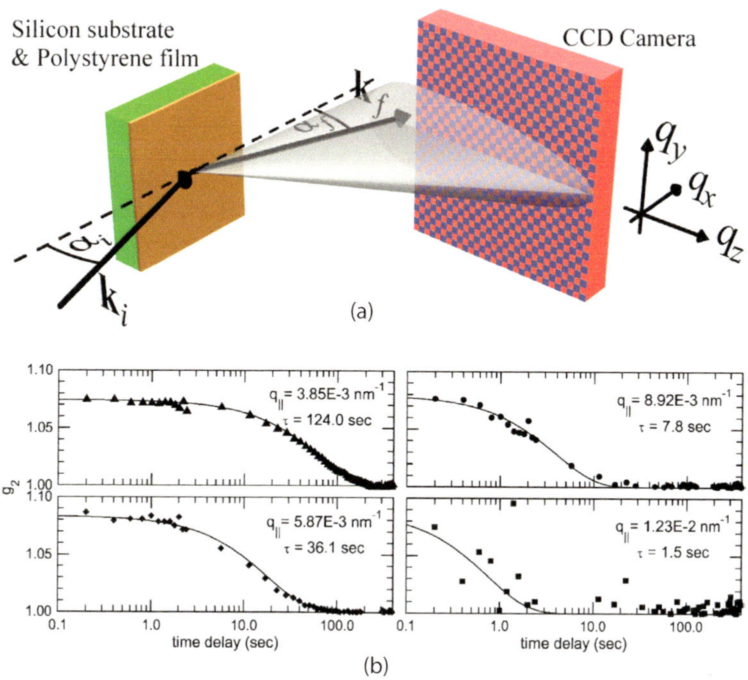

Figure 13.3 (a) Scheme of the experimental setup for XPCS in reflectivity geometry. The CCD collects the diffuse scattering. (b) Autocorrelations obtained at four different in-plane wave vectors measured on a sample thickness of 84 nm at 160°C (symbols), compared with single-exponential fits (lines). The time constant at each q value is also displayed. Figure from Ref. [41].

The intensity ACF is then calculated for each wave vector transfer $Q_{||}$. Figure 13.3b shows the ACFs obtained for a 84-nm thick film at temperatures well above the glass transition. All correlation functions can be fitted using a single exponential function with a corresponding relaxation time

$$\tau = \frac{2\eta H}{\gamma \; Q_{||}F}, \quad (13.15)$$

where η is the viscosity, γ is the surface tension, and F and H are defined as follows: $F = \sinh(Q_{||}h)\cosh(Q_{||}h) - Q_{||}h$ and $H = \cosh^2(Q_{||}h) + (Q_{||}h)^2$ with h denoting the film thickness.

Kim et al. show that the ratio τ/h is a function of $Q_{||}h$ only and directly proportional to η/γ. This scaling behavior is confirmed by the measured relaxation time shown in Fig. 13.4. A single fit parameter η/γ delivers a fit-for-all data set and the measured film viscosities were found in good agreement with theory.

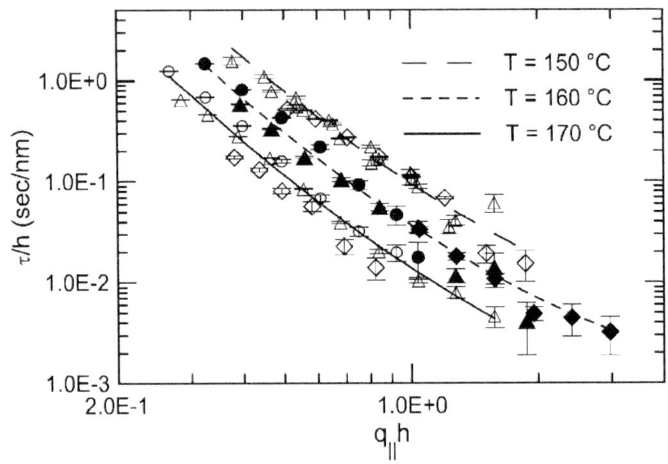

Figure 13.4 τ/h vs. $q_{||}h$ for film thicknesses 84 nm (circles), 170, 177 nm (triangles), and 312.5, 318, 333 nm (diamonds)]. Lines represent least-squares fits based on Eq. (13.15). Figure from Ref. [41].

This first experiment concentrated on films where the thickness h was much larger than the radius of gyration R_g of the polymer (i.e. $h > 8\,R_g$). In a second experiment, the same group studied the surface dynamics of films with much stronger confinement, that is, $h < 8\,R_g$, and they found that in the regime of film thicknesses of $2R_g$ the surface dynamics differs substantially from those of a simple viscous liquid [51]. For film thicknesses, $h = R_g$, no surface dynamics can be observed. This effect is due to pinning of the polymer chains on the solid substrate. This pinning leads to an effective shear modulus (elasticity) of the film, which alters the dispersion relation of the capillary waves. The relaxation time of a viscoelastic film with shear modulus μ is $\tau = \tau_0(Q_{||})/[1 + \tau_0(Q_{||})(\mu/\eta)]$, where τ_0 is the relaxation time given by Eq. (13.15). There are two asymptotic cases: (i) a small value of μ/η reduces the relaxation time to the value of the viscous model, whereas (ii) larger values of μ/η lead to a wave vector independent relaxation time of μ/η, which is the manifestation of viscoelasticity.

Figure 13.5 displays the $Q_{||}$ dependence of the relaxation time for different temperatures and radii of gyration for a film of thickness $2R_g$. The solid line presents the viscoelastic model and the dashed line the viscous model. The observed bending of the dispersion curve is in good agreement with the viscoelastic model. Jiang et al. found that it is not the absolute thickness of the film but rather the ratio of the thickness to R_g, which determines whether the elastic shear modulus is of importance in the dynamics of the thin film.

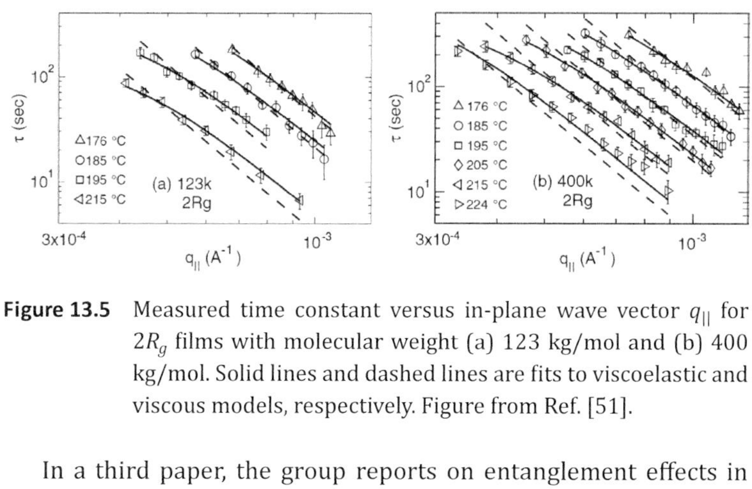

Figure 13.5 Measured time constant versus in-plane wave vector $q_{||}$ for $2R_g$ films with molecular weight (a) 123 kg/mol and (b) 400 kg/mol. Solid lines and dashed lines are fits to viscoelastic and viscous models, respectively. Figure from Ref. [51].

In a third paper, the group reports on entanglement effects in thin polymer films close to the glass transition temperature [47]. Here, the surface dynamics of polymer films with different molecular weights have been investigated. They prepared spin-cast PS-films of M_w ranging from 11 to 900 kg/mol and thicknesses ranging from 76 to 300 nm. Surprisingly, the XPCS results revealed a much more complicated surface dynamic behavior than expected.

Figure 13.6 shows the correlation functions obtained from a film of M_w = 129 kg/mol and thickness 160 nm at different temperatures. From 101°C to 111°C the curves are well described by single exponential correlation functions. A further increase in temperature results in a strongly stretched exponential decay of the g_2 function corresponding to stretching exponents of 0.1 and 0.26 at 116°C and 122°C, respectively.

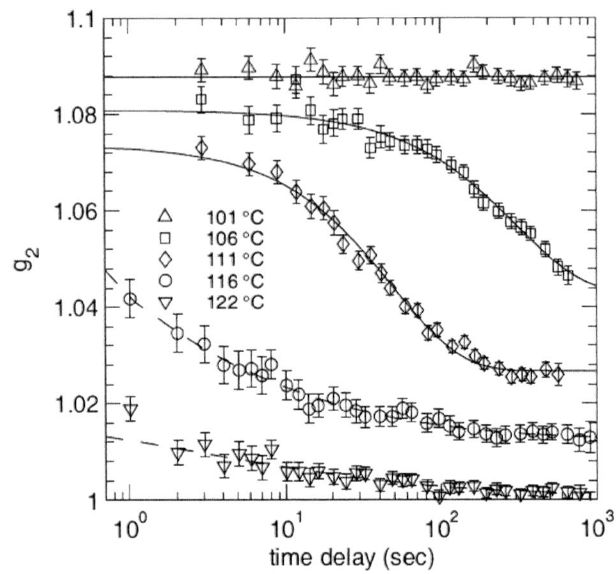

Figure 13.6 The g_2 functions obtained from a polystyrene film of M_w = 129 kg/mol and thickness d = 160 nm at various temperatures and Q = 1.9 nm^{-1}. Solid lines denote the best fits to a single exponential behavior and dotted lines denote the best fits to a stretching exponential behavior. The curves are vertically shifted for clarity. Figure from Ref. [47].

Once the temperature is increased to values well above T_G, the g_2 functions display again conventional single exponential decays. This behavior was observed for all films above M_w > 30 kg/mol. For all other films with M_w < 30 kg/mol, a single exponential decay was found at all temperatures.

Figure 13.7 shows a M_w – T diagram arising based on those measurements. The diagram is subdivided into several dynamic regions: The region SI denotes the single exponential decay dominated by chain segments of critical entanglement while region SII is the single exponential decay dominated by the full chains. In between, there is a region ST of highly stretched exponential behavior arising from the dynamics of chain segments of lengths ranging from the critical entanglement length to the full chain length. This distribution of time scales results in a highly stretched exponential correlation function.

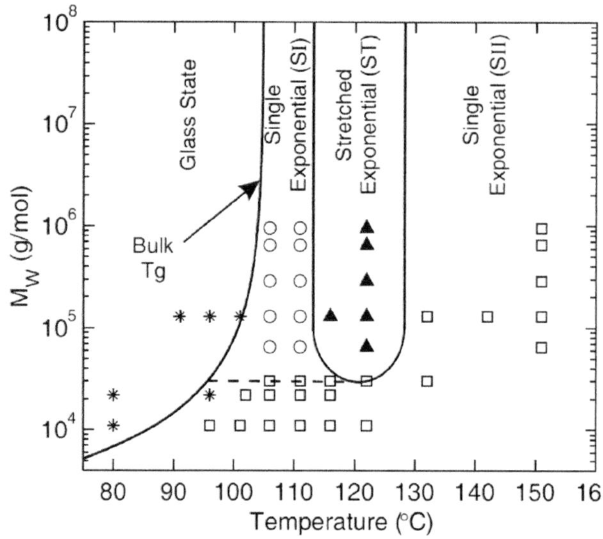

Figure 13.7 $M_w - T$ diagram from XPCS data on surface dynamics of supported PS films. Experiments are marked by symbols. The bulk T_g is shown as the leftmost solid line. The dashed line indicates the critical M_w for entanglement. In the glassy state there is no relaxation. Regions SI and SII denote the single exponential behavior dominated by chain segments of critical entanglement length and full chains. ST (N) denotes stretched exponential behavior for surface dynamics arising from the dynamics of chain segments between the critical entanglement length and the full chain length. Figure from Ref. [52].

13.7.2 Measuring Atomic Diffusion with Coherent X-Rays

The diffusion of atoms in solids determines the rates and pathways of countless chemical reactions and structural transformations. Our understanding of the mechanisms of these nonequilibrium processes, and thus our ability to synthesize new materials with better properties and stabilities, is typically built on models that describe the fundamental diffusion processes that take place in solids even at equilibrium. Whereas standard time-resolved scattering techniques can be used to observe the changes in the average structure that occur during nonequilibrium processes, it is more difficult to study the underlying equilibrium diffusion processes. From a macroscopic perspective, the equilibrium system is static, and the internal dynamics can only be revealed by a coherent, atomic-scale probe. Leitner *et al.* used XPCS to observe the dynamics of diffusing atoms [53]. The sample studied was a single crystal of $Cu_{90}\,Au_{10}$ at temperatures around 540 K, where the system is a substitutional solid solution, in which the Au atoms statistically occupy sites in the Cu fcc lattice. Intensity ACFs have been measured at Q values sufficiently large to detect diffusion on atomic length scales.

Figure 13.8 shows ACFs for fixed scattering angle and varying azimuthal angle. The coherence contrast at those large Q vectors is small but a correlation function is clearly visible. The decay times are on the order of 54 and 28 minutes proving very slow atomic diffusion in the crystalline solid. The relaxation time of a dilute substitutional alloy relaxing on a Bravais lattice is described by

$$\tau(Q) = \tau_0 \frac{I_{SRO}(Q)}{1-\sum_i p_i \cos(s_i Q)}, \qquad (13.16)$$

with a mean residence time τ_0 between the jumps of the solute atoms, s_i are possible jump vectors to the sites i, and p_i are the probabilities of the jumps to occur. $I_{SRO}(Q)$ takes short-range ordering effects (de Gennes narrowing) into account, where the local environment of an atom starts to influence the jump behavior. Values with high diffuse scattering intensities $I(Q)$ correspond to energetically favored and therefore long-lived fluctuations.

Figure 13.9 shows the extracted time constants as a function of the azimuthal angle and scattering angle. The solid lines are fits with Eq. (13.16) using a single free-fitting parameter τ_0. The model with nearest-neighbor exchange only (blue line) reproduces the data very well, whereas the second-nearest neighbor jumps (green line) cannot explain the data.

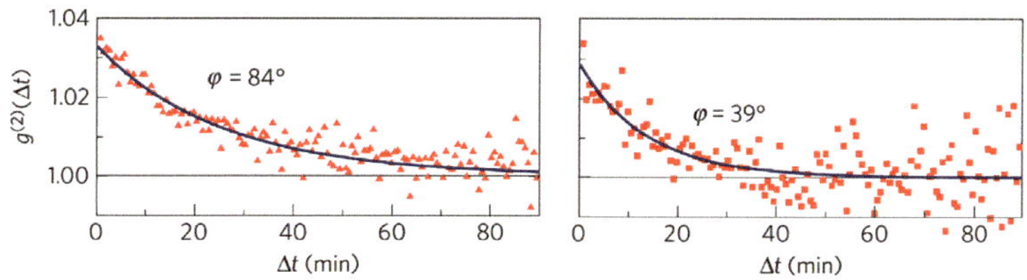

Figure 13.8 Variation of temporal intensity autocorrelations with position in reciprocal space. Shown are measurements at a temperature of 543 K, scattering angle 2θ = 25° and varying azimuthal angles φ. Although the modulus of the scattering vector is the same, the fitted correlation times (52 ± 5 min for φ = 84° and 28 ± 4 min for φ = 39°) differ by a factor of two. Figure from Ref. [53].

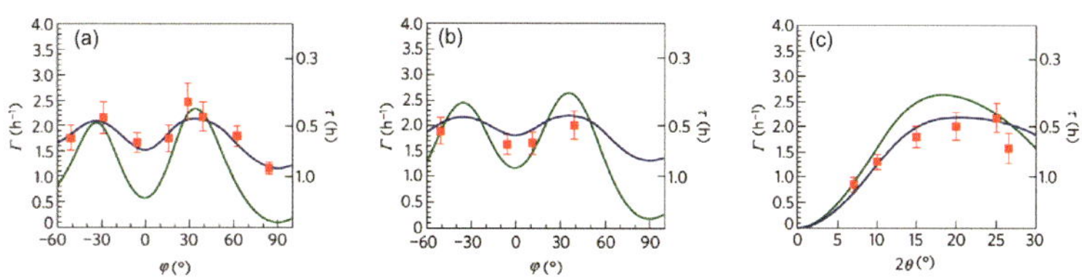

Figure 13.9 Experimental correlation times and with fits for three one-dimensional scans. (a) Azimuthal scan with fixed scattering angle 2θ = 25°, corresponding to Q = 1.75 Å^{-1}. (b) Azimuthal scan with fixed scattering angle 2θ = 20°, corresponding to Q = 1.41 Å^{-1}. (c) 2θ-scan with fixed azimuthal angle. The solid lines are one-parameter fits according to equation (13.16), with only the factor τ_0 free. A model with nearest-neighbor exchanges (blue line) reproduces the measurements well, whereas second-nearest-neighbour jumps (green line) cannot explain the data. Figure from Ref. [53].

The results of Leitner et al. show that the short-range order of the alloy leads to a diffusion dynamics different from the free diffusion in a disordered liquid. Instead of atoms moving freely to any lattice site, the dynamics of atoms with short-range order is dictated by the local neighborhood: an atom occupying a given lattice site influences the occupancies in its surrounding to minimize the configurational energy. If this atom leaves its site, this site will preferably be re-occupied by an atom of the same species—either the same atom moves back or another atom hops in as this is the energetically favorable case.

13.7.3 Antiferromagnetic Domain Wall Fluctuations

O. Shpyrko made an effort to observe fluctuations of antiferromagnetic domain walls with XPCS [54]. This large-angle XCPS experiment investigated speckles at the charge density wave (CDW) peaks in the vicinity of the (200) Bragg reflection.

The element chromium is an antiferromagnet. Below the Néel temperature of 311 K, three types of magnetic domains are formed spontaneously in a single crystal of chromium. Each magnetic domain is characterized by its spin density wave (SDW) direction along one of the unit cell vectors of this body-centered cubic metal. The SDW wavelength is of the order of λ_{SDW} = 6...8 nm and is accompanied by a CDW (see Fig. 13.10).

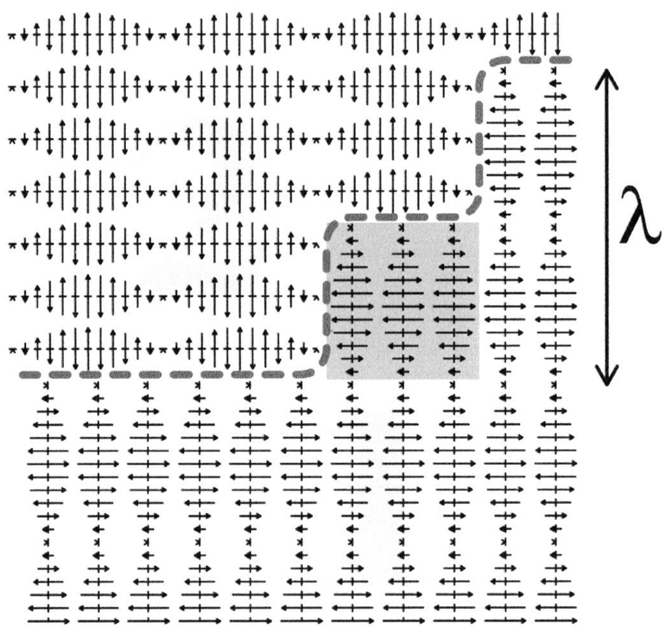

Figure 13.10 SDW domain wall (dashed line) in chromium. The two domains 1 and 2 (see numbers in the background) differ in the orientation of the SDW. The domain wall runs along the nodes with local magnetization 0. The shaded area represents one domain unit cell with lattice constant $\lambda/2$, where λ is the SDW period. Flipping of the shaded area changes the ratio between the two domains. This causes fluctuations at the according super lattice Bragg points. Figure from Ref. [54].

The intensity due to the CDW of a magnetic domain is directed to its corresponding pair of CDW satellite peaks. The spatial arrangement of the speckles of one of the CDW satellites thus represents the spatial arrangement of magnetic domains of a specific SDW direction. If the wall of one of these magnetic domains is modified by flipping the SDW direction of a portion of the domain near its edge, than the speckle arrangement will change accordingly (see Fig. 13.10). Thus, XPCS could be ideally suited to study the phenomena of antiferromagnetic domain wall fluctuations directly.

To perform a large-angle XPCS experiment, it is mandatory to show that the experiment is not disturbed by the long time stability of the setup. This experiment used the speckle pattern arising from the partial coherence of the X-ray beam of the (200) Bragg peak for this purpose. This speckle pattern was static over 5 h providing a long time limit for the actual experiment. The speckle pattern of the $[2 - 2\delta,0,0]$ CDW satellite peak was studied for temperatures ranging from 4 to 150 K. Time series of images at each temperature were analyzed and they revealed de-correlation times much faster than the long time limit, indicating that XPCS is indeed sensitive to study magnetic domain wall fluctuations.

Upon cooling, the correlation functions shift to larger time scales until the correlation functions start to overlap for temperatures below 40 K (see Fig. 13.11). Most data sets revealed a two-step correlation function. Thus, a double-stretched exponential form was used to model the data. It is visible that the observed dynamics are dominated by the slower fluctuations. The value of the stretching exponent was greater than 1. This behavior has been observed in many soft matter systems exploiting the "jamming" transition. It could indicate that blocks of spins are elastically coupled. The temperature dependence of the slow relaxation time τ_S shows that a plateau value $\tau_S \approx 5000$ s is reached below 40 K. A standard thermal activation model expects that the temperature dependence of the relaxation time τ_S is inversely proportional to $\tau_S^{-1} \approx \exp(\Delta E/k_B T)$. However, this model clearly fails for the low temperature data and an additional temperature-independent component is required (see Fig. 13.12).

By combining the standard thermal activation model with a quantum tunneling contribution represented by a temperature-independent residence time τ_{QT} to $\tau_S^{-1} = \tau_{QT}^{-1} + \tau_R^{-1}\exp(\Delta E/k_B T)$, the group of experimentalists were able to explain the data. It could

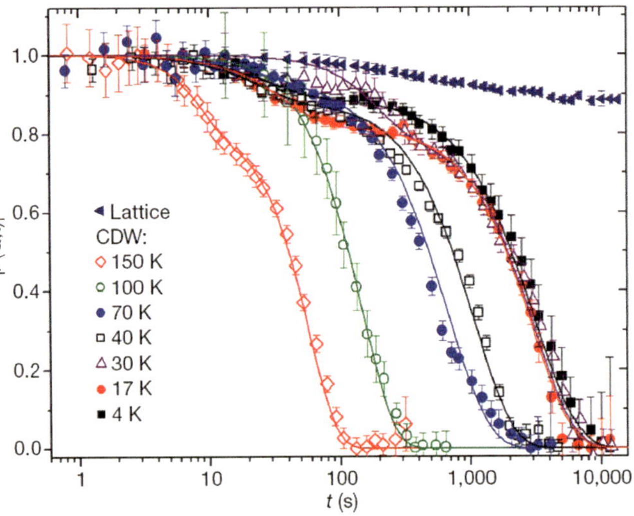

Figure 13.11 Intensity autocorrelation data for the [200] lattice Bragg peak, and for the CDW super lattice [2 – 2δ, 0, 0] peak, at different T. Two distinct timescales are clearly present in the CDW autocorrelation functions, visible as a kink. Figure from Ref. [54].

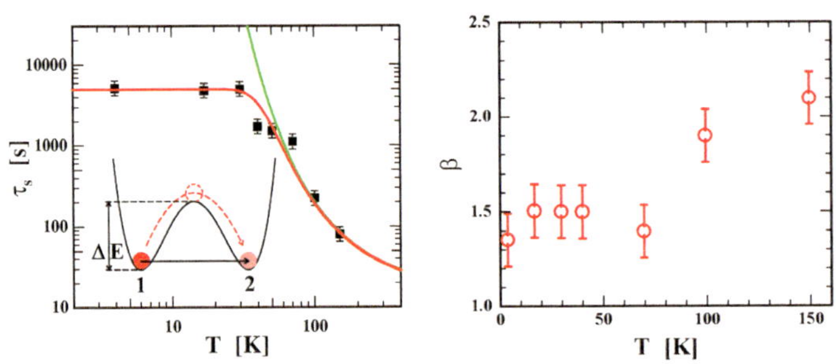

Figure 13.12 Temperature-dependent domain wall dynamics. Left: Characteristic slow fluctuation timescale τ_S obtained from fits to autocorrelation function data shown in Fig. 13.11, compared to a classical Arrhenius model (green line) and a model that also includes a temperature-independent switching rate term (red line). Inset: potential energy surface including thermally activated (red dashed arrow) and quantum tunneling (black arrow) mechanisms of the transition between the two low-energy domain configurations 1 and 2. Right: Values of stretching exponent β for various temperatures. Figure after Ref. [54].

be shown that the obtained results of $\tau_{QT} = 5000 \pm 1000$ s and $\tau_R = 4...60$ s are physically meaningful if applied to an "elemental" domain building block of volume $(\lambda_{SDW}/2)^3$ and the domain walls follow along the planes of minimized local magnetization. They are also consistent with other experimental observations of element chromium.

13.7.4 Reentrant Glassy Behavior

Mode coupling theory (MCT) is capable of predicting quantitatively collective dynamics near the glass transition. For example, the dynamical arrest in concentrated suspensions of hard spheres, where each particle is trapped by a cage of its surrounding neighbors is predicted quite well. Recently, MCT calculations predicted that it is possible for certain colloidal systems of particles with short-range attractive interaction as well as a hard core repulsion to enter the glassy state from the liquid state by heating. It predicted even that it is possible to transform an attraction-dominated glassy state (AG) via a liquid state to a repulsion-dominated glassy state (RG) purely by controlling the attraction strength via temperature.

A research team led by S. Mochrie investigated a colloidal system of silica spheres of 195 nm radius (volume fraction $\Theta \approx 0.52$) in a near-critical water-Lutedine binary mixture (Lutedine concentration $c = 0.24$) using XPCS in SAXS geometry [55]. This state-of-the-art XPCS experiment, carried out at 8ID-I/XOR, observed this "reentrant" transition in detail. Several timeseries measurements were made between 33.0 and 33.6°C. The data were taken using the fast SMD detector and consisted of up to several 10.000 frames per series to cover a sufficient time window of over 4½ decades.

Figure 13.13 shows the normalized intermediate scattering functions displaying a rich dynamical behavior over this small temperature range. At the low temperature end, the intermediate scattering function starts with a rapid decay (faster than the detector framing rate) before a plateau value of ~0.5 is reached. The plateau value stays constant over several time decades before the correlation is finally lost at timescales much greater than 100 s. Such a behavior has been previously observed for repulsive hard sphere glasses. The plateau value is called the 'nonergodicity' parameter and interpreted in terms of caging.

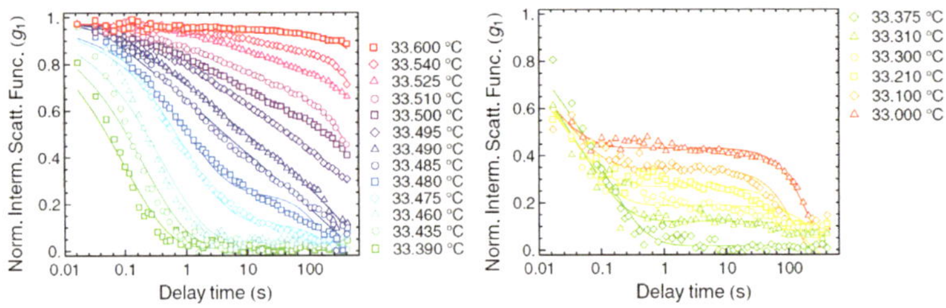

Figure 13.13 Normalized intermediate scattering function (g_1) versus delay time. For clarity, temperature ranges are separated. Left: temperatures from 33.000 to 33.375°C, Right: from 33.390 to 33.600°C. Figure taken from Ref. [55].

Upon heating the value of the plateau drops continuously to zero. At 33.375°C, only a single decay exists (delay time < 1 s) indicating that the system evolved from a RG to a liquid state. The single decay shifts back to longer timescales by increasing the temperature further. In addition, the shape of the intermediate scattering function changes. Around 33.5°C, the decay is nearly linear on the semi-logarithmic plot, indicating (nearly) logarithmic relaxations as predicted from MCT for the liquid to AG transition. The slope of the decay decreases as the temperature further increases; until near 33.6°C the intermediate scattering function is basically constant indicating that an attraction-dominated glassy state is entered.

13.7.5 Dynamical Heterogeneity in an Aging Colloidal Gel

Until now, the equilibrium nature of the dynamics allowed to use the time-averaged Siegert relation for quantifying the sample dynamics. However, many samples display strong out-of equilibrium behavior, which necessitates the use of two-time correlation functions and its fluctuations. The potential of this correlation method is illustrated with an example of an aging colloidal gel [22].

The colloidal gel consists out of silica nanoparticles (radius 22 nm) at a volume fraction of 0.33 in decalin (also bicyclo-decane, an organic solvent). Such gels display a thixotropic response to stress, that is, they are strongly viscous under normal conditions but become more fluid like and flow under stress. The recovery of the gel from the fluidizing shear can be monitored using XPCS. The evolution in the microscopic dynamics following the shear is visible in the two-time correlation function in Fig. 13.1 at a wave vector transfer close to the structure peak of the suspension. The measurement started after a waiting time of t_d = 700 s so that the sample age at any point in Fig. 13.1 can be calculated as $t_{\text{age}} = t_d + (t_1 + t_2)/2$. The slowing down of the microscopic dynamics, after the shear is ceased, becomes visible by the increasing broadening of the two-time correlation function with sample age along the diagonal from left (small t_{age}) to right (large t_{age}). Figure 13.14a shows correlation functions g_2, which have been obtained by averaging short time intervals, thus representing the evolution of the typical correlation times. The slowing down of the dynamics with aging time is now clearly visible. The correlation functions show the peculiar compressed exponential form typical for stress-mediated dynamics. More information can be obtained by calculating the variance of the two-time correlation functions. Figure 13.14b shows $\chi(Q, t)$ as a function of delay time t at the same waiting times and the same Q value as in Fig. 13.14a.

It is important to remove the contribution of measurement noise due to the finite number of CCD pixels by means of established procedures. The variance at each waiting time displays a pronounced peak at timescales that corresponds to the typical decorrelation times of the intensity correlation function g_2, illustrating the close correspondence between the maximum of the temporal heterogeneity and the correlation time. As can already be seen from Fig. 13.14b and as confirmed by further measurements, the maximum of the variance, which is the maximum value of the temporal heterogeneity, shows no systematic variation with waiting time. This suggests that the degree of heterogeneity of the dynamics remains largely unchanged as the system ages despite the dramatically slowing down of the dynamics.

The observations lead to the following interpretation of the dynamic processes. The increasing correlation time with age indicates a decrease of the characteristic strain velocity with time, triggered by the stop of the externally applied stress. The constant value of the temporal heterogeneity implies that the density of stress dipole sites—which are responsible for the dynamics—stays constant with aging time. However, the growth rate of the stress is decreasing resulting in the aging of the typical time constants in the system. Thus, careful measurements of the nonequilibrium dynamics allow extracting details of the nonequilibrium dynamics of soft matter yielding unique insights into this complex dynamical and structural behavior.

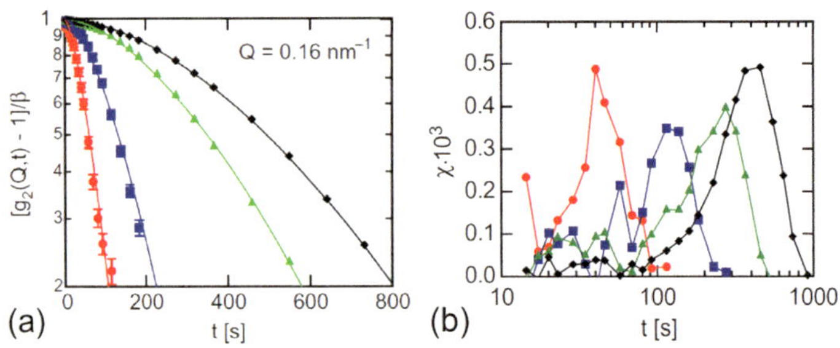

Figure 13.14 Intensity autocorrelation function for the concentrated gel at $Q = 0.16\ \text{nm}^{-1}$ and at different waiting times: $t_d = 1400$ s (red circles), 3,400 s (blue squares), 7,000 s (green triangles), and 14,000 s (black diamonds). The solid lines display the results of fits to a compressed exponential form with exponent 3/2. (b) Normalized variance of the instantaneous correlations as a function of delay time for waiting times t_d = 1,400 s (red circles), 3,400s (blue squares), 7,000s (green triangles), and 14,000s (black diamonds). Figure from Ref. [23].

13.7.6 Local Bond Order in Colloidal Glasses

Disordered matter, such as glasses and liquids, does not exhibit translational symmetry but able to accommodate different local symmetries in the same system, among them the icosahedral local order, which belongs to the forbidden motifs in periodic structures. This mysterious and so far experimentally inaccessible localized order within disorder has been fascinating scientists for many decades, because it is held responsible for the undercooling of liquids and the existence of the glass state. On the contrary, nonperiodic materials have always attracted the attention of materials scientists, because they do carry a unique potential to display novel smart functions [56].

Our experimental and theoretical approach to solve this age-old problem has followed the guiding principle that first the intrinsic spatial (and temporal) averaging mechanism performed in conventional (i.e., partially coherent) diffraction has to be eliminated experimentally. Then, a properly defined higher-order angular correlation function has to be devised and employed by hand to data in order to disclose the hidden local symmetries of disordered matter. The first results obtained from different amorphous colloidal structures deliver new structural details, which could already shed a fascinating new light on the origin of glasses and on the glass-forming mechanism.

For the experiment, the coherent X-ray beam prepared at station ID10A of the ESRF hits a colloidal glass sample and produces a full 2π speckle pattern at the 2D-detector covering a Q-range up to 0.2 nm^{-1}. The typical data set exhibits the expected isotropic granular "intensity rings," which are mediated by the random local order within the system (see Fig. 13.15a,b). The deduced angular averaged structure factor shows the standard radial intensity distribution. For going beyond this information, we must unravel correlations in the angular distribution of the X-ray speckles. This is done by constructing a first simple subset of these new types of higher-order correlations, that is, the local angular correlations with respect to a given azimuth Δ (see Eq. (13.12)).

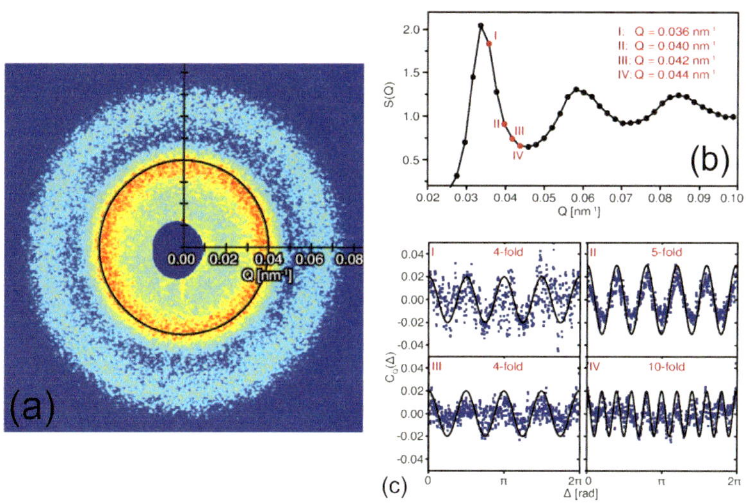

Figure 13.15 Experimental data and XCCA results. (a) CCD image shows a typical intensity spectrum with speckle structure; (b) Angular averaged structure factor of the image in (a), which is the standard radial intensity distribution; (c) Experimental results after applying the cross-correlator $C_Q(\Delta)$ to the data in (a) at different Q values. Solid lines are guide to the eyes. Figure from Ref. [34].

Figure 13.15c shows the result after applying $C_Q(\Delta)$ (see Eq. (13.3)) to data Fig. 13.15a, specifically for the intensity ring associated with $Q = 0.04\ \text{nm}^{-1}$. Most fascinating, $C_Q(\Delta)$ clearly reveals very pronounced anisotropies pointing toward hidden symmetrics in the colloidal glass.

We have made several systematic observations in all colloidal glass systems. The emerging local symmetries are coupled to distinct Q-vectors, which do not coincide with the maxima of the amorphous structure factor. A further most fascinating observation is associated with the temporal relaxation behavior of $C_Q(\Delta)$ (Fig. 13.16). Our new 4-point correlator uncovers a continuous change of the LSFs within the first 200s from initially 6-fold symmetry to 5-fold symmetry.

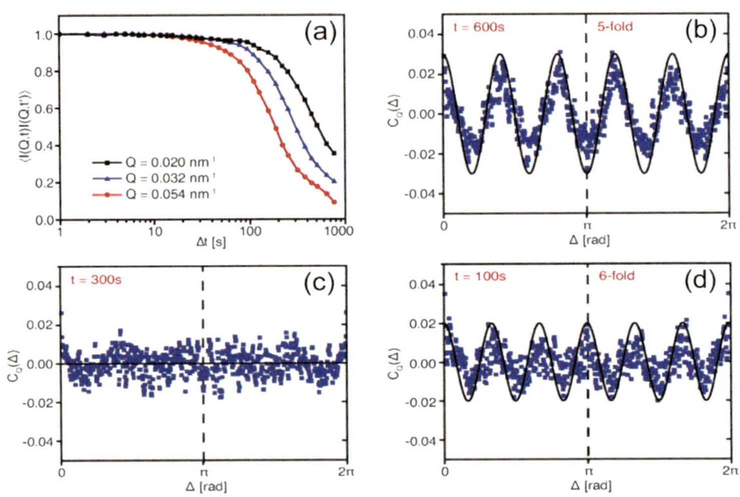

Figure 13.16 Temporal relaxation behavior of $C_Q(\Delta)$ and $< I(Q, t)\ I(Q, t')>$. (a) Normalized temporal intensity autocorrelation function $< I(Q, t)\ I(Q, t')>$ for different Q values as printed in the graph with $\Delta t = t' - t$. Solid lines are guide to the eyes. (b) The cross-correlation function $C_Q(\Delta)$ at $Q = 0.04\ \text{nm}^{-1}$ evolves from an initially 6-fold to a 5-fold symmetry. The curves at 600s and 300s are averaged over the subsequent 100s interval; the one at 100s is averaged over a 50s interval. Figure from Ref. [34].

Apparently, this involves a breaking and forming of bonds, thereby icosahedral clusters reorganize in a different orientation, transform to a local nanocrystal (hexagonal/fcc) or disorder. Such behavior is known from molecular dynamics simulations as "dynamical heterogeneity." Within the observed time frame, the temporal autocorrelation drops to 50 percent. This new XCCA concept reveals now that the processes, which are responsible for these relaxations, are accompanied by distinct changes of the local structure.

The availability of short-pulse XFEL radiation in the 0.1 nm regime and with 100 fs pulse length will open up the fascinating option to analyze the local structure of liquids (in particular also water) by applying the new concept of XCCA to single laser shot speckle diffraction pattern.

13.7.7 Summary

The basic concepts of XPCS were outlined. Standard XPCS is a versatile technique that allows to measure equilibrium dynamics in the time domain (g_2 analysis). Examples of standard XPCS experiments cover a broad range of problems from soft matter systems to hard condensed matter. However, using area detectors and sophisticated analysis schemes, XPCS can do even more. Nonequilibrium dynamics and local structural information become accessible. This allows it to study the dynamic processes during which the state of matter is changing, for example, from a liquid to a glass. Such studies are becoming increasingly important and XPCS can help to provide answers.

References

1. Sutton M, Mochrie SGJ, Greytak T, et al. (1991) Observation of speckle by diffraction with coherent X-rays, *Nature*, **352**, 608–610.
2. Sinha SK, Tolan M, Gibaud A (1998) Effects of partial coherence on the scattering of x rays by matter, *Phys Rev B*, **57**, 2740.
3. Berne BJ, Pecora R (1976) *Dynamic Light Scattering*, Wiley, New York.
4. Grübel G, Als-Nielsen J, Abernathy D, et al. (1994) Scattering with coherent x-rays *ESRF Newslett*, **20**, 14–15.
5. Grübel G, Abernathy D, Stephenson GB, et al. (1995) Intensity fluctuation spectroscopy using coherent X-rays, *ESRF Newslett*, **23**, 14–16.
6. Grübel G, Zontone Fe (2004) Correlation spectroscopy with coherent X-rays, *J Alloys Comp*, **362**, 3–11.

7. Grübel G, Madsen A, Robert A (2008) Photon Correlation Spectroscopy, in *Soft Matter Characterization* (Borsali R, Pecora B, ed), Springer, Heidelberg, pp. 954–995.
8. Sutton M (2002) Coherent X-ray diffraction, in *Third Generation Hard X-Ray Synchrotron Radiation Sources* (Mills DM, ed), John Wiley & Sons, New York, pp. 101–124.
9. Mochrie SGJ, Lurio LB, Rühm A, et al. (2003) X-ray photon correlation spectroscopy studies of colloidal diffusion and the capillary modes of polymer films, *Physica β*, **336**, 173–180.
10. Livet F (2007) Diffraction with a coherent X-ray beam: dynamics and imaging, *Acta Cryst A*, **63**, 87–107.
11. Konings S, Schüßler-Langeheine C, Ott H, et al. (2011) Magnetic domain fluctuations in an antiferromagnetic film observed with coherent resonant soft X-ray scattering, *Phys Rev Lett*, **106**, 077402.
12. Seu KA, Roy S, Turner JJ, Park S, Falco CM, Kevan SD (2010) Cone phase and magnetization fluctuations in Au/Co/Au thin films near the spin-reorientation transition, *Phys Rev B*, **82**, 012404.
13. Altarelli M, et al. (2006) *XFEL Technical Design Report*, DESY, Hamburg.
14. Ackermann W, et al. (2007) Operation of a free-electron laser from the extreme ultraviolet to the water window, *Nat Phot*, **1**, 336–342.
15. Emma P, et al. (2010) First lasing and operation of an Ångstrom-wavelength free-electron laser, *Nat Phot*, **4**, 641–647.
16. Gutt C, Ghaderi T, Tolan M, Sinha SK, Grübel G (2008) Effects of partial coherence on correlation functions measured by X-ray photon correlation spectroscopy, *Phys Rev B*, **77**, 094133.
17. Abernathy DL, Grübel G, Brauer S, et al. (1998) Small-angle X-ray scattering using coherent undulator radiation at the ESRF, *J Synchrotron Rad*, **5**, 37–47.
18. Sandy AR, Lurio JB, Mochrie SGJ, et al. (1999) Design and characterization of an undulator beamline optimized for small-angle coherent X-ray scattering at the advanced photon source, *J Synchrotron Rad*, **6**, 1174–1184.
19. Lumma D, Lurio LB, Mochrie SGJ, Sutton M (2000) Area detector based photon correlation in the regime of short data batches: data reduction for dynamic X-ray scattering, *Rev Sci Instrum*, **71**, 3274.
20. Malik A, Sandy AR, Lurio LB, et al. (1998) Coherent X-ray study of fluctuations during domain coarsening, *Phys Rev Lett*, **81**, 5832–5835.

21. Cipelletti L, Weitz DA (1999) Ultralow-angle dynamic light scattering with a charge coupled device camera based multispeckle multitau correlator, *Rev Sci Instrum*, **70**, 3214–3121.
22. Chung B, Ramakrishnan S, Bandyopadhyay R, et al. (2006) Microscopic dynamics of recovery in sheared depletion gels, *Phys Rev Lett*, **96**, 228301.
23. Madsen A, Leheny RL, Guo H, Sprung M, Czakkel O (2010) Beyond simple exponential correlation functions and equilibrium dynamics in X-ray photon correlation spectroscopy, *New J Phys*, **12**, 055001.
24. Trappe V, Pitard E, Ramos L, Robert A, Bissig H, Cipelletti L (2007) Investigation of q-dependent dynamical heterogeneity in a colloidal gel by X-ray photon correlation spectroscopy, *Phys Rev E*, **76**, 051404.
25. Duri A, Bissig H, Trappe V, Cipelletti L (2005) Time-resolved-correlation measurements of temporally heterogeneous dynamics, *Phys Rev E*, **72**, 051401.
26. Robert A, Wandersman E, Dubois E, Dupuis V, Perzynski R (2006) Glassy dynamics and aging in a dense ferrofluid, *Europhys Lett*, **75**, 764–770.
27. Grübel G, Stephenson GB, Gutt C, Sinn H, Tschentscher Th (2007) XPCS at the European X-ray free electron laser facility, *Nucl Instrum Meth B*, **262**, 357–367.
28. Roseker W, Franz H, Schulte-Schrepping H, et al. (2011) Development of a hard X-ray delay line for X-ray photon correlation spectroscopy and jitter-free pump-probe experiments at X-ray free-electron laser sources, *J Synchrotron Rad*, **18**, 481–491.
29. Roseker W, Franz H, Schulte-Schrepping H, et al. (2009) Performance of a picosecond X-ray delay line unit at 8.39 keV, *Opt Lett*, **34**, 1768–1770.
30. Gutt C, Stadler L-M, Duri A, et al. (2009) Measuring temporal speckle correlations at ultrafast X-ray sources, *Opt Express*, **17**, 55–61.
31. Frank FC (1952) Supercooling of liquids, *Proc R Soc A*, **215**, 43–46.
32. Bernal JD (1960) Geometry of the structure of monatomic liquids, *Nature*, **185**, 68 70.
33. Steinhardt PJ, Nelson DR, Ronchetti M (1983) Bond-orientational order in liquids and glasses, *Phys Rev B*, **28**, 784–805.
34. Wochner P, Gutt C, Autenrith T, et al. (2009) X-ray cross correlation analysis uncovers hidden local symmetries in disordered matter, *Proc Natl Acad Sci*, **106**, 11511–11514.

35. Clark NA, Ackerson BJ, Hurd AJ (1983) Multidetector scattering as a probe of local structure in disordered phases, *Phys Rev Lett*, **50**, 1459–1462.

36. Ackerson BJ, Taylor TW, Clark NA (1985) Characterization of the local structure of fluids by apertured cross-correlation functions, *Phys Rev A*, **31**, 3183–3193.

37. Falus P, Borthwick MA, Mochrie SGJ (2004) Fast CCD camera for X-ray photon correlation spectroscopy and time-resolved X-ray scattering and imaging, *Rev Sci Instrum*, **75**, 4383–4400.

38. Attwood D, Halbach K, Kim K-J (1985) Tunable coherent X-rays, *Science*, **228**, 1265–1272.

39. Narayanan S, Sandy A, Shu D, Sprung M, Preissner C, Sullivan J (2008) Design and performance of an ultra-high-vacuum-compatible artificial channel-cut monochromator, *J Synchrotron Rad*, **15**, 12–18.

40. Buff FP, Lovett RA, Stillinger FH (1965) Interfacial density profile for fluids in the critical region, *Phys Rev Lett*, **15**, 621–623.

41. Kim H, Rühm A, Lurio LB, et al. (2003) Surface dynamics of polymer films, *Phys Rev Lett*, **90**, 068302.

42. Seydel T, Madsen A, Tolan M, Grübel G, Press W (2001) Capillary waves in slow motion, *Phys Rev B*, **63**, 073409.

43. Sikharulidze I, Dolbnya IP, Fera A, Madsen A, Ostrovskii BI, de Jeu WH (2002) Smectic membranes in motion: approaching the fast limits of X-ray photon correlation spectroscopy, *Phys Rev Lett*, **88**, 115503.

44. Madsen A, Seydel T, Sprung M, Gutt C, Tolan M, Grübel G (2004) Capillary waves at the transition from propagating to overdamped behavior, *Phys Rev Lett*, **92**, 096104.

45. Sikorski M, Gutt C, Chushkin Y, Lippmann M, Franz H (2010) Dynamics at the liquid-vapor interface of a supercooled organic glass former, *Phys Rev Lett*, **105**, 215701.

46. Jiang Z, Kim H, Jiao X, et al. (2007) Evidence for viscoelastic effects in surface capillary waves of molten polymer films, *Phys Rev Lett*, **98**, 227801.

47. Jiang Z, Mukhopadhyay MK, Song S, et al. (2008) Entanglement effects in capillary waves on liquid polymer films, *Phys Rev Lett*, **101**, 246104.

48. Jiang Z, Kim H, Morchie SGJ, Lurio LB, Sinha SK (2006) Surface and interfacial dynamics of polymeric bilayer films, *Phys Rev E*, **74**, 011603.

49. Gutt C, Ghaderi T, Chamard V, et al. (2003) Observation of heterodyne mixing in surface X-ray photon correlation spectroscopy experiments, *Phys Rev Lett*, **91**, 076104.

50. Gutt C, Sprung M, Fendt R, Madsen A, Sinha SK, Tolan M (2007) Partially wetting thin liquid films: structure and dynamics studied with coherent X-rays, *Phys Rev Lett*, **99**, 096104.

51. Zhang J, Kim H, Jiao X, et al. (2007) Evidence for viscoelastic effects in surface capillary waves of molten polymer film, *Phys Rev Lett*, **98**, 227801.

52. Jiang Z (2007) *X-Ray Scattering Studies of Structure and Dynamics of Surfaces and Interfaces of Polymeric Liquids*, UC San Diego Electronic Theses and Dissertations, University of California San Diego, San Diego.

53. Leitner M, Sepiol B, Stadler L-M, Pfau B, Vogl G (2009) Atomic diffusion studied with coherent X-rays, *Nat Mater*, **8**, 717–720.

54. Shpyrko OG, Isaacs ED, Logan JM, et al. (2007) Direct measurement of antiferromagnetic domain fluctuations, *Nature*, **447**, 68–71.

55. Lu X, Mochrie SGJ, Narayanan S, Sandy AR, Sprung M (2008) How a liquid becomes a glass both on cooling and on heating, *Phys Rev Lett*, **100**, 045701.

56. Hufnagel TC (2004) Amorphous materials: finding order in disorder, *Nat Mater*, **3**, 666–667.

Index

For Product Safety Concerns and Information please contact our EU representative GPSR@taylorandfrancis.com Taylor & Francis Verlag GmbH, Kaufingerstraße 24, 80331 München, Germany